The Minerals, Metals & Materials Series

The Minerals, Metals & Materials Series publications connect the global minerals, metals, and materials communities. They provide an opportunity to learn about the latest developments in the field and engage researchers, professionals, and students in discussions leading to further discovery. The series covers a full range of topics from metals to photonics and from material properties and structures to potential applications.

Mohammad Shamsuddin

Thermodynamic Measurement Techniques

Mohammad Shamsuddin
Indian Institute of Technology
Varanasi, India

ISSN 2367-1181 ISSN 2367-1696 (electronic)
The Minerals, Metals & Materials Series
ISBN 978-3-031-47120-9 ISBN 978-3-031-47118-6 (eBook)
https://doi.org/10.1007/978-3-031-47118-6

This Springer imprint is published by the registered company Springer Nature Switzerland AG
The registered company address is: Gewerbestrasse 11, 6330 Cham, Switzerland

Preface

Currently, when so many textbooks on thermodynamics are available, it would be an act of ignorance to write another text on this subject. This availability is due to the fact that the subject is very basic to all physical, chemical, and biological scientists, and engineers. The subject is tough to start with and a person feels easy after being exposed several times or unless he starts teaching the subject. Thermodynamics is so diversified that it attracts a variety of person right from experimentalists to theoreticians. Hence, it is natural for a number of teachers to have a desire of conveying their understanding and enthusiasm to others in the form of a book.

I had been planning to write a book on metallurgical thermodynamics based on my experiences of teaching this subject as well as the courses based on thermodynamics to graduate and undergraduate students at the Department of Materials Science and Engineering, Massachusetts Institute of Technology and my association/interaction with the faculty members of the Department of Metallurgy and Metallurgical Engineering, University of Utah, Salt Lake City, during my 3 years (1978–1981) of visit to the United States, in addition to my long association as a faculty member with the Department of Metallurgical Engineering, Banaras Hindu University. But it was delayed due my sudden involvement in preparation of the manuscript for the second edition of the book *Physical Chemistry of Metallurgical Processes*, which was published in June 2021 (by Springer Nature and TMS).

Furthermore, while taking account of a number of books on chemical and metallurgical thermodynamics, I was in a dilemma regarding the extent of coverage of different topics in my book. After spending a lot of time on books of some prominent authors, I realized that there is very large coverage on thermodynamic principles but the thermodynamic measurement techniques have not been systematically discussed. This prompted me to write a book giving more emphasis on Thermodynamic Measurement Techniques, namely calorimetry, chemical equilibria, vapor pressure measurement, and electrochemical technique. In some books, these techniques have been either covered briefly or in very much detail. I have taken a balanced view. In addition, I have added a brief account on evaluation of thermodynamic quantities from phase diagram analysis. A chapter has been devoted to the combination of two techniques, either by coupling two thermodynamic techniques, viz. calorimetry and chemical equilibrium for simultaneous measurement of partial molar enthalpy and free energy. Vapor pressure has been coupled with electrical conductivity to study the effect of ambient pressure of tellurium on the mode of conduction in a large band gap compound semiconductor. Chapters 4, 5, 6, 7, 8, and 9 have been devoted to these techniques. A brief account has been included on the significance of various thermodynamic quantities in Chap. 1.

Currently, university courses provide inadequate background in chemical/metallurgical thermodynamics. At the majority of institutions, thermodynamics courses are formal. Often teachers feel satisfied by solving a few problems by plugging data in the thermodynamic expressions derived in the class. In this book, over 100 problems related to classical as well as solution thermodynamics and measurement techniques have been worked out in Chap. 10. The thermodynamic interrelationships concerning the problems have been summarized in Chaps. 2 and 3. For clarity the thermodynamics quantities have been defined together with an explanation of their physical significance. I am not consistent in using the SI unit throughout the book. Based on my long teaching experience of about four decades, I strongly feel that the use of different units will make students mature with regard to the conversion from one unit to another. The principal objective of the book is to enlighten students of chemistry, metallurgy, materials science and engineering, and chemical engineering with various thermodynamic measurement techniques.

I am thankful to a number of friends, colleagues, and library staff for necessary help in the preparation of this manuscript. It is not possible to mention all, but I shall be failing in my duty if I do not thank late Professor Somnath Misra, Professor Vasudeo B. Tare, and Professor K. K. Prasad of the Department of Metallurgical Engineering, Banaras Hindu University; late Professor K. P. Abraham of the Indian Institute of Science, Bengaluru; late Professor O. J. Kleppa of the James Franck Institute, University of Chicago; late Professor J. F. Elliott of the Department of Materials Science and Engineering, Massachusetts Institute of Technology; and late Dr. M. Nagamori of the Noranda Research Center, Canada for valuable discussions and suggestions during my association and private communications with them. Professor Abu Nasar, Chairman, Department of Applied Chemistry, Aligarh Muslim University deserves a special mention in this list for his willing assistance in drawing my attention to a number recent research papers on thermodynamic measurements based on fused salt electrolytes. Extensive thermodynamic data on thermodynamics of semiconducting intermetallics generated by him during his association (1985–1999) with the Department of Metallurgical Engineering, Banaras Hindu University has been frequently quoted in this book. Although due acknowledgments have been given to authors at appropriate places in the text for adapting their figures and tables published in various books and journals, I take this opportunity to thank the publishers (authors as well) listed below:

I. American Chemical Society

1. H.L. Johnston, A.L. Marshall, Langmuir type apparatus for determining reactive vapours. J. Am. Chem. Soc. **62**, 1382 (1940), for Fig. 6.4
2. O.J. Kleppa, Schematic diagram of Calvet microcalorimeter. J. Phys. Chem. **64**, 1937 (1960), for Fig. 4.5
3. W. Chen, A.J. Haslam, A. Macey, U.V. Shah, C. Brechtelsbauer, Schematic diagram of isoteniscope. J. Chem. Educ. (2016), https://doi.org/10.1021/acs.jchemed.5b00990, for Fig. 6.2
4. E. Veleckis, R.K. Edwards, Thermodynamic properties in the systems vanadium-hydrogen, niobium-hydrogen, and tantalum-hydrogen. J. Phys. Chem. **73**, 683(1969), for Fig. 9.1

II. AJP Publishing

1. M. Shamsuddin, O.J. Kleppa Thermodynamics of solution of hydrogen and deuterium in palladium-gold alloys at 555 K and 700 K. J. Chem. Phys. **71**, 5154 (1979), for Figs. 9.2 and 9.4
2. M. Shamsuddin, A. Nasar, V.B. Tare Electrical conductivity of tellurium deficient cadmium telliride. J. Appl. Phys. **74**, 6208 (1993), for Figs. 9.5, 9.6, and 9.7

III. Elsevier

1. M. Shamsuddin Hydrogen interaction in palladium alloys. J. Less Common Metals. **154**, 285 (1989), for Figs. 9.3(a), 9.3(b) and Tables 9.1 and 9.2

IV. Journal of Iron and Steel Institute

1. J.B. Bookey, Reduction equilibria involving phosphates. J. Iron Steel. **172**, 61 (1952), for Fig. 5.1
2. E.T. Turkdogan, S. Ignotowics, J. Pearson, The solubility of sulfur in iron and iron-manganese alloys. J. Iron Steel Inst. **180**, 349 (1955), for Fig. 5.2

V. Springer Nature

1. M. Shamsuddin, S. Misra, A thermoanalytical investigation of the thermodynamic properties of tin telluride. J. Therm. Anal. **7**, 310 (1975), for Figs. 4.9 and 4.10
2. O.J. Kleppa, Systematic aspects of the high temperature thermochemistry of binary alloys and related compounds. J. Phase Equilibria. **15**, 240 (1994), for Figs. 4.6 and 4.7.

VI. The Minerals, Metals & Materials Society (TMS)

1. H. Larson, J. Chipman, Fe^{2+}- Fe^{3+} equilibrium studies in CaO – FeO – Fe_2O_3 melts. Trans. Am. Inst. Min. Metall. Eng. **197**, 1089 (1953), for Fig. 5.3
2. M. Shamsuddin, *Physical Chemistry of Metallurgical Processes*, Second Edition (TMS, Springer, 2021) for Figs. 3.1, 3.2, 3.3, and 3.8
3. L.S. Darken, Thermodynamics of binary metallic solutions. Trans. Met. Soc. AIME. **239**, 80 (1967), for Figs. 3.9, 3.10, 3. 11, 3.12, 3.13, 3.14, and Table 3.2
4. M. Shamsuddin, A. Nasar, Thermodynamic investigations of Te-saturated solid CdSe-CdTe alloys. Metall. Trans. **23 B**, 467 (1992), for Figs. 3.15, 3.16, and 3.17
5. M. Shamsuddin, A. Nasar, On the thermodynamic behavior of Cadmium in Te-saturated HgTe-CdTe and CdSe-CdTe solid alloys. Metall. Mater. Trans. **26B**, 569 (1995), for Figs. 3.19 and 3.20
6. M. Shamsuddin, S. Misra, in *Chemical Metallurgy – A Tribute to Carl Wagner* (AIME, Chicago, Feb. 23–25, 1981), for Fig. 4.2.

VII. The Indian Institute of Metals

1. M. Shamsuddin, S. Misra, Constant temperature gradient calorimetry. Trans. Inst. Metals. **27**, 282 (1974), for Fig. 4.8
2. M. Shamsuddin, Thermodynamic properties of germanium telluride. Trans. Indian Inst. Metals. **28**, 303 (1975), for Fig. 4.3

VIII. Wiley & Sons, Inc.

1. R. Hultgren, R.L. Orr, P.D. Anderson, K.K. Kelley, *Selected Values of Thermodynamic Properties of Metals and Alloys* (John Wiley, New York, 1963), for Figs. 8.1 and 8.2.

Many people love thermodynamics and are fascinated by it and many more do not like it. I am in the first group; if the reader joins me there, I feel this book is a success.

The writing of any book consumes a lot of time which would otherwise have been spent with the family. I am extremely grateful to Mrs. Abida Khatoon for her understanding, taking care of the family, and for offering moral support and patience for my long hours spent working on the book.

Varanasi, India Mohammad Shamsuddin

Contents

Abbreviations

List of Symbols

A	Area of surface
a	Raoultian activity, Sievert's constant, cross sectional area
atm	Atmosphere (pressure)
c	Concentration
C	Heat capacity, component, constant
C_p	Heat capacity at constant pressure
C_v	Heat capacity at constant volume
D	Diffusion coefficient
d	Distance
e	Electron charge, interaction coefficient
E	Energy, electrode potential, emf, electron field mobility
f	Henrian activity coefficient
F	Degree of freedom, Faraday constant
(g)	Gaseous phase
G	Free energy
h	Planck constant
H	Enthalpy
I	Emergent intensity
I_o	Incident intensity
J	Joule
k	Sievert's constant, Boltzmann constant, kilo
K	Equilibrium constant, absorption coefficient
K_c	Concentration equilibrium constant
K_p	Pressure equilibrium constant
K_n	Molal equilibrium constant
l	Length
L	Latent heat of transformation, liter

L^e Latent heat of evaporation
L^f Latent heat of fusion
(l) Liquid state, liter
ln Naperian logarithm (loge)
m Mass
M Atomic/molecular weight
n Number of moles/atoms, number of electrons involved in an electrolytic cell, difference in number of moles of product and reactants in a chemical reaction
N Avogadro number, total number of molecules of reactants and products in a chemical reaction
p Partial pressure
P Total pressure
p_i Partial pressure of the component i
p^o Partial pressure of the pure component
q Quantity of heat
Q' Total extensive thermodynamic quantity
Q Molar extensive thermodynamic quantity
r Radius
R Gas constant, resistance
(s) Solid phase
S Entropy
t Time
T Absolute temperature (Kelvin), temperature °C
T^e Temperature of evaporation
T^f Temperature of fusion
U Internal energy
V Volume
W Thermodynamic probability
ω Weight, work done
x Atom/mole fraction, ionic fraction
z Valency

Greek Symbols

α A function factor in Gibbs–Duhem integration, Guggenheim interchange energy
β Degree of dissociation
γ Raoultian activity coefficient
γ_i^o Raoultian activity coefficient at infinite dilution
σ Specific conductivity
ε Interaction parameter
θ Angular deflection
μ Chemical potential (partial molar free energy), mobility
$\varnothing$ Fugacity
τ Torsion constant

Prefixes

Δ Change in any extensive thermodynamic property (between product and reactant), e.g., ΔG, ΔH, ΔS, ΔU, ΔV

d Very small change in any thermodynamic variable, e.g., dG, dT, dP

Suffixes[1]

aq Dissolved in water at infinitely dilute concentration, e.g., M^{2+} (aq)

g Gaseous, e.g., Cl_2 (g)

l Liquid, e.g., H_2O (*l*)

s Solid, e.g., SiO_2 (s)

Subscripts

cell E_{cell}, ΔG_{cell}

f Formation, e.g., ΔG_f

M Metal, e.g., E_M

sol Solution, e.g., ΔG_{sol}

vol Volume, e.g., ΔG_{vol}, ΔH_{vol}

Superscripts

– Partial molar functions, e.g., $\overline{G_i}$, e^-

\+ Electron hole concentration, e.g., h^+

o Thermodynamic standard state, e.g., ΔG^o

id Ideal thermodynamic functions, e.g., $\Delta G^{M,\ id}$

M Mixing, e.g., ΔG^M, ΔH^M

xs Excess thermodynamic functions, e.g., G^{xs}

Additional Symbols

( ) Solute in slag phase

{ } Gaseous phase

[ ] Solute in metallic phase, e.g., [*S*], concentration of a species in solution, e.g., [*A*]

[1] The physical state of a substance is indicated by the following symbols, placed in bracket ( ) after the chemical formulae of the substance.

List of Figures

List of Tables

Chapter 1
Introduction

Thermodynamics deals with the interconversion of heat and mechanical work which manifests itself into displacement of a body under mechanical force as noticed in heat engines. These studies evolved a particular branch of thermodynamics, known as mechanical engineering thermodynamics.

Heat may be evolved or absorbed on the formation of new molecules when substances react chemically. In this formation of molecules the atoms of the reacting molecules work against the binding forces in it on account of some energy available in the system comprising the reactants together with release of some energy while getting bound into the new molecules. The end result is that some chemical work is performed. Studies of such changes in the process is termed as chemical thermodynamics. In a similar way studies dealing with energy changes in reactions involving metals, and their oxides, sulfides and other compounds is named as metallurgical thermodynamics. Theoretically, chemical and metallurgical thermodynamics have many common concepts. However, they differ in ranges of working temperature and pressure. The former is concerned with relatively lower temperature and higher pressure often up to hundreds of atmosphere whereas metal production is generally carried out at lower pressure and higher temperature.

Thermodynamic laws have been formulated at macro level considering matters made up of atoms and molecules. In recent years scientists started looking into the micro-states within the systems and thus a new branch of thermodynamics called statistical thermodynamics has emerged, which is playing major role in interpretation of thermodynamic properties of micro constituent of the system.

The science dealing with the basic concepts, the thermodynamic laws and their interrelationships is known as the classical thermodynamics. The entire foundation of classical thermodynamics was laid down during the first half of the nineteenth century. Hess [1], Carnot [2], Joule [3], Clausius [4], Kelvin [5] and Helmholtz [6] were the pioneers in establishing the basic principles of the theory of energy. Following these principles thermodynamic theorems were developed by many scientists working in this field. Among them J. Willard Gibbs [7], K. van't Hoff [8], and H. Le Chatelier [9] are worthy of mention. The concepts initiated and

M. Shamsuddin, *Thermodynamic Measurement Techniques*, The Minerals, Metals & Materials Series, https://doi.org/10.1007/978-3-031-47118-6_1

perfected by them have important bearing in designing and developing many chemical and metallurgical processes for production of chemicals and extraction and refining of metals. The importance of their significant contributions can be easily judged by the following brief account on the Pidgeon process for the production of magnesium:

Since the free energy change for the following reaction under standard conditions is positive ($\Delta G^0 = +280$ kJ at 1200 °C) the reduction of MgO with silicon is not feasible, i.e. the following reaction will not proceed in the forward direction.

$$2\text{MgO}\ (s) + \text{Si}\ (\text{s}) = 2\ \text{Mg}\ (g) + \text{SiO}_2\ (s)$$

According to Le Chatelier principle the reduction of pressure will favor the reaction in the forward direction because in the above reaction two moles of magnesium gas are produced at 1200 °C. It is also important to note that the van't Hoff isotherm is more useful in judging the thermodynamic feasibility of reactions where ever activities of reactants and products differ significantly from unity. The silicothermic reduction of MgO (Pidgeon Process) sets a good example in understanding the problem of deviation from unit activity.

Making use of the van't Hoff isotherm the actual free energy change for above reaction may be expressed as

$$\Delta G = \Delta G^o + RT \ln K = \Delta G^o + RT\ \ ln\ \left(\frac{p^2_{Mg}.a_{SiO_2}}{a^2_{MgO}.a_{Si}}\right)$$

ΔG becomes negative, even if ΔG^o is positive, at reduced pressure of magnesium and at lower activity of SiO_2. The Pidgeon process operates at a reduced pressure and the activity of silica is brought down to <0.0001 by using calcined dolomite (CaO·MgO) instead of MgO. Thus calcined dolomite (CaO·MgO) obtained by the decomposition of dolomite ($CaCO_3$·$MgCO_3$) is reduced with silicon 1200 °C in a retort at a reduced pressure of about 10^{-4} atm according to the reaction:

$$2(\text{CaO} \cdot \text{MgO})(s) + \text{Si}(s) = 2\text{Mg}(\text{g}) + 2\ \ \text{CaO} \cdot \text{SiO}_2(l)$$

The Pidgeon process is much simpler compared to the reduction of MgO with carbon at 1900 °C (MgO + C = Mg + CO). In the carbothermic reduction process magnesium vapors get reoxidized with CO on cooling. In order to prevent the formation of finely divided pyrophoric magnesium powder magnesium vapors need to be quenched in cold hydrogen. There are many such examples in extraction of metals where these principles have been made use of in development of new processes.

The third stage of development has been in advancement of thermodynamic techniques to generate and accumulate accurate data useful in analysis of chemical and metallurgical processes. Haber [10] in Germany was the first scientist to publish the book based on the systematic study of the thermodynamic data required for the calculation of changes in free energy in some important chemical reactions. Nernst [11] made everlasting contributions in experimental and theoretical application of thermodynamics to chemistry and metallurgy. Clausius [4] introduced the concept of entropy. Further studies on entropy led Walter Nernst [11] to develop the concept of the third law of thermodynamics in 1906. This was under debate for some time and was settled in 1912 by Plank [12].

In the third part of the development G. N. Lewis [13] also contributed significantly. He successfully demonstrated through his numerous investigations as how to apply thermodynamics to various chemical systems. He made a major contribution in collection of necessary data for common chemical substances. Free energy and electrode potential tables developed by him occupied places in basic chemical handbooks. Significant contributions of Kleppa [14] in the development of a number calorimeters and microcalorimeters for determination of thermodynamic properties of salts and alloys, intermetallics and minerals occupy a high rating in the history of calorimetric measurements.

In the late nineteen fifties, Wagner [15] showed the utility of oxide solid electrolytes for thermodynamic measurements at high temperature. Initially solid glass and halide electrolytes were used, but it was established that oxide electrolytes were more promising. The most widely used solid electrolyte is zirconia doped with calcia. A large number of metallic and oxide systems have been studied since Wagner demonstrated the role of oxide electrolytes for thermodynamic measurements.

Thus thermodynamics originating from Physical Chemistry has played extremely important role in development of the existing processes as well as in inventing new processes in the fields of chemistry, extractive metallurgy and chemical engineering during the past century. It is going to continue further in years to come. Perhaps the best recommendation about the importance of thermodynamics could be given by the following quotation from Einstein [16]:

"A theory is the more impressive the greater the simplicity of its premises is, the more different kind of things it relates, and the extended is its area of applicability. Therefore the deep impression which classical thermodynamics made upon me-it is the only physical theory of universal content concerning which I am convinced that, within the framework of the applicability of its basic concepts, it will never be overthrown" [16].

It is now an accepted fact that for modification, replacement or development of various chemical and metallurgical processes, knowledge of the accurate thermodynamic data is very essential. With the help of this data one can at least say that the reaction is feasible (although whether the actual reaction takes place or not will depend upon kinetic factors). During the past few decades large number of thermodynamic data have been collected and compiled. These data have been continuously revised and replaced by more accurate data due to the development of better and precision experimental techniques. A chemist/metallurgist often needs the following information to study the feasibility of a particular process:

1. Molar heat capacities, standard molar entropies, heats of allotropic transformation, fusion and vaporization, partial and integral heats of mixing and heats of reaction.
2. Vapor pressures of elements and compounds.
3. Partial molar free energy of mixing and activity of components in binary and multicomponent systems.
4. Standard free energy changes in reactions involving metals, metal oxides, halides, sulfides, nitrides, carbides etc.
5. Equilibrium between different phases.

1.1 Significance of Thermodynamic Quantities

The concept of various thermodynamic quantities and their inter-relations will be discussed in Chaps. 2 and 3. In this Section a brief account is given about the significance of these quantities.

Information about thermodynamic properties of materials may contribute to an understanding about their nature. The free energy of formation of the compound is an absolute measure of its stability. In a multicomponent system a compound is stable if its free energy is lower than that of other competitive phases or phase mixtures. The heat of formation of a compound throws light on the type of bonding. A large value of exothermic heat of formation [17] of a compound indicates that the nature of its bonding is different from the bonding in the constituent elements, for example, when metallic elements form a compound in which the bonding is ionic or covalent. The exothermicity increases with the degree of ionicity of the bond and for intermetallic compounds, the heats of formation range between −83 and −335 kJ g atom^{-1}, approximately. On the other hand the heat of formation of covalent compounds is generally smaller than those of ionic compounds because of smaller energy change due to the sharing of the electrons between the atoms in the covalent bond. The heat of formation of most covalent intermetallic compounds range from −25 to −70 kJ g atom^{-1}. The heats of formation of compounds with metallic bonding are mainly due to the difference in coordination of the atoms in the component metals and the compounds. The heats of formation of most of these compounds are numerically smaller than −65 kJ g atom^{-1}. In addition to this the heat of formation of a compound is the measure of its stability in a condensed system, where its entropy of formation is small.

The entropy of formation reflects changes in the vibrational behavior and configurational arrangements of atoms on formation of the compound. The total entropy change on formation of a compound is the sum total of entropy changes due to changes in vibrational entropy, configurational entropy, electronic entropy and magnetic entropy. The change in vibrational entropy is due to the difference between the thermal vibrations of the atoms in the compound and in the component elements [18]. This difference is caused by changes in the bond strength and in the crystallographic arrangement of the atoms. A stronger bond in the compound reduces the vibrational entropy. The configurational entropy term tells about the degree of disorder, it decreases with decreasing the degree of disorder. Thus a value of zero indicates that the compound is perfectly ordered. Electronic and magnetic entropy effects are generally negligible in the formation of a compound but may contribute when the compound contains a transition element [19].

The heat capacity and the absolute entropy also reflect the vibrational behavior. The heat capacity of the solid just below the melting point sheds light on pre-melting phenomena. According to Kopp-Neumann rule, the heat capacity of an intermetallic compound is equal to the sum of the heat capacities of the component elements [20]. This rule is obeyed by compounds in which the bonding and crystallographic arrangements are similar to those in the component elements. Deviation from Kopp-Neumann rule of the heat capacities is expected when order-disorder or magnetic transitions occur in a compound. The heat capacity of solids increases appreciably just below the melting point which may be attributed [20] to the weakening of bonds, accelerated generation of point defects and accelerated disordering of an ordered compound.

The heat of fusion of a solid is the sum total of energy required to change the periodic arrangement of atoms in the solid into the less rigid arrangement of the liquid state and the energy required for the destruction of any order present in the solid at the melting point and for changes in the type of bonding accompanying the fusion process. The entropy of fusion is associated with the increase in vibrational freedom on melting and any disordering of an ordered compound.

Thermodynamic properties can be sensitively correlated to other physical properties. For example, Ettenberg et al. [21] have shown a linear correlation between the heat of formation and intrinsic degree of disorder in aluminides of transition elements. Chang et al. [22] have further correlated the degree of disorder to the entropy of formation. The abnormal change in the Darken's thermodynamic stability parameter [23] with composition in the intermediate composition ranges has been correlated to the change in bonding type and appearance of intermediate phases at lower temperatures. Shamsuddin and Misra [24] have correlated thermodynamic parameters with structural and electronic properties in semiconducting chalcogenides. Nasar and Shamsuddin [25] have correlated thermodynamic properties of II–VI compounds with melting point, bond length, energy gap, ionicity and molecular weight.

1.2 Measurement Techniques

From the above discussion it is clear that the knowledge of a number of thermodynamic properties is essential to understand the physical nature of compounds. When the data are not available with sufficient degree of accuracy one would, to start with, search for a suitable method to determine these experimentally. Confidence in and reliability of the data are considerably increased if the quantities are measured by two or more independent techniques. Several independent techniques may be mutually complementary. Independent measurements often yields reliable values of thermodynamic properties. The choice of the technique depends on the accuracy desired, the system to be investigated, container materials, and the temperature of interest. The usual thermodynamic quantities needed for material characterization and thermodynamic calculations have been listed above.

The different techniques available for determination of thermodynamic quantities can be broadly classified into seven main groups, namely, calorimetry, chemical equilibria, vapor pressure, electrochemical, computational, spectroscesopic, and phase diagram analyses. In this presentation the first five techniques will be discussed in detail in Chaps. 4, 5, 6, 7, and 8, respectively.

References

1. G. H. Hess, Recherches thermochimiques, Bull. Scientifique Academie imperial des sciences, **8** (1840) 257, St. Petersburg. https://www.abebooks.com/first-edition/Thermochemische-Untersuchungen-G.H-Hess-1839-1842-1839-1842/30250017161/bd
2. S. Carnot, The Reflections on the Motive Power of Fire (Paris, 1824) (quoted in Second Law of Thermodynamics by W. F. Magie (Ed), Harper & Brothers, New York, 1899)
3. J. Joule, Phil. Mag. **23**, 263 (1813)
4. R. Clausius, Ann. Phys. **79**, 368 (1850). **125**, 353 (1865)
5. W. T. Kelvin, Nature, **9**, 441 (1874); Proc. Royal Soc. Edin. **8**, 325 (1875); Phil. Mag. **33**, 291 (1892)
6. H. V. Helmholtz, *Die Thermodynamik chemischer Vorgange*, 1882
7. J. Willard Gibbs, Trans. Conn. Acad. Sci. **3**, 228 (1876)
8. K. van't Hoff, Svensk. Vet. Akad. Handl., 21 (1886) 1
9. H. Le Chatelier, Ann. Mines **13**, 157 (1888)
10. F. Haber, *Thermodynamik der technischen Gas Reaktionen* (R. Oldenbourg -Verlag, Munich, 1905) (trans. By A. B. Lamb, Longmans, Green & Co., Inc., New York, 1908)
11. W.H. Nernst, *Experimental and Theoretical Application of Thermodynamics to Chemistry* (Charles Scribner Sons, New York, 1907)
12. M. Plank, *Treatise on Thermodynamics.*, (trans. A. Ogg) (Longmans Green & Co. Inc., London, 1927)
13. G.N. Lewis, M. Randall, *Thermodynamics*, 1st edn. (McGraw-Hill Co. Ltd., New York, 1923)
14. O.J. Kleppa, J. Phase Equilibria **15**, 240 (1994)
15. K. Kiukkola, C. Wagner, J. Electrochem. Soc. **104**(308), 379 (1957)
16. Albert Einstein, in Albert Einstein: Philosopher – Scientist (Tuder Publishing Company, New York, 1949). Ann., Phys. **17**, 132 (1905)
17. P.M. Robinson, M.B. Bever, in *Intermetallic Compounds*, ed. by J.H. Westbrook, (John Wiley, New York, 1967)
18. J. Lumsden, *Thermodynamics of Alloys* (Institute of Metals, London, 1952)
19. O. Kubaschewski, *The Physical Chemistry of Solutions and Intermetallic Compounds*, National Physical Laboratory, Symposium No. 9, vol I & II (HMSO, London, 1959)
20. O. Kubaschewski, C.B. Alcock, *Metallurgical Thermochemistry*, 4th edn. (Pergamon, Oxford, 1979)
21. M. Ettenberg, K.L. Komarek, E. Miller, in *Ordered Alloys*, ed. by K.H. Kear et al., (Claitors, Baton Rouge, 1970)
22. Y.A. Chang, I. Cyuk, J. Franks, Acta Metall. **19**, 939 (1967)
23. L.S. Darken, Trans. Met. Soc. AIME **239**, 80 (1967)
24. M. Shamsuddin, S. Misra, *Materials Science Symposium on Structure Property Correlation* (Oct. 24–26, REC, Roukela, 1977)
25. A. Nasar, M. Shamsuddin, Def. Sci. J. **50**, 289 (2000)

Chapter 2
Classical Thermodynamics

Thermodynamics is concerned with the quantitative aspects of interconversion of heat and energy. The science dealing with the basic concepts, the thermodynamic laws and their interrelationships is known as the classical thermodynamics. For thermodynamic studies a system as any matter consisting of a definite amount of substance or substances is specified. The system exchanges heat with its surroundings. The system and surrounding together are considered as an isolated system. For example, a steel billet together with the reheating furnace housing the billet form a system. The surroundings of the billet would include the furnace, atmosphere inside and outside the furnace, the foundation on which the furnace has been built, and any other materials which can exchange a measurable amount of energy with the billet. The energy of a body is defined as its ability to do work. It can exist in various forms such as kinetic, potential, thermal, mechanical, chemical, etc. Each form of energy can be converted into another. Mechanical energy is transformed into electrical energy in a generator and vice versa in an electric motor. Chemical energy is transformed into electrical energy in a galvanic cell and vice versa in an electrolytic cell. Mechanical energy is transformed into heat energy in grinding processes. Although conversion of heat into other forms of energy is limited by the second law of thermodynamics, any form of energy can be completely converted into heat energy. Thus energy may be defined as heat or anything that can be converted into heat. All forms of energy can be considered as the product of an intensive property (e.g. pressure, temperature, etc.) and its conjugate extensive property (e.g. volume, entropy, etc.). Thus, energy $= p.\ dv$, $(= T.\ dS)$.

If the total quantity of energy of a system, whether electrical, kinetic, rotational, vibrational or any other form of energy but not the energy due to its position in space is called the internal or intrinsic energy of the system. It is not possible to measure quantitatively all different forms of energy possessed by the system but it does not matter because we are concerned with the measurement of energy change.

M. Shamsuddin, *Thermodynamic Measurement Techniques*, The Minerals, Metals & Materials Series, https://doi.org/10.1007/978-3-031-47118-6_2

2.1 Concept of Temperature and Zeroeth Law of Thermodynamics

The heat is a form of energy which is transferred due to temperature difference. Hence, the concept of temperature and changes therein are important for heat energy studies. We cannot define temperature quantitatively but we can discuss it as a level of heat just as the level of liquid in a vessel. We experimentally notice that liquid flows from a vessel with high level to another vessel at lower level when the two vessels are connected with a tube. In a similar manner we observe that heat flows from an object maintained at higher temperature to another object at lower temperature on bringing them in thermal contact. However, temperature can be measured and reported quantitatively using various scales such as Celsius, Fahrenheit, or Kelvin.

Based on our routine observation we find that when two bodies are in equilibrium with a third body, all the three are in thermal equilibrium with each other. This fact has been termed as the "Zeroeth Law of Thermodynamics."

2.2 The Concept of State

The properties of a system is defined by the state of the system. The relationship of volume, V of the system with pressure, P and temperature, T is known as equation of state of the system. In case of a definite amount of a gas if two of the three variables are fixed the third one is automatically fixed. This can be expressed as exact differential:

$$dV = \left(\frac{\partial V}{\partial P}\right)_T dP + \left(\frac{\partial V}{\partial T}\right)_P dT \tag{2.1}$$

Thus in transferring the gas from state 1 to state 2 the volume change depends on the volumes of the gas in two states. The change is independent of the path followed by the gas between the two states because the volume of the gas is a state function and Eq. 2.1 is an exact differential.

Robert Boyle established that pressure of the gas is inversely proportion to the volume at constant temperature, T (i.e. $P \propto \frac{1}{V}$) or $P_1V_1 = P_2V_2$, whereas according to Charles $P \propto T$ (i.e. $V_1/T_1 = V_2/T_2$) at constant pressure. Based on both these laws we can write as

$$\frac{P_1 V_1}{T_1} = \frac{P_2 V_2}{T_2} \tag{2.2}$$

Since according to Avogadro Hypothesis one g mole of an ideal gas occupies 22.414 liters at 0 °C and 1 atm pressure the constant for Eq. 2.2 can be calculated as

$$\frac{P_1 V_1}{T_1} = \frac{1 \times 22.414}{273.15} = 0.082057 \text{ liter atm deg}^{-1} \text{mol}^{-1}$$

This constant known as the universal gas constant, denoted by R, is applicable to all gases. Hence, for one mole of gas, Eq. 2.2 is universally written as

$$\mathrm{PV} = \mathrm{RT} \tag{2.3}$$

Equation 2.3 known as the equation of state for 1 mole of gas defines the ideal gas law. In all thermodynamic discussions the equation of state is frequently used.

Properties, relevant to the thermodynamic studies are known as thermodynamic properties. These are classified into two groups: extensive and intensive properties. The former depends upon the quantity of matter of the system, such as volume, mass, energy. Intensive property is independent of mass of the system, such as temperature, pressure, surface tension, density, refractive index, viscosity, etc.

2.3 The First Law of Thermodynamics

The First Law of thermodynamics is concerned with the quantitative aspects of interconversion of energies. It states that energy can neither be produced nor destroyed in a system of constant mass, although it can be converted from one form to another. Irrespective of the mode of transformation, a given amount of work always produces the same amount of heat within the experimental error of the technique. This establishes that heat and mechanical work are equivalent and it is possible to measure both in one single unit of energy and this forms the basis of the first law of thermodynamics.

Until the beginning of the nineteenth century heat was regarded as fluid called "caloric" by majority of scientists. This caloric was assumed to pass from one body to another without actually being lost. Thompson [1] disproved this concept as he noticed that a tremendous amount of heat was generated during boring of metal canons. This observation led him to believe that heat was generated at expend of mechanical work for boring. Subsequently, he determined the amount of heat produced by work. He found that 940 ft lb of work was needed to raise the temperature of 1 lb of water by 1°F (correct value 980 ft lb) and thus he initiated the interconvertability of heat and energy.

For the next 50 years there was lot of conflict over the nature of heat and the correct value of the mechanical equivalent of heat, i.e. the amount of work equivalent to 1 calorie. Finally Thompson's hypothesis that heat is energy, was proved by Joule [2]. According to Joule 10^7 ergs of work expended by paddles in water contained in a calorimeter produced 0.241 calorie (present value 0.239). Based on this observation, Joule then systematically converted electrical and mechanical work into heat in a variety of ways such as electric heating, mechanical stirring, compression of gases and friction. He found nearly the same value of the conversion factor in

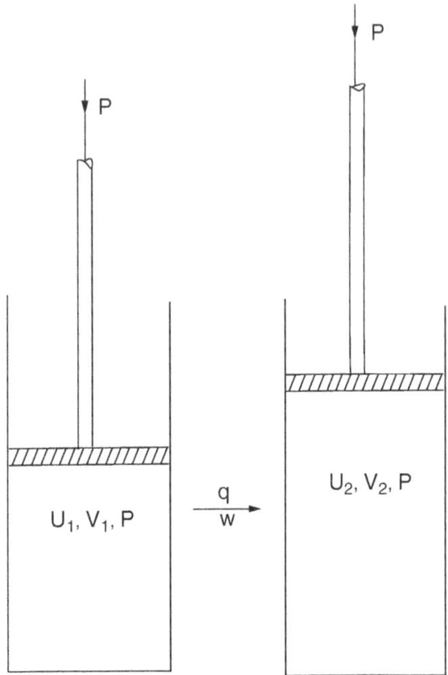

Fig. 2.1 The first law of thermodynamics

every case of conversion. In this way he for the first time established that a given amount of work always produced the same amount of heat within the experimental error of the same technique, whatever may be the mode of transformation.

To formulate a mathematical expression based on this law consider one mole of an ideal gas contained in a cylinder fitted with a frictionless piston (Fig. 2.1) at pressure, P and temperature, T, and let the internal energy of this gaseous system is denoted by U_1. Suppose external energy, q in the form of heat is supplied to the system and the system while absorbing this energy, performs the amount of work, w and attains the final energy, U_2. In this process the piston having an area, A moves by a distance dx.

$$\text{Work done by the gas} = \text{force} \times \text{distance} = (P.A).\ dx = P\ (A.dx) = P\ dV$$

Hence, the increase in energy of this system $= U_2 - U_1 = \Delta U$ (change in internal energy) $= q - w$. Change in internal energy of an ideal gas at constant volume, i.e. when $dV = 0$, $w = 0$, and $\Delta U = q_v$ (heat absorbed at constant volume). The change in internal energy of a gaseous system is due to changes in the translational, rotational and vibrational energies of the gaseous molecules.

If the initial energy of the given mass of gas is U_1 and volume V_1, on introducing q calories of heat in to the system there will be an increase in temperature and volume to keep the internal pressure of the gas constant. If V_2 and U_2 are respectively the final volume and energy, the change in internal energy at constant pressure can be expressed as follows:

$$\Delta U = U_2 - U_1 = q_p - w = q_p - PdV$$
$$\text{Or} \quad \Delta U = U_2 - U_1 = q_p - P\ (V_2 - V_1)$$
$$\therefore\ (U_2 + PV_2) - (U_1 + PV_1) = q_p$$

Hence, $(U + PV) = q_p$

The term $(U + PV)$ on the left hand side of the equation occurs often in thermodynamics. In order to simplify our calculation at constant pressure, it is given a symbol H and is called enthalpy; i.e. total heat content.

$$\text{Thus,} \quad H = U + PV \tag{2.4}$$

∴ Heat content (enthalpy) = internal energy + energy term dependent on the state of the system.

An increase in enthalpy therefore denotes the heat absorbed during a process at constant pressure.

The absolute value of H of a system cannot be defined, hence for convenience, H for pure elements at atmospheric pressure and 25 °C (298 K) is assumed as zero.

2.4 Internal Energy, Enthalpy and Heat Capacity

When heat is supplied to the system consisting of 1 mole of an ideal gas at constant volume (Fig. 2.1) there is an increase in temperature. Correspondingly, there will be an increase in internal energy, dU. Consequently, if this gas loses energy by doing work on the surroundings (e.g. by adiabatic expansion) temperature decreases. This change in internal energy dU varies with respect to the temperature change of the system, i.e., $dU \propto dT$

$$\text{or} \quad dU = C_v dT$$
$$\therefore C_v = \left(\frac{dU}{dT}\right)_v \tag{2.5}$$

Where C_v is a constant of proportionality, known as heat capacity at constant volume.

On the other hand if on supplying dq amount of heat energy to the above system at constant pressure, there is an increase in temperature, dT the heat capacity, C_p at constant pressure is expressed as:

$$dq = C_p dT \tag{2.6}$$

$$\therefore C_p = \left(\frac{dq}{dT}\right)_p \tag{2.7}$$

2.5 Expansion of an Ideal Gas Under Isothermal and Adiabatic Conditions

In the above case (Fig. 2.1) if the gas expands at constant pressure on supplying heat energy, dq we have:

$$dq = dU + PdV \tag{2.8}$$

From Eqs. 2.5, 2.6 and 2.8 we get:

$$C_p dT = C_v dT + PdV \tag{2.9}$$

$$\begin{aligned} \text{or} \quad C_p &= C_v + \frac{PdV}{dT} = C_v + d(PV)/dT \ \text{(since P is constant)} \\ &= C_v + d(RT)/dT \end{aligned} \tag{2.10}$$

$$\text{or} \quad C_p = C_v + R \tag{2.11}$$

Hence for an ideal gas:

$$C_p - C_v = R \tag{2.12}$$

In Fig. 2.1 if dq amount of heat is supplied and the system is maintained at a constant temperature, i.e., $dT = 0$, the expansion of the gas is considered to be under isothermal condition and hence, $dU = 0$, and thus from Eq. 2.8, we get:

$$dq = PdV \tag{2.13}$$

If the gas expands from the initial volume, V_1 to occupy the final volume V_2 with the corresponding decrease of pressure from P_1 *to* P_2 on supplying q amount of heat, integration of Eq. 2.13 gives:

$$q = \int_{V_1}^{V_2} PdV = \int_{V_1}^{V_2} \frac{RT}{V} dV = RT\ [ln\ V]_{V_1}^{V_2} \tag{2.14}$$

$$\text{Hence,} \quad q = RT \ln \left(\frac{V_2}{V_1}\right) \tag{2.15}$$

$$\text{Since} \quad P_1 V_1 = P_2 V_2,$$

$$q = RT \ln \left(\frac{P_1}{P_2}\right) \tag{2.16}$$

If gas expands adiabatically, $dq = 0$

$$\text{Hence,} \quad dU = -PdV$$

$$\text{or} \quad C_v dT = -PdV = -\left(\frac{RT}{V}\right) dV$$

$$\text{or} \quad C_v \left(\frac{dT}{T}\right) + R \left(\frac{dV}{V}\right) = 0$$

$$\text{or} \quad C_v \left(\frac{dT}{T}\right) + (C_p - C_v) \left(\frac{dV}{V}\right) = 0$$

$$\text{or} \quad \left(\frac{dT}{T}\right) + (\gamma - 1) \left(\frac{dV}{V}\right) = 0 \quad (\gamma = C_p/C_v)$$

$$\text{or} \quad ln\, T + (\gamma - 1) \ \ln V = constant$$

$$\text{or} \quad T \ V^{(\gamma - 1)} = constant \quad \left(T = \frac{PV}{R}\right) \tag{2.17a}$$

$$\text{or} \quad \frac{PV^\gamma}{R} = constant$$

$$\text{or} \quad PV^\gamma = constant \tag{2.17}$$

2.6 Thermochemistry

Thermochemistry being a part of thermodynamics deals with the heat changes associated with chemical reactions. Since temperature, pressure and the states of the reactants and products affect the enthalpy change of a reaction, all of these must be specified in the normal balanced chemical equation. For example, the heat evolved during the formation of water by the reaction between hydrogen and oxygen depends on the state of water:

$$2H_2 \ (g) + O_2 \ (g) \rightarrow 2H_2O \ (l) \quad \Delta H^0_{298} = -571.6 \text{ kJ}$$

$$2H_2\ (g) + O_2\ (g) \rightarrow 2H_2O\ (g) \quad \Delta H^0_{298} = -483.6\ \text{kJ}$$

The amounts of heat liberated differs in the two reactions due to the heat of vaporization of water at 298 K. The heat of vaporization of water: H_2O (l) → H_2O (g) at 298 K (= 44.0 kJ) can be calculated from ΔH^0_{298} values of the two chemical reactions.

2.6.1 *Heat of Reaction*

When the reactants in moles taking part in the chemical reaction, presented by the balanced equation react completely, the enthalpy change of the reaction, ΔH is known as the heat of reaction. For example, 46.7 kJ mol^{-1} is evolved when 3 moles of hematite are reduced with 1 mole of carbon monoxide to form 2 moles of magnetite and 1 mole of carbon dioxide at 727 °C.

$$3Fe_2O_3(s) + CO\ (g) = 2Fe_3O_4(s) + CO_2(g) + 46.7\ \text{kJ}$$

Reactions with evolution of heat are known as exothermic reactions and heat energy change accompanying such reactions will have negative sign, for example, in the above case: $\Delta H = -46.7$ kJ mol^{-1}. On the other hand reaction which absorbs heat, that is where heat energy is supplied from external source for completion of the reaction, is called endothermic reaction. In such cases, the heat change, ΔH will have a positive sign.

2.6.2 *Heat of Formation*

The enthalpy change, ΔH on formation of one mole of the compound from its constituent elements in their stable forms at 25 °C and 1 atm pressure is known as heat of formation. For example, the heat of formation of zinc sulfide is $\Delta H^0_{298} = -201.7$ kJ mol^{-1}. The subscript 298 refers to the temperature of measurement, 298 K.

$$C\ (s) + O_2\ (g) = CO_2\ (g) + 393.51\ \text{kJ}$$

Since in the above reaction 393.51 kJ is evolved on formation of one mole of CO_2 gas (i.e. it is lost to the surroundings) the heat of formation of CO_2, ΔH is -393.51 kJ mol^{-1}.

On the other hand, if a finite amount of heat is given to the system to carry out the reaction, ΔH will have a positive sign.

2.6.3 Heat of Combustion

It is the amount of heat evolved on complete combustion of one mole of an element or a compound at a given temperature and one atmosphere pressure. For example,

$$C\ (s) + O_2\ (g) = CO_2\ (g), \quad \Delta H^0_{298} = -394\ \text{kJ}$$

$$Mg\ (s) + \tfrac{1}{2}\ O_2\ (g) = MgO\ (s) \quad \Delta H^0_{298} = -603\ \text{kJ}$$

The above figures for the heat of combustion of carbon and magnesium are also heats of formation of CO_2 and MgO respectively.

2.6.4 Heat of Transformation

This refers to the change in enthalpy when 1 mole of a substance undergoes a physical change, namely, allotropic modification, melting, and evaporation. In general, a symbol L^t is given to such changes but it is commonly denoted according to the particular transformation.

2.6.5 Heat of Solution

The change in enthalpy on dissolution of one mole of the solute to form a solution of a particular concentration is known as the heat of solution.

2.6.6 Hess' Law of Constant Heat Summation

According to the First Law of Thermodynamics the total energy of a system and surroundings remains constant. The thermodynamic variables like enthalpy, entropy and free energy changes accompanying a reaction depend only on the initial and final states, not on the path chosen. This happens to be the basis of **Hess' law of constant heat summation**. The law states that the overall heat change of a chemical reaction is the same whether it takes place in one or several stages, provided the temperature and either the pressure or the volume remains constant.

2.7 Effect of Temperature on Heat of Reaction

A discussion on the effect of temperature on heat of reaction requires data about heat capacities of reactants and products taking part in a chemical reaction. Heat capacity, C is defined as the amount of heat required to raise the temperature of a substance by a certain amount. In other words, it is the ratio of the heat, Q absorbed by a system to the resulting increase in temperature $(T_2 - T_1)$, that is, ΔT. Since the heat capacity usually varies with temperature,

$$C = \lim_{T_1 \to T_2} \frac{Q}{\Delta T} = \frac{q}{dT} \tag{2.18}$$

Where q = quantity of heat change, dT = small rise in temperature.

At constant volume, $q_v = \Delta U_v$

$$\therefore C_v = \frac{dU_v}{dT} = \left[\frac{\partial U}{\partial T}\right]_v \tag{2.19}$$

Hence, the heat capacity of a system at constant volume is equal to the rate of increase of internal energy content with temperature at constant volume.

Similarly, at constant pressure,

$$C_p = \frac{q_p}{dT} = \left[\frac{\partial H}{\partial T}\right]_p \tag{2.20}$$

Thus, the heat capacity of a system at constant pressure is consequently equal to the rate of the increase of heat content with temperature at constant pressure.

An expression for the variation of the heat of reaction with temperature can be derived in a simple manner. If H_A is the total heat content of the reactants and H_B is that of the products at the same temperature and pressure then the heat change:

$$\Delta H = H_B - H_A \tag{2.21}$$

All the quantities refer to the same pressure. Differentiating Eq. 2.21 with respect to temperature:

$$\left[\frac{\partial (\Delta H)}{\partial T}\right]_P = \left[\frac{\partial H_B}{\partial T}\right]_P - \left[\frac{\partial H_A}{\partial T}\right]_P \tag{2.22}$$

By definition $\left[\frac{\partial H}{\partial T}\right]_p = C_p$

$$\text{Hence,} \quad \left[\frac{\partial(\Delta H)}{\partial T}\right]_P = (C_p)_B - (C_p)_A \tag{2.23}$$

Where $(C_P)_A$ and $(C_P)_B$ are the total heat capacities of the reactants and products, respectively at the given constant pressure. The right hand side of Eq. 2.23 is the increase in the heat capacity of the system accompanying the chemical reaction and so it may be represented by ΔC_P, then Eq. 2.23 takes the form:

$$\left[\frac{\partial(\Delta H)}{\partial T}\right]_P = \Delta C_P \tag{2.24}$$

Thus the rate of variation of the heat of reaction with temperature, at constant pressure, is equal to the increase in the heat capacity accompanying the reaction. Equation 2.24 is known as Kirchhoff's equation [3–5]. The equation is useful in estimation of the heat of reaction at one temperature, if that is known at another temperature. This involves the integration of the Eq. 2.24 between the temperature limits of T_1 and T_2. The variation of the heat of reaction with temperature can be expressed as:

$$\Delta H_2 - \Delta H_1 = \int_{T_1}^{T_2} \Delta C_p.\ dT \tag{2.25}$$

Where ΔH_1 and ΔH_2 are the heat of reaction at temperatures T_1 and T_2, respectively and ΔC_p is the difference in the total heat capacities of the reactants and products taking part in the reaction. In case reactant(s) and/or product(s) undergo any transformation at T^t the above equation is modified as:

$$\begin{aligned} \Delta H_2 - \Delta H_1 = & \int_{T_1}^{T^t} \Delta C_p\ (T_1 \leftrightarrow T^t).d\mathrm{T} \pm L^t \\ & + \int_{T^t}^{T_2} \Delta C_p\ (T^t \leftrightarrow T_2).d\mathrm{T}\ \ (T_1 \leq T^t \leq T_2) \end{aligned} \tag{2.26}$$

2.8 The Second Law of Thermodynamics

The equality between mechanical work and energy produced (or heat lost) or vice versa is valid for mechanical systems, e.g. heat engines. But it is not useful in which a chemist or metallurgist is interested. The concept was first used to predict maximum efficiency of heat engines. Later on the same concept was used to predict the maximum yield in a chemical reaction at equilibrium. The central problem in chemistry is control and understanding of a chemical reaction (change). In a chemical change we have to consider:

(i) Whether two substances will react on mixing?
(ii) Whether the reaction will be accompanied by energy release, if at all it reacts?
(iii) At what composition(s) of reactant/product, will reaction cease and equilibrium be established, if it begins?
(iv) What is the rate of reaction?

Chemical thermodynamics is concerned with the first three questions. It does not predict the rate of a chemical reaction. It does however provide us with an understanding of the factors that determine whether a given mixture of substances has a spontaneous tendency to react to form another substance. It tells us the conditions that prevail when chemical changes no longer occur and chemical equilibrium has been established. No information on atomic/molecular level, capable of formulating necessary conditions nor sufficient conditions, nothing about the rate of reaction, no statement can be made about the actual yield obtainable under a given set of conditions only maximum yield may be predicated. In spite of these limitations chemical thermodynamics is quite useful.

The First Law of thermodynamics is concerned with the quantitative aspects of interconversion of energies. This law neither allows us to predict the direction of conversion nor the efficiency of conversion when heat energy is converted into mechanical energy. In the last half of the nineteenth century many scientists put their concentrated efforts to apply the first law of thermodynamics to the calculation of maximum work obtained from a perfect engine and prediction of feasibility of a reaction in a desired direction. These considerations led to the development of the Second Law of thermodynamics which has had a far reaching influence on the subsequent development of science and technology. The Carnot cycle made it possible to assess the efficiency of engines and it also showed that under normal conditions all the heat supplied to the system cannot be converted into work even by perfect engines. Theoretically speaking, a perfect heat engine would convert all the heat supplied into work, if the heat sink (lower temperature of the process) can be maintained at absolute zero of temperature. However, this is an impossible situation, specially, because any heat sink on receiving some heat cannot remain at absolute zero of temperature.

2.8.1 The Carnot Cycle

The Carnot cycle is the name for a series of four changes that a substance (in principle, any substance) may undergo. In the forward order these are isothermal expansion, adiabatic expansion, isothermal compression and adiabatic compression, in that order. At the end of this series of changes, the substance must return to its initial state.

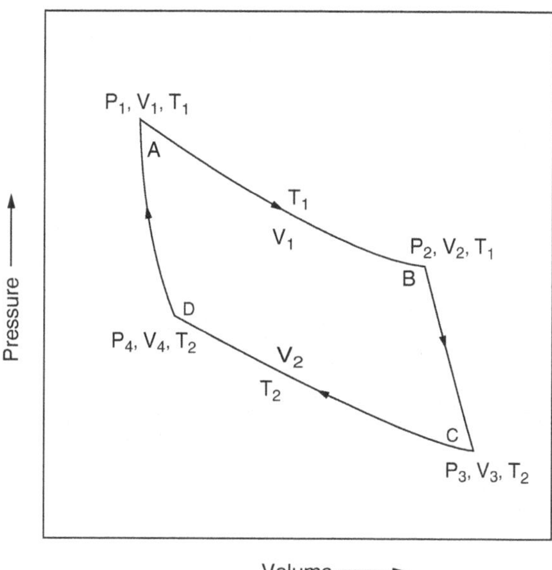

Fig. 2.2 Carnot's cycle

Carnot calculated the theoretical efficiency of such an ideal gas engine. For convenience he used in his cyclic process 1 g mole of a perfect gas as a working substance, since its behavior with changes of temperature and pressure is described by the ideal gas equation.

Let P_1 and V_1 are respectively, the initial pressure and volume of 1 g mole of the perfect gas at temperature T_1 (starting point A) in Fig. 2.2. It is possible to complete the full cycle of operations so as to bring the working substance back to its original condition at A in the following stages:

Stage I: From point A the gas is allowed to expand isothermally and reversibly to B where it occupies volume V_2 and attains pressure P_2. Since the internal energy of the given mass of gas remains constant at constant temperature, the gas must have gained heat q_1 from the surroundings, which must be equal to the work done by the gas, w_1.

The work done under isothermal conditions,

$$w_1 = RT_1 \ln \frac{V_2}{V_1} = q_1 \tag{2.27}$$

Stage II: The gas is then allowed to expand under adiabatic conditions to the point C to occupy volume V_3 and attain pressure P_3. The temperature falls to T_2 due to decrease in kinetic energy of gaseous molecules. Since heat cannot enter or leave the system, heat gained during the process, $q = 0$ and the work w_2 is performed at the expense of the internal energy of the system which is related to the molar heat capacity at constant volume, C_v and may be expressed as:

$$w_2 = C_v(T_1 - T_2) \tag{2.28}$$

Stage III: The gas is then compressed isothermally to D at temperature T_2, until its volume becomes V_4 and pressure P_4. In this process work is done on the gas, that is converted into heat energy q_2, which is lost to the surroundings. As the process is isothermal:

$$w_3 = RT_2 \ln \frac{V_4}{V_3} = q_2 \tag{2.29}$$

Since $V_3 > V_4$, w_3 will be negative because work is done on the gas.

Stage IV: Finally the gas is compressed under adiabatic condition until it reaches its original temperature and volume, i.e. the point A, $q = 0$ (since heat is neither gained or lost by the gas)

$$\text{Work done by the gas} = w_4 = C_v(T_2 - T_1) \tag{2.30}$$

Total work done by this cyclic process: $w = w_1 + w_2 + w_3 + w_4$

$$\begin{aligned} w &= RT_1 \; ln \frac{V_2}{V_1} + C_v(T_1 - T_2) + RT_2 \; \ln \frac{V_4}{V_3} + C_v(T_2 - T_1) \\ &= RT_1 \; ln \frac{V_2}{V_1} + RT_2 \; \ln \frac{V_4}{V_3} \end{aligned} \tag{2.31}$$

Note V_2 and V_3 are end points of the adiabatic process in stage II, whereas V_4 and V_1 are initial and final volume in stage IV of the adiabatic process. According to Eq. (2.16) the volume and temperature at any point in an adiabatic process are related as

$$T_1 V_2^{\gamma - 1} = T_2 V_3^{\gamma - 1} \tag{2.32}$$

$$T_1 V_1^{\gamma - 1} = T_2 V_4^{\gamma - 1} \tag{2.33}$$

From Eqs. (2.32 and 2.33) we get:

$$\frac{V_2}{V_1} = \frac{V_3}{V_4}$$

$$\begin{aligned} \text{Hence} \quad w &= RT_1 \; ln \frac{V_2}{V_1} + RT_2 \ln \frac{V_4}{V_3} \\ &= RT_1 \; ln \frac{V_2}{V_1} - RT_2 \ln \frac{V_2}{V_1} \\ &= R \; (T_1 - T_2) \; ln \frac{V_2}{V_1} \end{aligned} \tag{2.34}$$

Where w is the total work done by the gas in a cycle.

In practice however, we are interested to know how much heat can be converted into useful work i.e. the efficiency of the process:

Efficiency = work done/heat supplied at the higher temperature

$$= \frac{w}{q_1} = \frac{R\,(T_1 - T_2)\quad ln\,\frac{V_2}{V_1}}{RT_1\quad ln\,\frac{V_2}{V_1}} = \frac{(T_1 - T_2)}{T_1} \tag{2.35}$$

From the above equation it is concluded that the efficiency of a perfect engine depends on the temperature difference between the two isothermal stages of the process. Thus there is an ideal upper limit to the quantity of work that can be obtained by a given input of heat energy. This establishes that under normal conditions all the heat supplied to the system cannot be converted into work even by perfect engines. In order to convert all the heat supplied into work by a perfect engine the lower temperature of the process must be made equal to zero on the absolute temperature scale, a condition which can never be fulfilled. This proof of practical impossibility of complete conversion of heat energy into mechanical work/ energy, was the starting point of the Second Law of Thermodyanics which is stated as:

Energy is transformed from a higher to a lower energy level. Although the mechanical energy in any process can be completely converted into heat energy, the heat energy can never be completely transformed into mechanical energy; unless the process is reversible and the part of the process is at the lowest temperature, i.e., at absolute zero (a condition never fulfilled in practice).

Let us see whether the heat absorbed at the higher temperature q_1 is equal to the heat evolved q_2 at the lower temperature.

From Eqs. 2.27, 2.28, 2.29, and 2.31 quantity of heat converted into work, w may be expressed as

$$w = q_1 + q_2$$

From Eq. (2.35) $\frac{w}{q_1} = \frac{(T_1 - T_2)}{T_1}$

$$\therefore \frac{w}{q_1} = \frac{q_1 + q_2}{q_1} = \frac{(T_1 - T_2)}{T_1}$$

$$\text{or}\quad 1 + \frac{q_2}{q_1} = 1 - \frac{T_2}{T_1}$$

$$\text{or}\quad \frac{q_2}{q_1} = -\frac{T_2}{T_1}$$

$$\text{or}\quad \frac{q_1}{T_1} + \frac{q_2}{T_2} = 0 \tag{2.36}$$

Since $T_1 > T_2$ the heats absorbed at various temperatures are not numerically equal. But the sum of the ratios of heats absorbed reversibly to temperature at the different temperatures is equal to zero.

Therefore it can be generalized as

$$\sum \frac{q_{rev}}{T} = 0 \tag{2.37}$$

This equation is true for any working substance whether solid, liquid or gas. According to Sadi Carnot [6] every perfect gas engine working reversibly between the same temperature limits has the same efficiency whatever may be the working substance.

2.9 Entropy

In Section 2.8.1, we have noticed that the heat absorbed or evolved during the isothermal steps in Carnot's cycle (Fig. 2.2) depends on the temperature at which the steps occurred and the numerical values of the ratios $\frac{q_{rev}}{T}$ were the same (e.g., $\frac{q_1}{T_1} = -\frac{q_2}{T_2}$ *etc.*). The sums of these terms for a complete cyclic process irrespective of the number of stages is zero. Clausius [7] was first to recognize the fact that the ratios $\frac{q_{rev}}{T}$ was characteristic (state property) of the system since this varies from temperature to temperature. Thomson [8] gave the standard notation of $\sum \frac{q_{rev}}{T} = 0$, for a number of steps involved in a cyclic process. Finally, Clausius called the ratio δq/T as entropy change and denoted this change by the symbol ΔS and considered it a state function. It is an extensive property of the system as it depends on the mass of the system, and is a thermodynamic variable. Since entropy = energy/ temperature, its unit would be: cal $\deg^{-1}$ mol^{-1} (e.u.) or J $\deg^{-1}$ mol^{-1}. As the system absorbs heat, its entropy increases, for example, during melting and boiling. Entropy depends on the state of a substance or system and not on its previous history, irrespective of whether the path is thermodynamically reversible or not and it is also independent of the substance involved.

2.9.1 Entropy Is a State Property

Consider a system which absorbs an infinitesimal quantity of heat under reversible conditions. According to Clausius the entropy change of the system will be given as $dS = \frac{dq_{rev}}{T}$. From Eq. 2.8 we have $dq = dU + PdV$ ($dU = C_v dT$ *and* $P = RT/V$)

$$\therefore\ dq = C_v dT + RT\frac{dV}{V}$$

$$\text{or}\quad \frac{dq}{T} = \frac{C_v dT}{T} + R\frac{dV}{V} \tag{2.38}$$

Since C_v *and* R are constants the right hand side of the equation is an exact differential, it can be integrated. Therefore, the figure $\frac{dq}{T}$ on the left hand side of the Eq. 2.38 is also an exact differential, it will depend only on state of the system and not on the path through which it is attained.

2.9.2 Entropy Change in Gases

Since $dS = \frac{dq}{T}$, from Eq. 2.38,

$$dS = C_V\frac{dT}{T} + R\frac{dV}{V} \tag{2.39}$$

The entropy change for an ideal gas can be calculated from Eq. 2.39. Consider an ideal gaseous system (Fig. 2.3) with initial entropy S_1 and volume V_1 at pressure P_1, and temperature T_1. After undergoing a change of temperature from T_1 to T_2 the system attains final values of entropy, pressure and volume as S_2, P_2, and V_2, respectively.

The total entropy change of the above gaseous system, $\Delta S = (S_1 - S_2)$ can be estimated by integrating Eq. 2.39.

$$\begin{aligned}\Delta S &= (S_2 - S_1) = C_V \int_{T_1}^{T_2}\frac{dT}{T} + R\int_{V_1}^{V_2}\frac{dV}{V} \\ &= C_V\, ln\,\frac{T_2}{T_1} + R\, ln\,\frac{V_2}{V_1}\end{aligned} \tag{2.40}$$

Thus entropy change of an ideal gas can be calculated on heating provided C_V is known. Equation 2.40 can be used to calculate the entropy changes at constant temperature and pressure. Since $P_1V_1 = RT_1$ and $P_2V_2 = RT_2$.

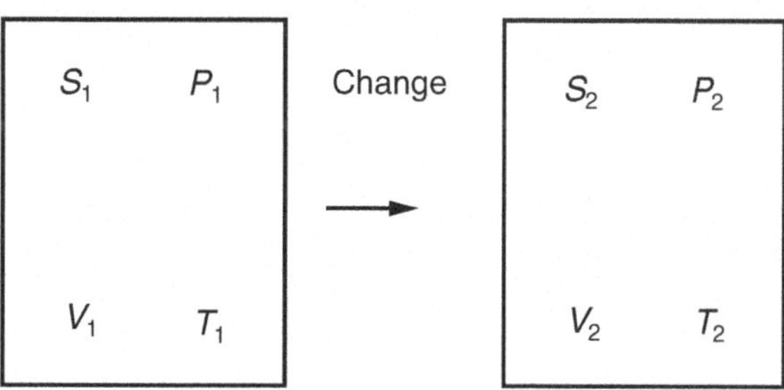

Fig. 2.3 Gaseous system undergoing changes of pressure, volume, temperature and entropy

$$\therefore \frac{V_2}{V_1} = \frac{T_2 P_1}{T_1 P_2} \tag{2.41}$$

From Eqs. 2.40, and 2.41, we get:

$$\begin{aligned} \Delta S &= C_V \ln(T_2/T_1) + R\ ln\ \ (T_2 P_1/T_1 P_2) \\ &= C_V \ln(T_2/T_1) + R\ ln\ (T_2/T_1) + R\ ln\ \ (P_1/P_2) \\ &= (C_V + R)\ \ ln\ (T_2/T_1) + R\ ln\ \ (P_1/P_2) \end{aligned}$$

$$\therefore \Delta S = C_P\ ln\ (T_2/T_1) + R\ ln\ \ (P_1/P_2) \tag{2.42}$$

Entropy change under isothermal conditions, i.e. at constant temperature: $T = T_1 = T_2$

$$\Delta S_T = R\ \ ln\ \ (P_1/P_2) = -R\ \ \ln\ \ (V_1/V_2) \tag{2.43}$$

For an isobaric process, $P_1 = P_2 = P$

$$\Delta S_P = C_P\ ln\ (T_2/T_1) \tag{2.44}$$

Since $dq = 0$, in an adiabatic process, $dS = 0$.

2.9.3 Calculation of Entropy Change from Heat Capacities

Clausius has mathematically defined entropy change as $\Delta S = q_{rev}/T$. Since heat capacity C_p has exactly the same units as entropy, that is, cal deg^{-1} mol^{-1}, for a limiting case of an infinitesimal change in the process, the entropy change can be expressed by the equation:

$$\frac{dq_{rev}}{dT} = C_p \tag{2.45}$$

and entropy change of an element

$$dS = \frac{dq_{rev}}{T} \tag{2.46}$$

From Eqs. 2.45, and 2.46 we get:

$$dS = \frac{C_p.dT}{T} \tag{2.47}$$

Equation 2.47 is a general differential equation for change in entropy and if we assume that entropy is zero at absolute zero, then the entropy of a substance at temperature T can be calculated according to the following equations:

$$S_T = S_0 + \int_0^T \frac{C_p.dT}{T} \tag{2.48}$$

$$\Delta S = S_T - S_0 = \int_0^T \frac{C_p.dT}{T} \tag{2.49}$$

Equation 2.49 is applicable in cases where there is no transformation from 0 to T K. In case of solid-solid, solid-liquid, and liquid-gas transformations (between 0 and T K), at T^t, T^f and T^e respectively, with the corresponding latent heat of transformation (L^t), heat of fusion (L^f), and heat of evaporation (L^e) Eq. 1.49 is modified as:

$$\Delta S = \int_0^{T^t} \frac{C_{p1}}{T}.dT + \frac{L^t}{T^t}. + \int_{T^t}^{T^f} \frac{C_{p2}}{T}.dT + \frac{L^f}{T^f} + \int_{T^f}^{T^e} \frac{C_{p3}}{T}.dT + \frac{L^e}{T^e} + \int_{T^e}^{T} \frac{C_{p4}}{T}.dT \tag{2.50}$$

Where $\frac{L^t}{T^t} = \Delta S^t$, $\frac{L^f}{T^f} = \Delta S^f$, $\frac{L^e}{T^e} = \Delta S^e$ and C_{p1}, C_{p2}, C_{p3} and C_{p4}, are respectively, the heat capacity of the substance in the temperature ranges ($0-T^t$), ($T^t - T^f$), ($T^f - T^e$) and ($T^e - T$).

2.9.4 *Entropy and Thermodynamic Absolute Temperature Scale*

This absolute temperature scale is based on the knowledge of coefficient of expansion of the perfect gas. Kelvin 1852 suggested that the Eq. 2.36 obtained by Carnot from his cycle can be used for estimating the absolute zero:

$$\frac{q_1}{T_1} + \frac{q_2}{T_2} = 0$$

Where q_1 and q_2 are reversible heats absorbed at temperature T_l and T_2 respectively. Kelvin defined the temperature intervals between the freezing point of water at 1 atm pressure and the normal boiling point of water to be equal to 100°.

Since $T_l = T_2 + 100$,

$$\frac{q_1}{T_2 + 100} + \frac{q_2}{T_2} = 0 \tag{2.51}$$

Where q_2 and q_1 are respectively the heat absorbed at the freezing point and at the boiling point of water. On rearranging Eq. 2.51:

$$-\frac{q_1}{q_2} = \frac{T_2 + 100}{T_2}$$

$-\frac{q_1}{q_2}$ was found to be 1.3661 by subjecting a body to a Carnot cycle between boiling water at 1 atm and freezing point of water at 1 atm (q_2 the heat given up to the ice).

$$\therefore \; T_2 = -273.15$$

This temperature is absolute since it is independent of the state of particular substances.

2.9.5 Entropy, Randomness and the Third Law of Thermodynamics

In Section 2.8.1 on Carnot cycle, from Eq. 2.36 we know that for a complete operation of a reversible heat engine: $\frac{q_1}{T_1} = -\frac{q_2}{T_2}$ where q_1 *and* q_2 are respectively, the quantity of heat given to the system under isothermal conditions at temperature T_1 and heat transfer out of the system at temperature T_2. Both the operations are performed under reversible conditions. It is to be noted that $\frac{q_1}{T_1}$ *and* $\frac{q_2}{T_2}$ are respectively, the integral of $\frac{dq_{rev}}{T}$ in the stage I and stage III of the cycle. Thus these values refer to entropy changes of the system in the first and third stages of a closed system. However, there is no entropy changes in stage II and IV of the cycle because they are adiabatic ($q = 0$) and reversible. The entropy change in the first and the third stages of the cycle are equal and opposite in sign. Thus the entropy change for the complete cycle is zero. This means the reversible engine takes up heat from the heat source, performs work and discards unutilized heat on completing the cycle of operation and there is no change in entropy.

The significance of the entropy change of a closed system can be demonstrated by considering the system (Fig. 2.4) having two identical copper vessels full of water, maintained at different temperatures, vessel 1 at T_1 and vessel 2 at T_2 ($T_1 > T_2$). The vessels are fully insulated and touch each other. Let us assume that the amount of heat flowing from higher to lower temperature is q cal. In this process vessel 1 loses heat and the other one, 2 gains. The temperature difference between the two vessels ($T_1 - T_2$) is not only a measure of the irreversibility of the process but also indicates about the amount of heat q that can be transferred from vessel 1 to vessel 2, is proportional to ($T_1 - T_2$). Using Clausius definition $\Delta S = {q_{rev}}/{T}$ we can express the entropy changes in the two vessels as:

Entropy change of the vessel 1, $\Delta S_1 = \frac{-q}{T_1}$ (q is negative since vessel 1 is releasing heat)

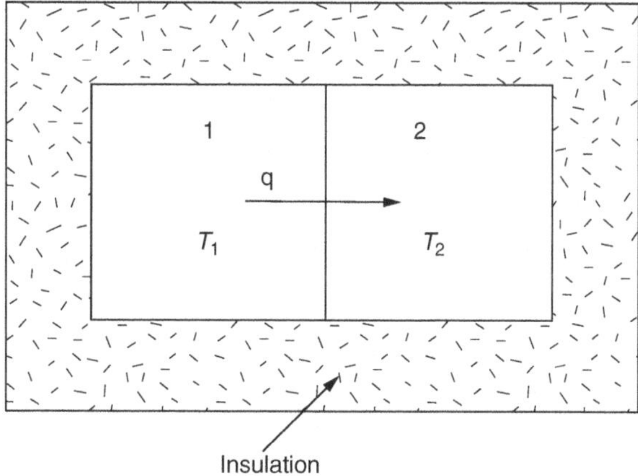

Fig 2.4 Second law of thermodynamics

Entropy change of the vessel 2, $\Delta S_2 = \frac{q}{T_1}$ (q is positive since vessel 2 is absorbing heat)

The total entropy change, i.e. the entropy change of the process = sum of the entropy changes of the two vessels:

$$\Delta S_{process} = \Delta S_1 + \Delta S_2 = \left(\frac{q}{T_2} - \frac{q}{T_1}\right) \tag{2.52}$$

$$\Delta S_{process} = \frac{q\,(T_1 - T_2)}{T_1 T_2} \tag{2.53}$$

Equation 2.53 allows us to draw the following important conclusions regarding the heat flow and entropy changes:

(1) Heat flows from higher temperature to lower temperature with entropy increase of the system, i.e., process is spontaneous, $\Delta S_{irr} > 0$ ($T_1 > T_2$).
(2) There is no heat transfer, i.e. a dynamic equilibrium exists, $\Delta S_{rev} = 0$ ($T_1 = T_2$).
(3) When $T_1 < T_2$, $\Delta S_{irr} < 0$, process is not feasible.

The above conclusions deduced from simple considerations of heat flow in a closed system of two vessels can be extended to the transfer of any other form of energy. Thus it follows that the entropy change of a closed system will be zero for a process at equilibrium, positive if the process proceeds in the desired direction, and negative in case of non-feasible process. Therefore, the Second Law of thermodynamics can be stated as "All spontaneous processes are accompanied by increase of entropy." This can be mathematically demonstrated by considering two copper billets (or any metallic billets) of the same weight and identical shape, placed in a furnace at different temperatures 400° and 200 °C. Assuming that there is no heat transfer from the billets to the materials of construction of the furnace, the heat content of copper billets can be calculated from the knowledge of the heat capacity value of copper, $C_p = 5.41 + 1.5 \times 10^{-3} T$ cal deg mol^{-1} [5]. Suppose both the billets after heat transfer attain equilibrium at 300 °C, the heat contents of the billets are given by the following expression (Eq 2.49):

$$\Delta S = \int_{T_1}^{T_2} \frac{C_p}{T} dT = \int_{T_1}^{T_2} \left(\frac{5.41}{T} + 1.5 \times 10^{-3} \right) dT$$

The entropy change from the hot billet:

$$\Delta S_h = \int_{673}^{573} \left(\frac{5.41}{T} + 1.5 \times 10^{-3} \right) dT = -0.921 \text{ cal deg mol}^{-1}$$

The entropy change from the cold billet:

$$\Delta S_c = \int_{473}^{573} \left(\frac{5.41}{T} + 1.5 \times 10^{-3} \right) dT = +1.188 \text{ cal deg mol}^{-1}$$

The total entropy change of the system and surroundings is given as

$$\Delta S_h + \Delta S_c = +0.267 \text{ cal deg mol}^{-1}$$

Since the entropy change is positive, the process of heat flow from the hot billet to the cold billet is spontaneous according to the Second Law of Thermodynamics.

Contrary to the above case, if the hot billet were to gain heat from the cold billet so that its temperature increases to 600 °C from 400 °C and that of the cold billet decreases from 200 °C to 0 °C, the total entropy change would be:

$$\begin{aligned} \Delta S_h + \Delta S_c &= \int_{673}^{873} \left(\frac{5.41}{T} + 1.5 \times 10^{-3} \right) dT + \int_{473}^{273} \left(\frac{5.41}{T} + 1.5 \times 10^{-3} \right) dT \\ &= 1.708 – 3.273 = -1.265 \text{ cal deg mol}^{-1} \end{aligned}$$

Since the entropy change is negative, total entropy of the system and the surroundings decreases, the process is not spontaneous. It is a matter of experience that heat does not flow from a colder body to a hotter body.

The entropy change for one molecule of a perfect gas under isothermal expansion (dT = 0) is given as $\Delta S_T = R \ln (V_2/V_1)$, where V_1 and V_2 are the initial and final volumes of the perfect gas. Since $V_2 > V_1$, ΔS_T is positive. Conversely it must have a negative value in compression when $V_2 < V_1$. The equation indicates that the expansion of gas is a natural spontaneous process because on expansion entropy increases and system becomes more disordered.

Imagine certain quantity of a gas contained in a container, is connected to an evacuated container by a tap [9]. On opening the tap the gas molecules will enter in the space available in the evacuated container due to their random motion until the number of molecules become the equal in both the containers. We know that the gas molecules on their own accord will never enter in the first container leaving the second container empty. In this way we see that in the second case which is more disordered is more probable. Thus a spontaneous process has taken place by increase in probability [9] and therefore with an increase in disorder of the system.

In another example, suppose there are number black and white balls placed on a flat table in a well arranged manner, i.e., the ordered state. Once all these balls are picked up and thrown on the table we find them placed in a disturbed manner, i.e., the disordered state. The second state is more probable than the first state. Thus a more disordered system is more probable than the ordered one and changes of the latter type is accompanied by an increase in disorder of the system.

Statistical mechanic principles are useful in establishing a relationship between entropy and thermodynamic probability. According to this theory a system undergoing a spontaneous process changes from a state of lesser to a state of greater probability. This change in the state of order is expressed as a change in property of the system, that is, entropy S. Thus entropy is a measure of disorder. Although entropy is associated with disorder, it is more accurately described as a measure of the number of microstates available to a system given its macroscopic properties. The entropy is related to the thermodynamic probability by the following expression:

$$S = k\,ln\,W + S_o \tag{2.54}$$

Where k is Boltzmann's constant and W the thermodynamic probability of the system, is defined as the ratio of the probability of the existing state of the system to the probability of the state of complete order of the system for the same energy and volume. The more randomly (orderly) the molecules are arranged, the larger (smaller) are W and S. If the system is perfectly ordered, $W = 1$ and $S = 0$.

The Third Law of Thermodynamics states that that the entropy of a pure crystalline substance approaches zero as the temperature approaches absolute zero. Thus a perfectly crystalline solid is supposed to be completely ordered at the absolute zero of temperature and should therefore, have zero entropy. In Eq. 2.54, the Third Law assumes a value of zero for S_o in perfectly ordered state, $W = 1$ and $S = S_o = 0$. Thus Eq. 2.54 is modified as:

$$S = k\,ln\,W \tag{2.55}$$

Thus the absolute value of entropy can be measured but we can only accurately measure the changes in internal energy and enthalpy, not the absolute value. In calculations we assume enthalpy of elements to be zero in a defined standard state but no atom or molecule can have zero heat/enthalpy at ambient conditions. Contrary to this a completely ordered (crystalline) solid will have zero entropy at temperature of absolute zero. From the above equation, it is evident that as the system becomes more disordered both *W and S* increase. The entropy of the system can be used as a measure of disorder, or probability of the system. For details readers are advised to refer books [10, 11]. Lewis and Randall [3] proposed the following statement of the third law: "If the entropy of each element in some crystalline state be taken as zero at absolute zero of temperature, every substance has a finite entropy; but at the absolute zero of temperature the entropy may become zero, and does so become in the case of perfectly crystalline substances."

2.9.5.1 Experimental Verification of the Third Law of Thermodynamics

The third law of thermodynamics is based upon experiments. It is not like the second law based on common experience that heat always flows from a hot body to a cold body. Since the third law is not so obvious it needs some convincing evidence for its acceptance. As it is not possible to carry out some experiment to measure the entropy change at absolute zero the verification of the third law involves the application of the second law making use of the principle that the net entropy change in a system undergoing a cyclic process is zero.

A cyclic process depicting the allotropic transformation of tin has been schematically presented in Fig. 2.5. The two crystallographic forms of tin, gray and white tin are respectively, stable below and above 19 °C (292 K) with the latent heat of transformation of 541 cal deg^{-1} g $atom^{-1}$[12]. White tin can be readily cooled below 19 °C without any transformation. The heat capacity of both the forms can be measured from low temperature to the transformation temperature. The starting point of the cycle is the system composed of gray tin at the absolute zero. In the first step of the cycle the gray tin is heated to the transformation temperature 292 K, where it is in equilibrium with the product white tin. In the second step the transformation is allowed to proceed to completion at 292 K. The separated product is cooled to the absolute zero in the third step. In the fourth step (imaginary) the reverse reaction is allowed to proceed.

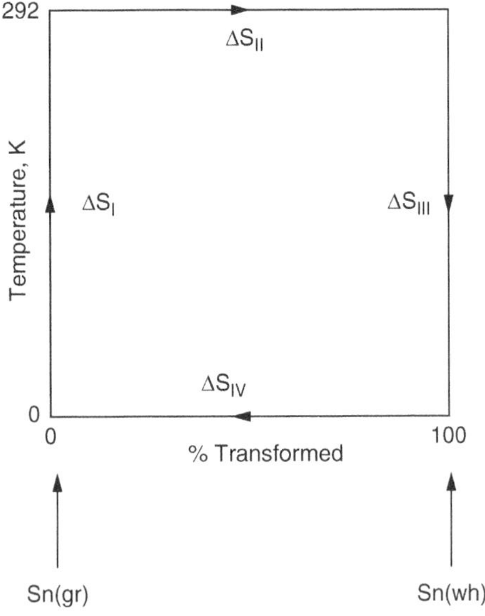

Fig. 2.5 A schematic diagram illustrating evaluation of the entropy change of a reaction at absolute zero (i.e., experimental verification of the third law of thermodynamics)

For the transformation reaction under consideration: Sn (gray) = Sn (white), according to the second law of thermodynamics for a cyclic process:

$$\Delta S_I + \Delta S_{II} + \Delta S_{III} + \Delta S_{IV} = 0$$

Where ΔS_I, ΔS_{II}, ΔS_{III}, *and* ΔS_{IV} respectively, designate the entropy change during each step of Fig. 2.5 by the corresponding subscript. Thus the entropy change at the absolute zero is obtained as:

$$\begin{aligned}\Delta S_o = \Delta S_{iv} &= -(\Delta S_I + \Delta S_{II} + \Delta S_{III}) \\ &= -\left[\int_0^{292} C_{p(gr)} d\ \ln T + \frac{q_r}{T}\left(=\frac{541}{292}\right) + \int_{292}^{0} C_{p(wh)} d\ \ln T\right] \\ &= -(9.11 + 1.85 - 11.04) \\ &= 0.08\ \text{eu}\ \left(\text{cal deg}^{-1}\ \text{g atom}^{-1}\right)\end{aligned}$$

Thus entropy change at absolute zero is zero within the experimental error.

A similar experimental verification can be demonstrated by considering the cyclic process in case of sulfur which has two allotropes, an orthorhombic form (stable below 95.5 °C) and a monoclinic form with a latent heat of transformation of 95.602 cal deg^{-1} g atom^{-1} at 95.5 °C.

2.9.6 Trouton's Rule

The rule [5] states that the ratio of the latent heat of evaporation, L^e to the normal boiling point, T^e is constant. The ratio known as the entropy of evaporation ($\Delta S^e = \frac{L^e}{T^e}$) is approximately 22.4 cal deg^{-1} mol^{-1} for a large number of substances (metals as well as salts) as depicted in Table 2.1.

Trouton's Rule fails for associated liquid e.g., water, alcohols etc. Based on the value of ΔS^e we can conversely estimate whether there is any association or polymerization in the liquid. In other words we can estimate the liquid structure from the entropy value.

2.9.7 Richard's Rule

The ratio of the latent heat of fusion, L^f to the normal melting point, T^f known as the entropy of fusion is not as constant as that of entropy of evaporation [5]. The change in state of order on melting is not as large as on evaporation. The state of order of a solid due to variety of the possible binding forces, has therefore a proportionately large effect on the entropy of fusion. Crompton and Richards [5] were the first to point that the entropy of fusion of metals should be nearly constant.

Table 2.1 Entropy of evaporation of some substances

Substance	Boiling point, K	Latent heat of evaporation L^e cal mol^{-1}	$\Delta S^e = \frac{L^e}{T^e}$
Hg	630	14100	22.4
Cd	1038	23900	22.8
Zn	1180	27300	24.4
Mg	1378	30500	22.2
Bi	1830	41100	22.5
Ag	2420	61600	25.4
Fe	3343	81300	24.2
KCl	1680	39000	23.2
NaCl	1738	40700	23.3
AgCl	1837	42520	23.2
$FeCl_2$	1299	30200	23.2
$MgCl_2$	1691	32700	19.3
$BiCl_3$	714	17300	24.2

The larger value of entropy of evaporation ($\Delta S^e = \frac{L^e}{T^e}$) approximately 22.4 cal deg^{-1} mol^{-1} as compared to the entropy of fusion ($\Delta S^e = \frac{L^e}{T^e} \sim 2$ cal deg^{-1} mol^{-1}) indicates that the change in state of order in liquid to gaseous transformation is much larger and the gaseous state is more disordered.

2.10 Feasibility of a Reaction

In daily life we observe objects falling from roof top to the ground because its potential energy is lowered in falling. An electric current flows along a wire because it is driven by electric energy. Similarly one can find many other examples establishing the fact that physical changes take place due to the loss of energy. These physical phenomena encourage us to think over as how does a chemical reaction take place? The search for reasons behind chemical reactions compels one to believe whether enthalpy change at constant pressure (ΔH) or change in intrinsic energy at constant volume (ΔU) accompanying a reaction is responsible for the feasibility of a reaction. Assuming that these would be a measure of the driving force behind chemical reactions, Berthelot [13] carried out successful determinations of enthalpy changes in the nineteenth century. Based on this argument it is concluded that if a system losses energy as a result of a chemical reaction, the exothermic reaction will take place spontaneously. And the greater the quantity of heat lost, the greater would be the driving force behind the reaction.

For example, carbon burns in air spontaneously with evolution of heat energy according to the reaction:

$$C\ (s) + O_2\ (g) = CO_2\ (g) + 94050\ \text{cal mol}^{-1} \tag{2.56}$$

$\Delta H^o_{298} = -94050$ cal mol^{-1}, is the driving force in this case.

The following two reactions take place spontaneously because both have negative heats of reaction at 1600 °C

$$2\ Fe\ (s) + O_2\ (g) = 2FeO\ (s) \tag{2.57}$$

$$3FeO\ (s) + 2\ Al\ (l) = 3Fe\ (l) + Al_2O_3(s) \tag{2.58}$$

On the other hand, if we consider the reduction of ZnO with carbon according to the reaction:

$$ZnO\ (s) + C\ (s) = Zn\ (g) + CO\ (g) \tag{2.59}$$

ΔH^o for the reaction at 1100 °C is 83295 cal mol^{-1} compared to 56812 cal mol^{-1} at 25 °C. The reaction will not take place at 25 °C but if the system is heated to 1100 °C' carbon will reduce zinc oxide to produce zinc metal.

Since endothermic reactions do take place spontaneously in addition to exothermic reactions, one cannot use heats of reaction as a criterion for feasibility of the reaction. This encourages us to look for a more consistent rule [9] for the driving force of a reaction. In order to sort out this problem let us analyze the above reaction at two different temperatures which produce in gaseous and solid zinc.

$$\text{At 1100 °C}, ZnO\ (s) + C\ (s) = Zn\ (g) + CO\ (g)$$

$$\text{At 25 °C}, \quad ZnO\ (s) + C\ (s) = Zn\ (s) + CO\ (g) \tag{2.60}$$

At 1100 °C (i.e., above the boiling point of Zn, 907 °C), the reaction produces gaseous zinc whereas at 25 °C, zinc is a solid (melting point 419.6 °C). Thus these reactions do not only differ in their temperature of operation but also differ in the physical state, that is, a difference in the state of order of a system and consequently in the entropy of the system. Since two molecules of gas are being produced at 1100 °C from two solid molecules, ΔS^o at 1100 °C should be much more than that at 25 °C because two solid molecules produce only one gaseous mole at 25 °C.

A solid to gas transformation takes place with a comparatively larger entropy increase which is evident from the following figures for the reduction of ZnO by C:

$$\Delta S^o_{298} = 46.1\ \text{cal deg}^{-1}\ \text{mol}^{-1}\ \text{and}\ \Delta S^o_{1373} = 68.12\ \text{cal deg}^{-1}\ \text{mol}^{-1}$$

The idea of using a positive entropy change of a reaction as a measure of the driving force behind a chemical reaction is immediately contradicted while considering the formation of the oxide of any metal:

$$Fe\ (s) + ½\ O_2\ (g) = FeO\ (s),\quad \Delta S^0_{298} = -16.97\ \text{cal deg}^{-1}\ \text{mol}^{-1} \qquad (2.61)$$

This reaction is spontaneous because FeO film is readily formed on iron at room temperature. Hence a positive entropy change is not the criterion of a chemical reaction.

The Second Law of thermodynamics states that a spontaneous process is always accompanied by an increase in entropy of the system and its surroundings. In the above case, the system consisting of Fe, O, and FeO, undergoes a loss of entropy but the surroundings gains heat by the heat evolved due the exothermic reaction, and will therefore undergo an increase in entropy.

If enthalpy change of the system at constant temperature and pressure is ΔH, the surroundings will receive a quantity of heat, and hence the enthalpy change of the surroundings = −ΔH.

Therefore, the entropy increase of the surroundings = −ΔH/T.

If ΔS is the entropy change of the system, total entropy change of the system and surroundings would be ΔS−ΔH/T.

According to the Second Law (ΔS−ΔH/T) must be positive for any chemical reaction to take place spontaneously. Since T is positive (ΔH−T ΔS) must always be negative in order that the total entropy change of the system and surroundings can be positive, when the reaction proceeds.

Let us examine the factor (ΔH−T ΔS) for the above three reactions under consideration:

For the reduction of ZnO by C at 25 °C.

$\Delta H^o_{298} - T\Delta S^o_{298} = 56892 - 298 \times 46.14 = 43080\ \text{cal mol}^{-1}$, the reaction cannot proceed because this factor is positive.

$$\text{At } 1100\ °\text{C},\ \Delta H^o_{1373} - T\Delta S^o_{1373} = 83295 - 1373 \times 68.12 = -10230\ \text{cal mol}^{-1},$$

Since this is negative the reaction will proceed spontaneously at 1100 °C. This explains why reduction of zinc oxide by carbon must be carried out at temperature of the order of 1100 °C.

For the oxidation of iron at 25 °C

$$\Delta H^o_{298} - T\Delta S^o_{298} = -63193 - 298 \times (-16.91) = -58154\ \ \text{cal mol}^{-1},$$

Again it is negative, and the reaction takes place spontaneously at 25 °C.

Consistent Rule [9]: The driving force of a reaction can be calculated as (ΔH−TΔS); the more negative this factor, the greater the driving force and if the factor is positive, the reaction will not proceed spontaneously.

2.10.1 Free Energy

The factor ($\Delta H - T\Delta S$) has dimensions of energy and it is known as the change in the "free energy" of the system. Free energy is a thermodynamic function of great significance. According to the first law of thermodynamics, the change in internal energy, ΔU for a system undergoing a thermodynamically reversible change at constant temperature and constant volume, is expressed as $\Delta U = q_{rev} - w$, where q_{rev} is the heat absorbed reversibly by the system at temperature T and w is the maximum work done by the system.

Since $\Delta S = \frac{q_r}{T}$

$$\Delta U = T\Delta S - w \tag{2.62}$$

$$\therefore -w = \Delta U - T\Delta S \tag{2.63}$$

In Eq. 2.63, $-w$ is the maximum work (whether mechanical or electrical) that can be obtained from the system. A thermodynamic function, A, the "work function" or Helmholtz free energy (after H. von Helmholtz) can be defined as:

$$-w = \Delta U - T\Delta S = \Delta A \tag{2.64}$$

A is a thermodynamic variable. It depends only on the state of the system, not on its history, because U, T, and S are all thermodynamic variables. The work done by the system at constant temperature and constant pressure ($P\Delta V$) as a result of a volume change is not a 'useful work'. The useful work will then be the maximum work, $-w$, less the energy lost due to volume change ($-P\Delta V$).

Hence, the 'useful work' is expressed as ΔG, the Gibbs free energy change of the system:

$$\Delta G = -w - (-P\Delta V) = \Delta A + P\Delta V \tag{2.65}$$

From Eqs. 2.64 and 2.65, we get:

$$\Delta G = \Delta U - T\Delta S + P\Delta V = (\Delta U + P\Delta V) - T\Delta S = \Delta H - T\Delta S \tag{2.66}$$

In Eq. 2.66, $\Delta G = \Delta H - T\Delta S$, G is the Gibbs free energy of the system (after J. Willard Gibbs), which depends only on the thermodynamic variables, H, T, and S. It is the maximum work available from a system at constant pressure other than that due to a volume change. In Section 2.10 the fundamental importance of the factor ($\Delta H - T\Delta S$), which is equal to ΔG, has been discussed. Hence, ΔG is a measure of feasibility of a chemical reaction. For a reaction to proceed spontaneously,the free energy change for the reaction must be negative; the more negative the value of ΔG, the greater will be the driving force.

2.10.2 Some More Thermodynamic Relationships

By definition $G = H - TS = U + PV - TS$.

On differentiation, we get:

$$dG = dU + PdV + VdP - TdS - SdT \tag{2.67}$$

For a reversible process involving work only due to expansion at constant pressure, according to the first law we know: $dU = dq - PdV$, and from the second law : $dq = TdS$.

$$\therefore\ dU = TdS - PdV \tag{2.68}$$

From Eqs. 2.67 and 2.68 we get:

$$dG = VdP - SdT \tag{2.69}$$

At constant pressure, $dP = 0$,

$$\therefore\ \left(\frac{\partial G}{\partial T}\right)_P = -S \tag{2.70}$$

At constant temperature, $dT = 0$, $dG = VdP$ and since for 1 mole of an ideal gas: $PV = RT$

$$\therefore\ dG = \frac{RT}{P}\,dP \tag{2.71}$$

If P_A and P_B respectively, denote the initial and final pressures of the system with corresponding free energies as G_A and G_B, integration of Eq. 2.71 between the limits P_A and P_B at constant temperature results:

$$\Delta G = G_B - G_A = RT\int_{P_A}^{P_B}\frac{dP}{P} = RT\ ln\,\frac{P_B}{P_A}$$

This corresponds to the maximum work done by a gaseous system at constant temperature when pressure changes from P_A to P_B, $-w = RT\ ln\,\frac{P_B}{P_A}$

$$\text{Since}\quad \Delta G = -w,\quad \therefore\ \Delta G = RT\ ln\,\frac{P_B}{P_A} = -RT\ ln\,\frac{P_A}{P_B} \tag{2.72}$$

When system undergoes a change at constant pressure, using Eq. 2.70, we can write:

$$dG_A = - \quad S_A.\mathrm{dT} \text{ and } dG_B = - \quad S_B.\mathrm{dT}$$

$$\text{Hence, } (G_B - G_A) = -(S_B - S_A)\, dT$$

$$\text{but } G_B - G_A = \Delta G \text{ and } S_B - S_A = \Delta S$$

$\therefore (\Delta G) = -\Delta S.\ dT$ and $\left(\frac{\partial \Delta G}{\partial T}\right)_P = -\Delta S$; on substitution in Eq. 2.66 we get:

$$\Delta G = \Delta H - T\Delta S = \Delta H + T\left(\frac{\partial \Delta G}{\partial T}\right)_{P,T} \tag{2.73}$$

Equation 2.73 is known as the Gibbs–Helmholtz equation.

Since free energy is an extensive thermodynamic quantity it can be added and subtracted in the same way as enthalpy changes. As values of ΔH and ΔS are usually tabulated at 298 K, ΔG_{298} can be calculated. ΔH and ΔS vary with temperature and this variation can be calculated from the following equations. Combining Eqs. 2.25, 2.49, and 2.66 we get:

$$\Delta G_T = \Delta H_{298} + \int_{298}^{T} \Delta C_P.dT - T\Delta S_{298} - T\int_{298}^{T} \frac{\Delta C_P}{T}\ .dT \tag{2.74}$$

Neglecting any heat of transformation this leads to a generalized formula of the type:

$$\Delta G_T = a + bT\, log\, T + cT^2 + eT^{-1} + fT \tag{2.75}$$

where a, b, c, e, and f are constants in ΔG versus T relationship of a particular system. However, experimental errors involved in the determination of the data do not often justify such complex formulae and normally two or three term formulae of the following types suffice:

$$\Delta G_T = a + bT \tag{2.76}$$

$$\Delta G_T = a + bT\, log\, T + cT \tag{2.77}$$

2.10.3 Maxwell's Relations

If x is a single-valued function of the variables y and z, e.g., a thermodynamic property of a closed system, the complete differential dx can be expressed as,

$$dx = M\, dy + N\, dz \tag{2.78}$$

where *M and N* are also functions of the variables. If z is constant, dz is zero, Eq. (1.78) gives

$$\left(\frac{\partial x}{\partial y}\right)_z = M \tag{2.79}$$

If y is constant, dy is zero,

$$\left(\frac{\partial x}{\partial z}\right)_y = N \tag{2.80}$$

On differentiating Eq. 2.79 with respect to z, keeping y constant, and 2.80 with respect to y, keeping z constant, we get:

$$\left(\frac{\partial M}{\partial z}\right)_y = \left(\frac{\partial N}{\partial y}\right)_z \tag{2.81}$$

The above result referred as the Euler criterion is also known as the reciprocity relationship. Eq. 2.81 can used to derive some useful thermodynamic relationships while applying to the expressions $dU = TdS - PdV$, $dH = TdS + PdV$, $dA = SdT - PdV$, and $dG = SdT + VdP$:

$$\left(\frac{\partial T}{\partial V}\right)_S = -\left(\frac{\partial P}{\partial S}\right)_V \tag{2.82}$$

$$\left(\frac{\partial T}{\partial P}\right)_S = \left(\frac{\partial V}{\partial S}\right)_P \tag{2.83}$$

$$\left(\frac{\partial S}{\partial V}\right)_T = \left(\frac{\partial P}{\partial T}\right)_V \tag{2.84}$$

$$\left(\frac{\partial S}{\partial P}\right)_T = -\left(\frac{\partial V}{\partial T}\right)_P \tag{2.85}$$

Equations 2.82, 2.83, 2.84, and 2.85 are known as Maxwell relations.

2.11 Thermodynamic Equilibrium

This topic is very important in understanding the usefulness of thermodynamics. Imagine puncturing of an inflated tube of an automobile vehicle in an open atmosphere. The pressurized air from the tube will leak out through the puncture like a jet till no extra air is left inside the tube.

What will happen if the same inflated tube is punctured in the chamber which has also been inflated to the same pressure? The tube will not get deflated. Since the pressure outside and the inside the tube are equal i.e. since mechanical potentials on either side of the tube are the same, there exist a pressure or 'mechanical equilibrium' in the system: the tube and its surrounding system.

When two objects at different temperature are physically brought together heat flows from the one at higher temperature to the other at the lower temperature. If the two were at the same temperature there is no heat flow. It is then said to be in 'thermal equilibrium'.

When a chemical reaction takes between two reactants in contact, if the rate of the forward reaction is equal to that of the backward reaction, it is a case of chemical equilibrium.

Complete thermodynamic equilibrium is thus that situation where the system in equilibrium with respect to all such potentials like mechanical, thermal, chemical etc. The term equilibrium is applied to a system not to a process.

A system in state of partial equilibrium means that the two phases involved in the system are in equilibrium with respect to any one of the parameters (thermal, chemical, pressure etc.) or a combination of these but not all of them at a time.

2.11.1 The Law of Mass Action and Eqilibrium Constant

The law of mass action proposed by Guldberg and Waage [14] states that the velocity of a reaction at a given temperature is proportional to the product of active masses of the reacting substances. In chemical reactions the reacting species are called the reactants whereas the resulting species products. In general, the extent to which reaction proceeds depends on temperature.

For example, if A and B as reactants give rise to the formation products C and D in a chemical reaction, the rate of the forward reaction, v_f is proportional to the product of active masses of A and B. Thus the rate of the forward reaction is given as:

$$v_f = k_1.\ [A][B]$$

Where $[A]$ and $[B]$ are respective concentrations of A and B in g moles per liter. Similarly the rate of the reverse reaction, v_r is

$$v_r = k_2.\ [C][D]$$

Where $[C]$ and $[D]$ are respective concentrations of C and D in g moles per liter. k_1 and k_2 are respectively, the rate constant for the forward and reverse reactions. At equilibrium the two rates become equal.

$$k_1.\ [A][B] = k_2.\ [C][D]$$

$$\therefore\ \frac{k_1}{k_2} = \frac{[C][D]}{[A][B]} = K$$

Where K is the equilibrium constant of the reaction at constant temperature. In general for any reversible reaction, $aA + bB \rightleftharpoons cC + dD$, the concentration equilibrium constant, K_c is expressed as:

$$K_c = \frac{[C]^c[D]^d}{[A]^a[B]^b} \tag{2.86}$$

The value of K_c depends on temperature. Three different equilibrium constants are in use depending on the way concentrations are expressed. These are namely (i) K_c the concentration equilibrium constant, (ii) K_p the pressure equilibrium constant, and (iii) K_n the molal equilibrium constant. All the three equilibrium constants may be used in gaseous reactions. The pressure equilibrium constant, K_p is expressed in terms of the partial pressure of the gaseous reactants and products. If a number of moles of gas A reacts with b number of moles of gas B to produce c and d moles of the product gases C and D, with their respective partial pressures $p_A, p_B, p_C\ and\ p_D$ at equilibrium K_p is related to the partial pressure according to the following expression, provided each gas obeys the ideal gas law:

$$K_p = \frac{p_C^c . p_D^d}{p_A^a . p_B^b} \tag{2.87}$$

A relation between K_c and K_p may developed for perfect gases making use of the equation of state: $pv = RT$ for 1 mole of gas, and noting that $C = 1/v$ (i.e. the number of moles per unit volume).

$$\therefore\ p \times \frac{1}{C} = RT \ \text{ Or } \ p = CRT$$

Where C is concentration of a gas in the gaseous mixture. The partial pressure of the individual gas in the mixture can be expressed as:

$$p_A = C_A RT, p_B = C_B RT, p_C = C_C RT, p_D = C_D RT$$

Substituting partial pressures terms in Eq. (2.87) we get:

$$K_p = \frac{(C_C RT)^c (C_D RT)^d}{(C_A RT)^a (C_B RT)^b} = \frac{C_C^c C_D^d}{C_A^a C_B^b} RT^{(c+d)-(a+b)} = \frac{C_C^c C_D^d}{C_A^a C_B^b} (RT)^n$$

$$\text{Where } n = (c + d) - (a + b)$$

$$\therefore \; K_p = K_c (RT)^n \tag{2.88}$$

From Eq. 2.88, it is clear that when the total number of moles of reactants and products are equal (i.e. $n = 0$), $K_p = K_c$.

If P is the total pressure of the gaseous mixture and N the total number of g moles of all the gases in the mixture the partial pressures p_A, p_B, p_C *and* p_D can be expressed as:

$$p_A = \frac{n_A}{N} P, \; p_B = \frac{n_B}{N} P, \; p_C = \frac{n_C}{N} P, \; p_D = \frac{n_D}{N} P$$

Where $N = n_A + n_B + n_C + n_D$ and n_A, n_B, n_C *and* n_D are number of moles of gases A, B, C, and D, respectively. Then for the reaction $aA + bB \rightleftharpoons cC + dD$,

$$K_p = \frac{p_C^c.\, p_D^d}{p_A^a.\, p_B^b} = \frac{n_C^c.\, n_D^d}{n_A^a.\, n_B^b} \cdot \frac{P^{(c+d)}}{P^{(a+b)}} = \frac{n_C^c.\, n_D^d}{n_A^a.\, n_B^b} . P^{(c+d) - (a+b)}$$

$$\therefore \; K_p = K_N \, P^n \tag{2.89}$$

Thus when number of molecules of reactants and products in a chemical reaction are equal $K_p = K_N = K_c$. The molal equilibrium constant for gases is the only constant [4] enables us to calculate the effect of pressure on the equilibrium at a constant temperature.

2.11.2 The van 't Hoff Isotherm

The van't Hoff isotherm corrects the Gibbs free energy change of a reaction at a particular temperature with the equilibrium constant. The expression relating ΔG and K, known as the van't Hoff isotherm is extremely useful to calculate the equilibrium constants when ΔG is known or ΔG from the equilibrium constant. Mackowiak [4] and Parker [9] may be referred for details on this topic. A brief discussion has been included in this section.

Consider a chemical reaction in a large reaction box where two ideal gases A and B are allowed to react at constant temperature to produce gases C and D according to the reaction: A + B = C + D. Let us assume that initial partial pressures, P_A and P_B of one mole of each of the reactant gases, A and B are changed from their respective pressures to the equilibrium pressures, p_A and p_B under isothermal and reversible conditions. The gases are transferred into a large box containing gases A, B, C, and D at equilibrium, with their respective partial pressures: p_A, p_B, p_C and p_D. Since the box contains a large amount of gas, the addition of 1 mole of A and B will not change partial pressures in the box.

The Gibbs free energy change for the isothermal process in which partial pressures of A and B, respectively change from, P_A to, p_A and P_B to p_B is given by

$$\begin{aligned}\Delta G_1 &= RT\ln\frac{p_A}{P_A} + RT\ln\frac{p_B}{P_B}\\ &= RT\ \ ln\frac{p_A p_B}{P_A P_B}\end{aligned}$$

There is no change in free energy when the gases are introduced into the equilibrium box because the system is at equilibrium.

The gases A and B react in the box to form gases C and D at partial pressures of p_C, and p_D, respectively. On removing 1 mole of each of gases C and D from the box their partial pressures change from p_C and p_D to P_C and P_D, respectively under isothermal and reversible conditions, the free energy change is given by

$$\begin{aligned}\Delta G_2 &= RT\ln\frac{P_C}{p_C} + RT\ln\frac{P_D}{p_D}\\ &= RT\ln\frac{P_C P_D}{p_C p_D}\end{aligned}$$

Thus after conducting the entire operation in which one mole of each A and B at pressures $P_{A,}$ and P_B have reacted to form one mole of each C and D at partial pressures of P_C and P_D, respectively the total free energy is expressed as

$$\begin{aligned}\Delta G &= RT\ ln\frac{p_A p_B}{P_A P_B} + RT\ ln\frac{P_C P_D}{p_C p_D}\\ &= RT\ ln\frac{p_A p_B}{p_C p_D} + RT\ ln\frac{P_C P_D}{P_A P_B}\end{aligned}$$

The equilibrium constant of the reaction: A + B + C + D is expressed as

$$K_p = \frac{p_C p_D}{p_A p_B}$$

$$\therefore\ \Delta G = -RT\ \ ln\, K_p + RT\ ln\frac{P_C P_D}{P_A P_B} \tag{2.90}$$

Equation 2.90 is known as the van't Hoff isotherm. It consists of two terms. The first term is constant because R, T, and K_p are constant. The second term varies depending on the initial pressures P_A *and* P_B and final pressures P_C *and* P_D (i. e. $\Delta G=$ a constant term + a variable term). Thus the value of ΔG at a constant temperature depends on the value of the second term. The value of the free energy change at temperature T for a reaction in which all the reactants and products are in their standard states (unit activity, and unit pressure) is known as the standard free energy change.

$$\therefore \ \Delta G^o = -RT \ \ ln\, K_p + RT \ln \frac{1 \times 1}{1 \times 1} = -RT \ \ ln\, K_p \tag{2.91}$$

Therefore, van't Hoff isotherm may be written as

$$\therefore \ \Delta G = \Delta G^o + RT \ \ ln \frac{P_C P_D}{P_A P_B} \tag{2.92}$$

(i.e. a constant term at constant temperature + a variable term called the correction factor)

For a reversible reaction in which different number of moles of reactants and products take part (e.g. a moles of A and b moles of B give rise to the formation of c moles of C and d moles of D) as for example:

$$aA + bB \rightleftharpoons cC + dD$$

$$\Delta G = \Delta G^o + RT \ \ ln \frac{P_C^c \cdot P_D^d}{P_A^a \cdot P_B^b} \tag{2.93}$$

The above relationship is known as the van't Hoff isochore [15].

In case of solid and liquid reactants and products, Eq 2.90 is accordingly modified in terms of concentrations rather than partial pressures:

$$\Delta G = -RT \ \ ln\, K_c + RT \ ln \frac{C_C C_D}{C_A C_B} \tag{2.90a}$$

The free energy change for the reaction: A + B = C + D, will depend on the concentrations of the reactants and products. Depending on whether reactants and products are in pure state or present as a dissolved state in dilute solutions, the free energy will vary and in such cases the approach of van't Hoff is extremely useful for correct assessment. This approach will be more clearly understood with the concept of activity (i.e. effective concentration), which will be discussed in Section 3.2. The silicothermic reduction of MgO [16, 17] (the Pidgeon process) sets a good example for understanding the problem of deviation from pure state (i.e. unit activity). Since the free energy change for the following reaction under standard conditions is positive (ΔG^o = + 280 kJ at 1200 °C), the reaction will not proceed in the forward direction.

$$2\,MgO\,(s) + Si\,(s) = 2\,Mg\,(g) + SiO_2(s)$$

Making use of the van't Hoff isotherm the actual free energy change for the reaction is given by

$$\Delta G = \Delta G^o + RT\ ln\ K = \Delta G^o + RT\ ln\ \left[\frac{p_{Mg}^2 . a_{SiO_2}}{a_{MgO}^2 . a_{Si}}\right]$$

ΔG, becomes negative, even if ΔG^o is positive, at reduced pressure of magnesium and at lower activity of SiO_2. The Pidgeon process thus operates at a reduced pressure and the activity (effective concentration) of silica is brought down to < 0.0001 by using calcined dolomite (CaO.MgO) instead of pure MgO.

Differentiation of Eq. 2.91, $\Delta G^o = -RTlnK_P$ with temperature at constant pressure gives:

$$\left[\frac{\partial\,(\Delta G^o)}{\partial T}\right]_P = -R\,ln\,K_P - RT\left[\frac{\partial\ ln\,K_P}{\partial T}\right]_P$$

On multiplying by T we get:

$$T\ \left[\frac{\partial\,(\Delta G^o)}{\partial T}\right]_P = -RT\ ln\,K_P - RT^2\left[\frac{\partial\ ln\,K_P}{\partial T}\right]_P = \Delta G^o - RT^2\left[\frac{\partial\ ln\,K_P}{\partial T}\right]_P$$

From the Gibbs–Helmholtz Equation ($\Delta G^o = \Delta H^o - T\Delta S^o$), where $\Delta S^o = -\left[\frac{\partial(\Delta G^o)}{\partial T}\right]$

$$\Delta G^o = \Delta H^o + T\ \left[\frac{\partial\,(\Delta G^o)}{\partial T}\right]$$

Hence, from the above equations, on comparison we get:

$$T\ \left[\frac{\partial\,(\Delta G^o)}{\partial T}\right] = \Delta G^o - \Delta H^o = \Delta G^o - RT^2\ \left[\frac{\partial\ ln\,K_P}{\partial T}\right]$$

$$\therefore\ \Delta H^o = RT^2\left[\frac{\partial lnK_P}{\partial T}\right]$$

$$\text{Or}\quad \frac{d\ ln\,K_P}{dT} = \frac{\Delta H^o}{RT^2} \tag{2.94}$$

The above relation, known as van't Hoff's equation, shows the effect of temperature on the equilibrium constant of a reaction. If K_1 and K_2 are equilibrium constants at temperatures T_1 and T_2 respectively, assuming enthalpy to be independent of temperature in the close temperature interval ($T_1 - T_2$), Eq. 2.94 can be integrated:

$$\int_{k_1}^{k_2} d\ ln\ K_P = \frac{\Delta H^o}{R}\ \int_{T_1}^{T_2}\frac{dT}{T^2}$$

$$\ln\left[\frac{K_2}{K_1}\right]=\frac{\Delta H^o}{R}\left[\frac{T_2-T_1}{T_1T_2}\right] \tag{2.95}$$

From $G = H - TS$,we can write: $G/_T = H/_T - S$

On differentiation with respect to T at constant pressure:

$$\left[\frac{\partial(G/_T)}{\partial T}\right]_P=\left[\frac{\partial(H/_T)}{\partial T}\right]_P-\left[\frac{\partial S}{\partial T}\right]_P=H\left[\frac{\partial(1/_T)}{\partial T}\right]_P+\frac{1}{T}\left[\frac{\partial H}{\partial T}\right]_P-\left[\frac{\partial S}{\partial T}\right]_P$$

Since $\left[\frac{\partial H}{\partial T}\right]_P = C_P$ (from Eq. 2.19) and $\left[\frac{\partial S}{\partial T}\right]_P = \frac{C_P}{T}$ (from Eq. 2.47)

$$\left[\frac{\partial(G/_T)}{\partial T}\right]_P=-\frac{H}{T^2}+\frac{C_P}{T}-\frac{C_P}{T}=-\frac{H}{T^2}$$

$$\text{Similarly}\quad\left[\frac{\partial(\Delta G/_T)}{\partial T}\right]_P=-\frac{\Delta H}{T^2} \tag{2.96}$$

This is known as the Gibbs–Helmholtz equation, useful in determining ΔH at a particular temperature if the variation of ΔG with T is known.

2.11.3 Le Chatelier Principle

Le Chatelier principle states that if a constraint is imposed on a system in equilibrium the system will undergo readjustments to nullify the constraint. The principle can be easily understood by considering the effect of temperature on the equilibrium constant of an exothermic reaction:

$$M+O_2=MO_2+q, \Delta H=-ve$$

On increasing the temperature of the above reaction (i.e. by adding heat into the reaction system) some of the MO_2 will decompose into M and O_2 in order to consume a part or the entire amount of heat supplied to the system. As a result a lesser amount of the product MO_2 will be produced by the increase of temperature. Hence the value of the equilibrium constant, K for the reaction decreases with the increase of temperature. This qualitative conclusion based on the Le Chatelier principle can be calculated quantitatively by the van't Hoff equation (Eq. 2.95):

$$\ln\left[\frac{K_2}{K_1}\right]=\frac{\Delta H^o}{R}\left[\frac{T_2-T_1}{T_1T_2}\right]$$

For exothermic reactions, where ΔH is negative, the higher the temperature the smaller the value of K. On the other hand for endothermic reactions i.e. when ΔH is positive K will increase with increase in temperature

Pressure/Volume considerations: Le Chatelier principle states that the magnitude of the total pressure of a system will not affect an equilibrium if the number of moles of gas on either side of the chemical equation are equal. If the numbers differ, an increase in total pressure will displace the equilibrium in favor of the side with the smaller number of moles and vice versa. For example,

1. $WO_3\ (s) + 3H_2\ (g) = W\ (s) + 3H_2O\ (g)$

This reaction is independent of total pressure of the system because number of moles of H_2 and H_2O are equal on both the sides of the chemical equation. (See problem 26 in Chapter 10.).

2. Ni (s) + 4CO (g) = $Ni(CO)_4$ (g)

This reaction will be favored in the forward direction with increase of pressure because there is decrease in the number of moles on the right hand side of the equation. (See problem 38 in Chapter 10.).

3. Nb_2O_5 (s) + 5C (s) = 2Nb (s) + 5CO (g)

In this case decrease of pressure (i.e. application of vacuum) will favor in the forward direction.

2.12 Clausius–Clapeyron Equation

The Clausius–Clapeyron equation is extremely useful in calculating the effects of temperature and pressure changes on the melting point of solids, boiling point of liquids, and any solid-solid phase transformations. It also enables us to quantify the effect of the changes in the concentration of solutions on their freezing points, and boiling temperatures and is therefore very useful for calculation of phase boundaries of systems, which are either immiscible or partially miscible in the solid state.

When a single solid material is in equilibrium with its own liquid at the temperature of fusion at one atmosphere of pressure there is a natural tendency for atoms/ molecules to pass from the solid into the liquid and vice versa. The number of atoms passing from one state to another will depend on the temperature and pressure. Suppose G_s and G_l are respectively, the molar free energy of the solid and liquid at constant temperature and pressure. Then depending on whether G_s is greater than, equal to, or less than G_l, the following situations arise:

(i) $G_s > G_l$, solid melts because $\Delta G = -$ve (solid transforms to liquid)
(ii) $G_s = G_l$, solid $\rightleftharpoons$ liquid, since $\Delta G = 0$ (equilibrium between the two phases)
(iii) $G_s < G_l$, liquid solidifies because $\Delta G = +$ve (liquid transforms to solid)

Thus for a dynamic equilibrium between the solid and the liquid phases the required condition is

$$G_s = G_l$$

and hence, $dG_s = \mathrm{d}G_l$

Since the change in free energy of the solid or liquid is due to the change in temperature and pressure of the phases. Hence,

$$dG = f(T, P)$$

We can mathematically express the effects of temperature and pressure as total differentials of dG_s and $\mathrm{d}G_l$ with respect to T and P. Thus

$$dG_s = \left(\frac{\partial G_s}{\partial T}\right)_P dT + \left(\frac{\partial G_s}{\partial P}\right)_T dP$$

$$dG_l = \left(\frac{\partial G_l}{\partial T}\right)_P dT + \left(\frac{\partial G_l}{\partial P}\right)_T dP$$

Since at equilibrium,

$$dG_s = \mathrm{d}G_l$$

$$\therefore \left(\frac{\partial G_s}{\partial T}\right)_P dT + \left(\frac{\partial G_s}{\partial P}\right)_T dP = \left(\frac{\partial G_l}{\partial T}\right)_P dT + \left(\frac{\partial G_l}{\partial P}\right)_T dP$$

$$\text{or} \quad -S_s dT + V_s dP = -S_l dT + V_l dP$$

$$\text{or} \quad (V_l - V_s) dP = (S_l - S_s) dT$$

$$\text{or} \quad \Delta v.\ dP = \Delta S dT$$

$$\text{or} \quad \frac{dP}{dT} = \frac{\Delta S}{\Delta v} \tag{2.97}$$

Since the fusion of a metal is a thermodynamically reversible process, entropy of fusion is given as $\Delta S^f = \frac{q_{rev}}{T} = \frac{L^f}{T^f}$, Eq. 2.97 gives

$$\therefore \frac{dP}{dT} = \frac{L^f}{T^f(V_l - V_s)} = \frac{L^f}{T^f \Delta v} \tag{2.98}$$

where q_{rev} is the quantity of heat absorbed during fusion of 1 g mol of the substance and L^f refers to the corresponding latent heat of fusion, and v_s and v_l are the volumes in solid and liquid states, respectively.

The above equation known as the Clausius–Clapeyron equation [18] has been derived for a single chemical substance undergoing a change from one phase to another. The use of this equation can be extended to liquid-vapor, solid-vapor and solid-solid transformations by incorporating the appropriate latent heat of transformation and the volume change in each case. In every case, L is the heat absorbed during the transformation and Δv is the accompanying volume change.

The Clausius–Clapeyron equation in the above form enables us to calculate the changes in either pressure with temperature or latent heat of solid-solid, solid-liquid, solid-gas, and liquid-gas transformation. The above Eq. 2.98 does not allow us to calculate the absolute value of pressure and temperature for a given system undergoing a transformation. The equation will be of much wider application when transformed into a form suitable for integration by making the following assumptions:

a. In the liquid-gas transformation, the volume of 1 g atom/mole of the gaseous phase is much larger compared to the volume of 1 g atom/mole of the liquid phase, that is, the volume of the liquid is negligible. This assumption is reasonable because in case of $Fe(l) = Fe(g)$, v (liquid iron) = 10 cc g atom^{-1} whereas v (iron vapor) = 22400 cc g atom^{-1}.
b. Since metallic vapors behave ideally, $PV = RT$.
c. The latent heat of transformation is constant over the range of pressure and temperature under consideration.

Thus, for liquid-gas transformation we have:

$$\frac{dP}{dT} = \frac{L^e}{T^e\left(v_g - v_l\right)} \quad \left(\text{since, } v_g \gg v_l\right)$$

$$\therefore \frac{dP}{dT} = \frac{L}{Tv_g}$$

From, the second assumption: $PV = RT$, $v_g = RT/P$

$$\therefore \frac{dP}{dT} = \frac{L\,P}{RT^2} \tag{2.99}$$

From, the third assumption L/R can be taken out of the integral,

$$\therefore \int_{p_1}^{p_2} \frac{dP}{P} = \frac{L}{R}\int_{T_1}^{T_2} \frac{dT}{T^2}$$

Where p_1 and p_2 are the initial and final pressures at temperatures T_1 and T_2, respectively.

On integration we get:

$$\ln\left[\frac{p_2}{p_1}\right] = \frac{L}{R}\left[\frac{T_2 - T_1}{T_1 T_2}\right] \tag{2.100}$$

This equation allows us to calculate:

1. The latent of transformation provided p_1 and p_2 and the corresponding T_1 and T_2 are known.
2. The variation in boiling point with change in pressure, provided p_1, p_2 and T_1 are known.
3. The value of p, provided, T_1, T_2 and L are known.

Integration of Eq. 2.99 may also be expressed as:

$$ln\,P = -\frac{L}{RT} + C \tag{2.101}$$

This form of equation is very useful because plot of *lnP* vs 1/*T* gives a straight line, the slope of which is, $-\frac{L}{R}$.

In some materials latent heat of transformation varies with temperature (i.e. assumption 3 is not valid) but assumptions 1 and 2 are valid like that in the previous case discussed above (i.e., $v_g \gg v_l$ and v_s and vapors behaving ideally). Suppose ΔH is the heat of transformation in such a case. According to the Kirchhoff's equation we have

$$\frac{\partial(\Delta H)}{\partial T} = \Delta C_p$$

$$\therefore\ \Delta H = \int \Delta C_p\ dT = \Delta C_p.T + A$$

Thus the Clausius–Clapeyron equation can be written as

$$\frac{dP}{dT} = \frac{L\,P}{RT^2} = \frac{\Delta H.P}{RT^2} \tag{2.102}$$

On integration we get:

$$\int \frac{dP}{P} = \int \frac{\Delta H}{RT^2} dT = \int \left(\frac{\Delta C_p.T}{RT^2} + \frac{A}{RT^2}\right)\ dT \tag{2.103}$$

$$\therefore\ \ln P = \frac{\Delta C_p}{R} \ln T - \frac{A}{RT} + a\ constant \tag{2.104}$$

2.13 Phase Rule

Thermodynamics has been found useful to scientists working in fields of chemistry, physics, and metallurgy on account of three reasons. Firstly, it allows them to predict the maximum number of possible phases in a given system. Secondly, it helps in establishment of simple equilibrium phase diagrams from a very limited thermodynamic data. Finally, it is useful in rechecking parts of phase diagrams when the conventional experimental methods for the determination of certain boundary conditions present widely different results. The phase rule mathematically relates phase, component and degree of freedom by means of a relation. Before deriving the phase rule equation these terms are defined below.

Phase A phase is defined as any homogeneous and physically distinct part of a system which is separated from other part of the system by a bounding surface.

For example, ice, water and water vapor can coexist in equilibrium at 273.15K. When ice exists in more than one crystalline form, each form will represent a separate phase because it is clearly distinguishable from each other. In general every solid in a system constitutes a separate phase but a homogeneous solid or liquid solution forms a single phase irrespective of number of chemical components present. However two immiscible liquids constitute two phases, since there is a boundary between them. Gases on the other hand pure or as a mixture always form one phase since their molecules are intimately mixed together and thus produce a homogeneous gas mixture (a phase).

Component The number of components in a system at equilibrium is the smallest number of independently variable constituents by means of which the composition of each phase present can be expressed directly or in the form of a chemical equation. As an example let us consider decomposition of calcium carbonate:

$$CaCO_3\ (s) = CaO\ (s) + CO_2\ (g)$$

According to the above definition, at equilibrium this system will consist of two components since the third one is fixed by the equilibrium conditions. Thus we have three phases—two solids ($CaCO_3$ and CaO) and a gas (CO_2) and the system has only two components. If CaO and CO_2 are taken, the composition of calcium carbonate phase can be expressed as x CaO + x CO_2 giving x $CaCO_3$ (by the chemical reaction). The composition of calcium oxide phase is y CaO + 0 CO_2 and that of the gaseous phase is 0 CaO + y CO_2. The composition of the three phases could be expressed equally by taking $CaCO_3$ and CaO or $CaCO_3$ and CO_2 as the components. In these cases it would have been necessary to use a minus sign in the chemical equation, e.g. y $CaCO_3$ – y CO_2 is clearly y CaO.

The dissociation of any carbonate, oxide or similar compounds involves two components; the same is true in the case of salt hydrate equilibrium, for example: $CuSO_4.5H_2O(s) \rightarrow CuSO_4.3H_2O(s) + 2H_2O\ (g)$ where the simplest components are evidently $CuSO_4$ and H_2O.

In the slightly more complicated equilibrium: Fe (s) + H_2O (g) = FeO (s) + H_2 (g) it is necessary to choose three components in order that the composition of each of the three phases can be expressed. The composition of the two solid phases could be given in terms of Fe and O, but these alone are insufficient to define the gaseous phase which is a mixture of hydrogen and water vapor, a third component, viz. H_2O is necessary. The water system for example consists of one component, viz. H_2O

each of the phases in equilibrium i.e. solid, liquid and vapor may be regarded as being made of this component only.

The same applies to acetic acid which consists of entirely of double molecules in the solid state, to a great extent in the liquid and partially in the vapor; the composition of each phase, however, be expressed in terms of $C_2H_4O_2$ and this is only the component. On the other hand if ammonium chloride is vaporized in a vacuum, the system consists of only one component, in spite of the dissociation of the vapor into ammonia and hydrogen chloride; the resultant composition of the gaseous phase still represented by NH_4Cl. If however, an excess ammonia or HCl gas is introduced, there would be two components; the vapor no longer has the same ultimate composition as the solid.

Degree of Freedom The number of degrees of freedom is the number of variable factors, such as temperature, pressure and concentration that need to be fixed in order that the condition of a system at equilibrium may be completely defined when referring to its equilibrium phase diagrams.

2.13.1 Derivation of the Phase Rule Equation

In order to derive the Phase Rule equation, consider a system consisting of C components distributed between P phases which will be influenced by only external factors such as temperature, pressure and internal concentrations. In such a system if we know (C−1) concentrations in each phase, the composition of each phase will be completely known because the last concentration can be calculated by difference. In addition to the composition, the temperature and pressure of the system have to be stated. If all the phases are assumed to be at the same temperature and pressure at equilibrium, the total number of variables need to be known = P (C−1) + 2. (I)

We can eliminate certain number of variables with the aid of thermodynamics. It is well established that at equilibrium the chemical potential of any one component must be the same in each of the phases, otherwise there would be a net transfer of this component from one phase to another. Suppose there are two immiscible solutions, L_1 and L_2 (Fig. 2.6), each containing a certain amount of the component species *i*. If the chemical potential of the component *i* in solutions L_1 and L_2 are respectively, μ_{i_1} and μ_{i_2}, then under equilibrium conditions at constant temperature and constant pressure, these two chemical potentials would be equal (i.e. $\mu_{i_1} = \mu_{i_2}$) since otherwise *i* would either pass from L_1 to L_2 or vice versa until its chemical potential becomes the same in each phase.

Hence, based on the above facts, in the system of C components and P phases we need to know only one concentration term for each component to be able to calculate all the concentrations of this component in the other phases from its chemical potential.

Thus, the number of concentration terms (variables) that are automatically fixed by the chemical potential = C (P−1) (II).

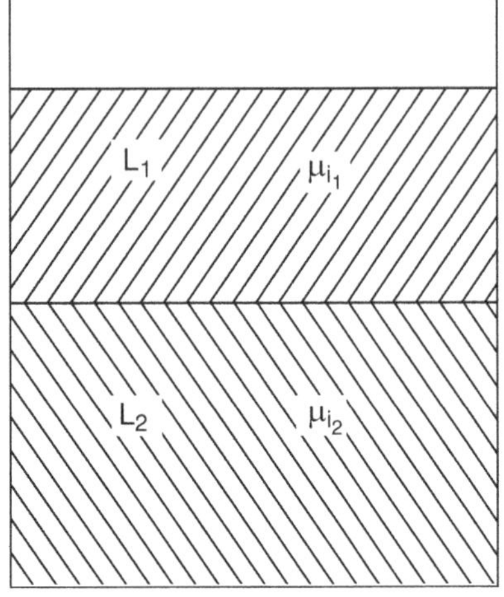

Fig. 2.6 Two immiscible solutions L_1 and L_2 consisting a certain amount of the component species *i* under equilibrium

Therefore, the total number of variables, i.e. the degree of freedom, F that must now be specified in order to define the system completely, is the difference between equations I and II.

$$F = [P\,(C - 1) + 2] - [C\,(P - 1)]$$

$$\text{or}\quad F = C - P + 2 \tag{2.105}$$

Equation 2.105 is known as the Gibbs Phase Rule. This is applicable to all macroscopic systems which are in state of heterogeneous equilibrium, and are influenced by temperature, pressure, and composition. The phase rule does not account for molecular complexity. In general, it is useful to specify the number of independent variables involved in a system at equilibrium.

2.13.2 Applications

2.13.2.1 Gibbs Phase Rule in Single Component System

Figure 2.7 shows the phase diagram of water where three phases, i.e., solid ice, liquid water and water vapor are in equilibrium depending on temperature and pressure. In the figure solid and vapor are separated by the line IO. OP separates the regions of liquid and vapor whereas solid and liquid regions are separated by LO.

At any point on the line IO, there are two phases, viz. ice and water vapor are in equilibrium. Since there is only one component, according to the phase rule, degree of freedom, $F = C - P + 2 = 1 - 2 + 2 = 1$. Thus any system represented on the line IO, OP or LO there has only one degree of freedom, i.e., only one variable temperature or pressure. If one fixes temperature, pressure is automatically read

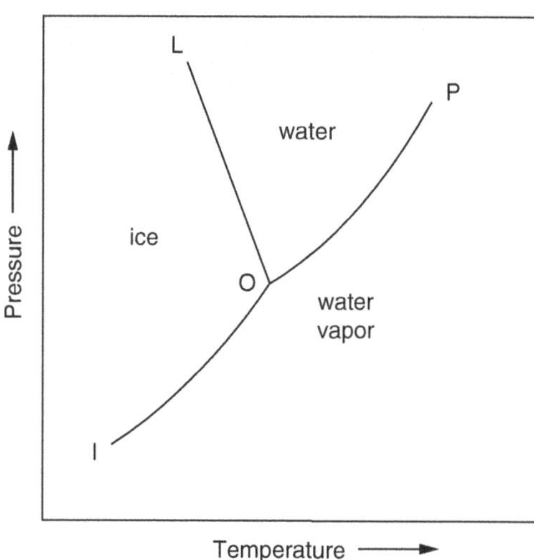

Fig. 2.7 Phase diagram of water

from the diagram. Additional change in either temperature or pressure will alter the number of phases. A higher pressure will convert all the water vapor into ice, and a lower pressure will convert all the ice into water vapor.

At point O, where all the three phases solid, liquid and vapor coexist, has zero degree of freedom ($F = 1 - 3 + 2 = 0$). Hence O is called the invariant or triple point. There is only one such point for any substance and at this point, both temperature and pressure are fixed. Any change in either temperature or pressure or in both together will cause disappearance of one phase.

Any other point outside IO, OP or LO represents existence of only one phase, i.e., either solid, liquid or vapor, hence according to the Gibbs Phase Rule, $F = C - P + 2 = 1 - 1 + 2 = 2$. This means system is bi-variant, since it has two degrees of freedom and therefore in order to locate the position of a point, representing the system in the diagram two variables, the temperature and the pressure of the system must be known.

Similarly single component system CO_2 can be discussed.

2.13.2.2 Phase Rule to Binary Eutectic Alloy

Figure 2.8 shows the phase diagram of the alloy system A – B. This is a two component system. At point 1, 2 and 3 we have respectively one, two and three phase(s) in equilibrium at a particular temperature. While considering effect of composition at different temperature, pressure is fixed. Hence the Phase Rule equation for a condensed system is modified as: $F = C - P + 1$. Thus the number of degree(s) of freedom at three different points may be evaluated as follows:

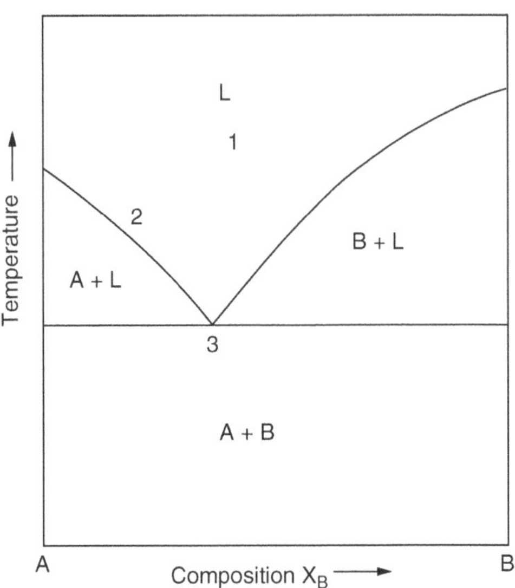

Fig. 2.8 Eutectic phase diagram

At point 1, P = 1, C = 2, F = 2 − 1 + 1 = 2 (temperature and composition)
At point 2, P = 2, C = 2, F = 2 − 2 + 1 = 1 (either temperature and composition)
At point 3, P = 3, C = 2, F = 2 − 3 + 1 = 0 (no degree of freedom i.e. invariant point)

2.13.2.3 Phase Rule in Reduction, Decomposition and Dehydration

(i) Consider reduction of FeO with CO under standard conditions, i.e., P = 1 atm.

$$FeO\ (s) + CO\ (g) = Fe\ (s) + CO_2\ \ (g)$$

In this system we have P = 3 (i.e. two solids FeO and Fe and a gaseous phase: CO + CO_2) and C = 3

$$\therefore\ F = C - P + 1 = 3\text{–}3 + 1 = 1$$

Thus the above system has only one degree of freedom, either temperature or pressure.

(ii) Decomposition of $CaCO_3$ may be represented as $CaCO_3$ (s) = CaO (s) + CO_2 (g).

In this system P = 3 (two solids $CaCO_3$ and CaO and a gas CO_2) and C = 2 (already discussed in the preceding paragraph.

$$\therefore\ F = C - P + 1 = 2\text{–}3 + 1 = 0$$

This means if pressure is fixed decomposition temperature gets automatically fixed. According to the equation F = C – P + 2, i.e., when T and P both are variables, F = 2 – 3 + 2 = 1, either temperature or pressure can be varied.

(iii) Dehydration of $CuSO_4$. 5 H_2O is illustrated as

$$CuSO_4.5H_2O \rightarrow CuSO_4.3H_2O \rightarrow CuSO_4.H_2O \rightarrow CuSO_4 + H_2O$$

In this case, $CuSO_4$ *and* H_2O are evidently the simplest components, C = 2, but P = 5. Hence, F = 2–5 + 2 = –1, that is, the system is over-constrained. This means it is impossible to achieve equilibrium in the dehydration of $CuSO_4$. 5 H_2O [19].

2.13.2.4 Phase Rule in Roasting of Sulfides

The necessary conditions for the formation of different products of roasting can be illustrated by the relationship between the equilibriums in any M–S–O system. In this system, we have three components and according to the phase rule ($P = C - F + 2$) there can be five phases (in simplest case), that is, four condensed phases and one gaseous phase ($P = 3 - 0 + 2 = 5$). If temperature is fixed, we have $P = 3 - 0 + 1 = 4$ (3 condensed phases and 1 gaseous phase). The gas phase normally contains SO_2 and O_2 but some SO_3 and even sulfur vapor (S_2) may be present. Among the gaseous compounds the following equilibria exist:

$$S_2\ (g) + 2O_2\ (g) = 2SO_2\ (g)$$

$$2SO_2\ (g) + O_2\ (g) = 2SO_3\ (g)$$

At any selected temperature the composition of the gas mixture is defined by the partial pressures of two of the gaseous components. Further, for any fixed gas composition, the composition of the condensed phase gets fixed. Thus, the phase relationship in the ternary system at constant temperature may be described in a two-dimensional diagram, where the two co-ordinates are the partial pressures of two of the gaseous components. Generally, p_{SO_2} and p_{O_2} are chosen because during roasting sulfide reacts with oxygen to produce SO_2 which is a predominant gas species in the flue. Such isothermal plots shown in Fig. 2.9, are called Predominance Area Diagrams.

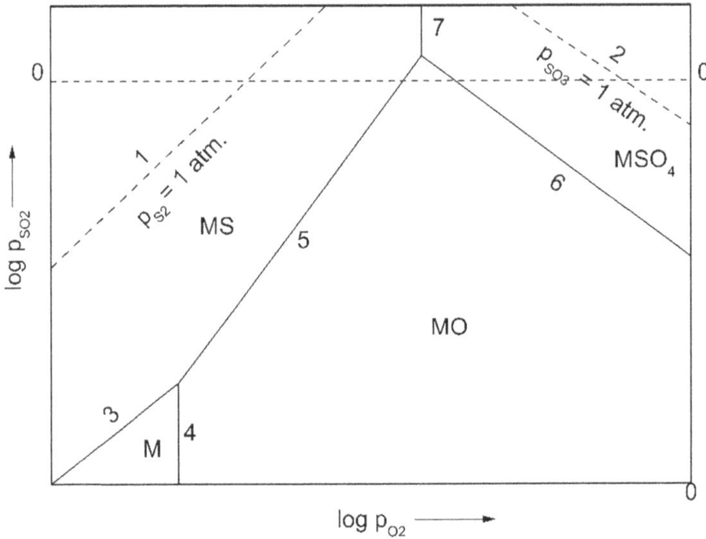

Fig. 2.9 Predominance area diagram of the M–S–O system. (Reproduced from M. Shamsuddin and H. Y. Sohn, [20] with the permission of The Minerals, Metals & Materials Society)

References

1. B. Thompson, Phil. Mag. **23**, 243 (1813)
2. J. Joule, Phil. Mag. **23**, 263 (1813)
3. G.N. Lewis, M. Randall, *Thermodynamics*, 1st edn. (McGraw-Hill Co. Ltd, New York, 1923)
4. J. Mackowiak, *Physical Chemistry for Metallurgists* (American Elsevier Publishing Co. Inc, New York, 1966)
5. O. Kubaschewki, C.B. Alcock, *Metallurgical Thermochemistry*, 5th edn. (Pergamon, Oxford, 1979)
6. S. Carnot, (quoted in Second Law of Thermodynamics by W. F. Magie (Ed), Harper & Brothers, New York, 1899).
7. R. Clausius, Ann. Phys. **79**, 368 (1850)
8. W. Thomson, Trans. Royal Soc. Edin. **21**, 123 (1854)
9. R.H. Parker, *An Introduction to Chemical Metallurgy*, 2nd edn. (Pergamon, Oxford, 1978)
10. R.A. Swalin, *Thermodynamics of Solids* (John Wiley & Sons, New York, 1964)
11. D.R. Gaskell, *Introduction to the Thermodynamics of Materials*, 4th edn. (Taylor & Francis, New York, 2003)
12. L.S. Darken, R.W. Gurry, *Physical Chemistry of Metals* (McGraw-Hill Co. Ltd, London, 1953)
13. M. Berthelot, (quoted in Metallurgical Thermochemistry by O. Kubaschewski, and C. B. Alcock, [9]).
14. C.M. Guldberg, P. Waage, *Etudes sur les affinities chimique, Brogger and Christie* (Christiania, 1867)
15. K. van't Hoff, Svensk Vet. Akad. Handl. **21**, 1 (1886)

16. M. Shamsuddin, *Physical Chemistry of Metallurgical Processes*, The Minerals, Metals & Materials Society, 2nd edn. (Springer Nature, Switzerland, 2021)
17. M. Shamsuddin, H.Y. Sohn, JOM **71**, 3266 (2019)
18. B.P.E. Clapeyron, J. de L'Ecole Polytechnique **14**, 153 (1834)
19. A.M. James, *A Dictionary of Thermodynamics* (John Wiley, New York, 1976)
20. M. Shamsuddin, H.Y. Sohn, JOM **71**, 3253 (2019)

Chapter 3
Solution Thermodynamics

A solution may be defined as a homogeneous phase composed of different chemical substances, whose concentration may be varied without the precipitation of a new phase. It differs from a mixture by its homogeneity and from a compound by being able to possess variable composition. Solution may be gaseous, liquid or solid. It may be classified as binary or ternary solution depending on whether it contains two or three components. A binary solution has two chemical substances (elements or compounds), e.g., molten cadmium and zinc miscible in all proportions. The major component of the solution is called solvent and the minor one as solute. Composition of solution may be expressed in a number of ways, e.g., weight, mole or atom percent. If w_1 and w_2 are weights of the solvent and solute, respectively in the solution:

$$\text{wt\%solute} = \left[\frac{w_2}{w_1 + w_2}\right] \times 100$$

On the other hand if w_A and w_B are the weights of components A and B in the solution, having atomic/molecular weights M_A and M_B, respectively, the atom percent (at %) or mole percent (mol %) is expressed as:

$$\text{at/mol\%A} = \left[\frac{w_A/M_A}{w_A/M_A + w_B/M_B}\right] \times 100$$

The atom/mole fraction (x), most widely used in thermodynamic equations, is defined as the number of atoms/moles of a substance divided by the total number of atoms/moles of all the substances present in the solution. If n_A number of moles of A and n_B number of moles of B form a solution: A–B, atom fractions of A and B are given as:

M. Shamsuddin, *Thermodynamic Measurement Techniques*, The Minerals, Metals & Materials Series, https://doi.org/10.1007/978-3-031-47118-6_3

$$x_A = \frac{n_A}{n_A + n_B} \text{ and } x_B = \frac{n_B}{n_A + n_B}, \text{ and } x_A + x_B = 1$$

Raoult's law [1–4] is useful in discussing the properties of solutions.

3.1 Raoult's Law

The law states that the relative lowering of the vapor pressure of a solvent due to the addition of a solute is equal to the mole fraction of the solute in the solution. Suppose x_A atom/mole fraction of A and x_B atom/mole fraction B form a solution, A–B, in which p_A and p_B are the partial pressures exerted by vapors of A and B, respectively, and P is the total pressure of the solution. If p_A^o and p_B^o are the partial pressure of pure A and pure B, respectively, at the same temperature at which solution exists, then according to the Raoult's law we have:

$$\frac{p_A^o - p_A}{p_A^o} = x_B \text{ and } \frac{p_B^o - p_B}{p_B^o} = x_A$$

$$\text{or } 1 - \frac{p_A}{p_A^o} = x_B \text{ and } 1 - \frac{p_B}{p_B^o} = x_A$$

$$\text{or } \frac{p_A}{p_A^o} = 1 - x_B = x_A \quad \text{and} \quad \frac{p_B}{p_B^o} = x_B$$

$$\therefore\ p_A = p_A^o \cdot x_A \quad \text{or} \quad p_A \propto x_A$$

$$\text{and } p_B = p_B^o \cdot x_B \quad \text{or } p_B \propto x_B$$

Therefore, in general, for the species i in a solution, we can write as:

$$p_i = p_i^o . x_i \tag{3.1}$$

This means if the solution obeys Raoult's law, the vapor pressure of the component is directly proportional to its atom/mole fraction in the solution. The constant of proportionality is the vapor pressure of the component in the pure state. A solution that obeys Raoult's law, is called an ideal solution. In order to form an ideal solution: A–B, the molecules of A and B must be of similar size and must attract one another with the same force, the molecules of A attract other molecules of A or molecules of B attract other molecules of B, and also the vapor should behave as an ideal gas. Thus, the solution obeying Raoult's law satisfies the following condition:

$$(A \leftrightarrow B) = (A \leftrightarrow A) = (B \leftrightarrow B) = \frac{1}{2}\ \{(A \leftrightarrow A) + (B \leftrightarrow B)\}$$

Where $(A \leftrightarrow A)$, $(B \leftrightarrow B)$ and $(A \leftrightarrow B)$ denote the bond energies between similar and dissimilar atoms.

3.1.1 *Ideal Solutions*

An ideal solution obeys Raoult's law, which may be represented by plotting vapor pressure against the mole fraction. This gives straight lines with p_A^o and p_B^o being the intersection of the line with the vapor pressure axes as shown in Figs. 3.1 and 3.2. In ideal solutions, gas pressures p_A and p_B obey the ideal gas equation: $PV = RT$ and their physical properties will be additive, that is, total pressure, $P = p_A + p_B$.

3.1.2 *Nonideal or Real Solutions*

Deviations from Raoult's law occur when the attractive forces (bond energies) between the molecules of components A and B of the solution are weaker or stronger than those existing between A and A or B and B in their pure states. For example, if the attractive force $(A \leftrightarrow B)$ between the components A and B in the solution is weaker than the mutual attraction [i.e. bond energy $(A \leftrightarrow A)$ and $(B \leftrightarrow B)$] between molecules of A in pure A and molecules of B in pure B, there would be a higher

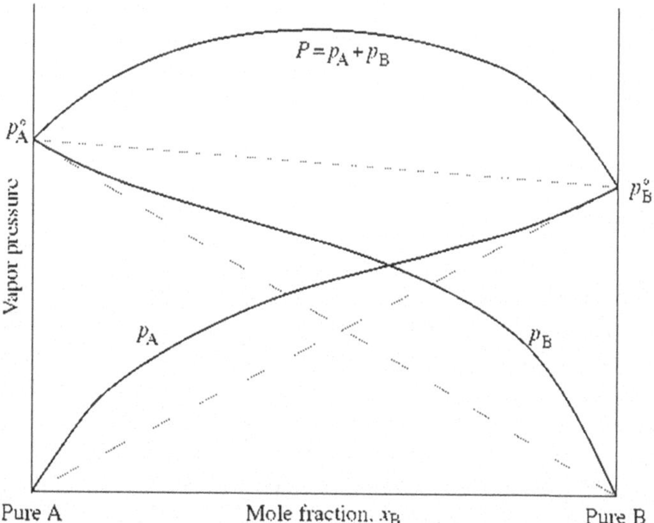

Fig. 3.1 Schematic representation of positive deviation from Raoult's law (broken lines represent ideal behavior). (Reproduced from *Physical Chemistry of Metallurgical Processes* by M. Shamsuddin, [5] with the permission of The Minerals, Metals & Materials Society)

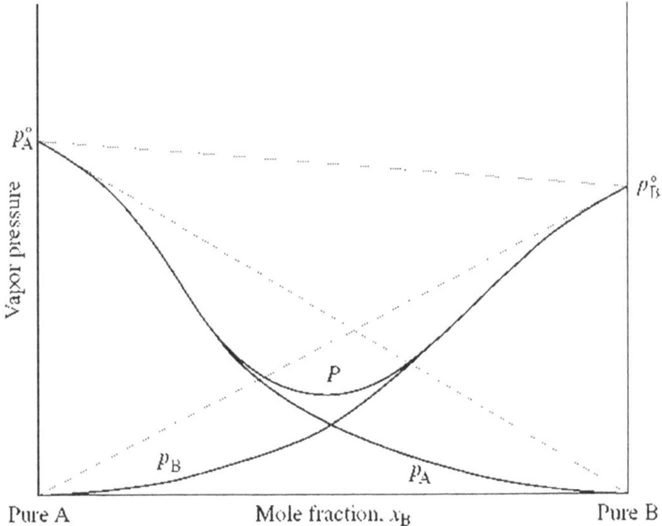

Fig. 3.2 Schematic representation of negative deviation from Raoult's law (broken lines represent ideal behavior). (Reproduced from *Physical Chemistry of Metallurgical Processes* by M. Shamsuddin, [5] with the permission of The Minerals, Metals & Materials Society)

tendency for these components to leave the solution. In this case the vapor pressure would be more than that predicted from Raoult's law. This is known as the positive deviation from Raoult's law (Fig. 3.1). In case of a stronger attraction between A and B as compared to A and A or B and B, vapor pressures of both A and B (separately) would be less than expected according to Raoult's law. This is known as the negative deviation from Raoultian behavior (Fig. 3.2). Systems with intermetallic compounds exhibit negative deviation since the attractive force between the components is large.

3.2 Activity

In Sect. 2.11.1 on The Law of Mass Action [2], it has been stated that the rate of a chemical reaction is proportional to the product of the active masses of the reacting substances. Active mass means the amount of substance available for the reaction or which is actively participating in the reaction. This approach is applicable under ideal conditions (solutions) when components of a chemical system act independently from each other. On the other hand, when constituents react with one another the availability of reactants for chemical reaction will not depend on the concentration of the components present in the system, instead only a fraction of concentration will take part in the chemical reaction. The fraction of that concentration is called activity, the effective concentration available for reaction.

In actual solution the vapor pressure of a component is not directly proportional to the mole fraction of that component. It may be either greater or lesser than that expected from the solution if it obeyed Raoult's law. We can now define "activity," a_A of the component A in the solution.

For ideal solutions, we have $p_A = p_A^o . x_A$

In case of nonideal solutions, $p_A \neq p_A^o . x_A$

Therefore, in order to maintain equality in case of nonideal solutions, we introduce a new term activity, a_A hence,

$$p_A = p_A^o \cdot a_A \quad \text{and} \quad p_B = p_B^o \cdot a_B \tag{3.2}$$

Thus, for an ideal solution, $a_A = x_A$. In order to account for any deviation from ideality we introduce a factor, γ_A and write: $a_A = \gamma_A . x_A$. γ_A is known as Raoultian activity coefficient. It may be greater or less than unity for a positive or negative deviation, respectively. For pure A: $x_A = 1$, $a_A = 1$ and hence, $\gamma_A = 1$. Thus, a pure substance having unit activity is said to be in its "standard state".

Since the vapor pressure of a substance is a measure of its attraction to the solution in which it exists, it is therefore a measure of its availability for reaction, perhaps with another phase. Thus, activity may be defined as that fraction of molar concentration, which is available for reaction.

In phase diagrams, compound formation indicates negative deviation whereas existence of immiscibility exhibits positive deviation from ideality. Since in all these cases the activity coefficient is related to the Raoultian ideal behavior line, it is called the "Raoultian activity coefficient". The activity of any component in a mixture depends upon the environment in which it is present. Forces of attraction or repulsion caused by the environment acting on the substances determine its activity. Let us now consider the physical significance of both Raoult's ideal and nonideal behavior in binary systems.

3.2.1 Raoult's Ideal Behavior

A solution A–B composed of components A and B will exhibit Raoult's ideal behavior if (i) the attractive force (bond energy) between A and B atoms $(A \leftrightarrow B)$ is of the same order as that between A and A $(A \leftrightarrow A)$ or between B and B $(B \leftrightarrow B)$ or else the attractive forces between dissimilar atoms A and B is the average of the attractive forces (bond energy) between the similar atoms A and A and between B and B, that is $[(A \leftrightarrow B) = \frac{1}{2}\left[(A \leftrightarrow A) + (B \leftrightarrow B)\right]$. Under this condition, the activities of A and B in the solution at all concentrations will be equal to their mole fractions and the solution is said to be ideal. The system Bi-Sn serves as an example of such a solution at a particular temperature. In this case the net attractive force between Bi and Sn in the solution can be represented as:

$$(Bi \leftrightarrow Sn) = \frac{1}{2}\ \{(Bi \leftrightarrow Bi) + (Sn \leftrightarrow Sn)\}$$

In general, the ideal behavior of any solution A–B, can be expressed as:

$$(A \leftrightarrow B) = (A \leftrightarrow A) = (B \leftrightarrow B) = \frac{1}{2}\ \{(A \leftrightarrow A) + (B \leftrightarrow B)\}$$

3.2.1.1 Positive Deviation

When the net attractive force (bond energy) between components A and B is less than the average those between A–A and B–B, the solution A–B, exhibits positive deviation from Raoult's law. In terms of the forces of attraction (bond energy) this can be represented as:

$$(A \leftrightarrow B) < \frac{1}{2}\ \{(A \leftrightarrow A) + (B \leftrightarrow B)\}$$

In this case the Raoultian activity coefficient is always greater than unity except when the concentration of the solvent A approaches unity (i.e. $x_A \rightarrow 1$). Pb-Zn liquid solutions show such a behavior at temperature above 1071 K. In general, the systems showing positive deviation are endothermic in nature.

3.2.1.2 Negative Deviations

Negative deviations occur when the attractive force between dissimilar components A and B is larger than the average those between A–A and B–B, that is, $(A \leftrightarrow B) > \frac{1}{2}\ \{(A \leftrightarrow A) + (B \leftrightarrow B)\}$. Negative deviations generally indicate a tendency for compound formation. For example, formation of Mg_3Bi_2 in Mg-Bi system shows such a behavior. The systems exhibiting negative deviation are usually exothermic.

Occasionally, both negative and positive deviations from Raoult's law occur in the same system. Zn-Sb, Cd-Bi, and Cd-Sb systems are outstanding examples of such combined deviations. Ideal, positive and negative behaviors are shown in Fig. 3.3.

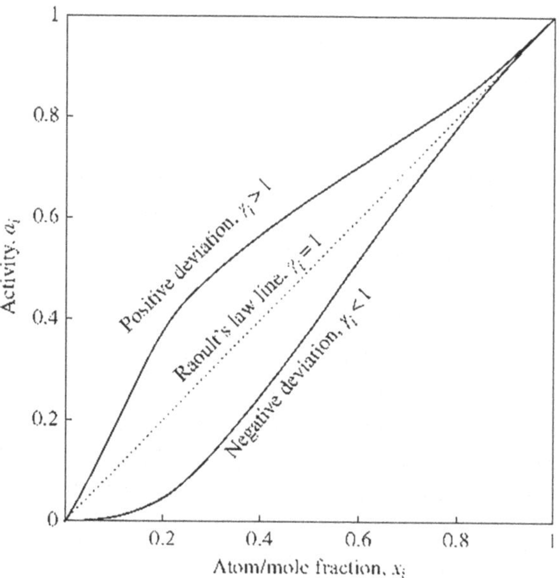

Fig. 3.3 Variation of activity with mole fraction of a component *i* in ideal and nonideal solutions showing positive and negative deviations from Raoult's ideal behavior. (Reproduced from *Physical Chemistry of Metallurgical Processes* by M. Shamsuddin, [5] with the permission of The Minerals, Metals & Materials Society)

3.3 Partial Molar Quantities

While discussing thermodynamics of solutions, we are frequently concerned with the thermodynamic property of a particular component in a solution. This points out towards a question as how to express the molar quantities [6] of a component in the solution. When two liquids are mixed, the total volume of the solution is often not equal to the sum of the individual volumes before mixing. This is due to the difference between the inter-atomic forces in the pure substance and in the solution. For example, on mixing liquid A and liquid B, if ΔV is the change in volume on formation of the solution, one is really confused in assessing the individual contribution made by A and B in the expansion or contraction that occurs on mixing. Similar problems may arise for other thermodynamic properties of the components in solutions. The problem can however be sorted out by the introduction of partial molar quantities. Since the same general treatment is applicable to any extensive thermodynamic quantity such as volume, energy, entropy, and free energy, we shall use the symbol Q to represent any one of these. Prime is used to indicate any arbitrary amount of solution rather than 1 mol, molar quantities are represented as unprimed.

Let us assume that Q' and Q represent the total quantity of solution and the molar solution, respectively. If n_1, n_2, n_3 ... are number of moles of components 1, 2, 3, ..., respectively in the solution, we can write:

$$Q = \frac{Q'}{n_1 + n_2 + n_3 + \ldots} \tag{3.3}$$

If an infinitesimal number of moles, dn_1, of component 1 is added to an arbitrary quantity of a solution at constant temperature and pressure without changing the amount of other constituents, the corresponding increment in the property Q' is noted as dQ', the ratio $\left[\frac{\partial Q'}{\partial n_1}\right]_{P,T,n_2,n_3}$ is known as the partial molar quantity of component, 1 and is designated as $\overline{Q}_1$. Thus,

$$\overline{Q}_1 = \left[\frac{\partial Q'}{\partial n_1}\right]_{P,T,n_2,n_3} \tag{3.4}$$

Analogous relations may be written for other components. $\overline{Q}_1$ may also be represented as the increment of Q' on addition of 1 mol of the first component to a very large quantity of the solution. For example, if the change in volume accompanying the addition of 1 g mol/atom of Cd to a large quantity of a liquid alloy is noted as 6.7 cc, the partial molar volume of cadmium in the alloy at the particular composition, temperature and pressure would be 6.7 cc. This is expressed as $\overline{V}_{Cd} = 6.7$ cc. From the fundamentals of partial differentiation, we have at constant pressure and temperature:

$$dQ' = \left[\frac{\partial Q'}{\partial n_1}\right]_{n_2,\ n_3} dn_1 + \left[\frac{\partial Q'}{\partial n_2}\right]_{n_1,\ n_3} dn_2 + \left[\frac{\partial Q'}{\partial n_3}\right]_{n_2,\ n_3} dn_3 + \ldots.. \tag{3.5}$$

$$\text{or } dQ' = \overline{Q}_1\, dn_1 + \overline{Q}_2\, dn_2 + \overline{Q}_3\, dn_3 + \ldots.. \tag{3.6}$$

If we add to a large quantity of solution n_1 moles of component 1, n_2 moles of component 2 and so on, the increment in Q' after mixing is given as $n_1\overline{Q}_1 + n_2\overline{Q}_2 + \ldots..$

If we now mechanically remove a portion containing $n_1 + n_2 + n_3$ moles, the extensive quantity Q' for the main body of solution is now decreased by $(n_1 + n_2 + n_3)\ Q$. Since at the end of these addition and removal steps, the main body of the solution is the same in composition and amount as it was initially, Q' has the same value finally as it did initially and the increment in Q' accompanying the individual addition is equal to the decrement accompanying their mass withdrawal, which can be expressed as:

$$(n_1 + n_2 + n_3 + \ldots)\ Q = n_1\overline{Q}_1 + n_2\overline{Q}_2 + n_3\overline{Q}_3 + \ldots$$

Dividing by $(n_1 + n_2 + n_3 + \ldots)$ and noting that $\left[\frac{n_i}{n_1+n_2+n_3+\ldots}\right] = x_i$, we get:

$$Q = x_1\overline{Q}_1 + x_2\overline{Q}_2 + x_3\overline{Q}_3 + \ldots. \tag{3.7}$$

On multiplying by $(n_1 + n_2 + n_3 + \ldots)$ we can write as:

$$Q' = n_1\overline{Q}_1 + n_2\overline{Q}_2 + n_3\overline{Q}_3 + \ldots. \tag{3.8}$$

For a binary solution, Eqs. 3.7 and 3.8, are expressed as:

$$Q = x_1\overline{Q}_1 + x_2\overline{Q}_2 \tag{3.9}$$

$$Q' = n_1\overline{Q}_1 + n_2\overline{Q}_2 \tag{3.10}$$

On differentiation of Eq. 3.10, we get:

$$dQ' = n_1 d\overline{Q}_1 + n_2 d\overline{Q}_2 + \overline{Q}_1 dn_1 + \overline{Q}_2 dn_2 \tag{3.11}$$

From Eqs. 3.6 and 3.11, for a binary solution we can write:

$$n_1 d\overline{Q}_1 + n_2 d\overline{Q}_2 = 0 \tag{3.12}$$

On dividing by $(n_1 + n_2)$ we get,

$$x_1 d\overline{Q}_1 + x_2 d\overline{Q}_2 = 0 \tag{3.13}$$

Equation 3.13 is one of the forms of the Gibbs–Duhem equation. This is the most important form of all the equations dealing with solutions, and a number of subsequent relations have been derived from this.

3.3.1 Methods of Estimating Partial Molar Quantities from Molar Quantities

If the experimental value of the molar quantity Q is known as a function of composition the partial molar quantities $\overline{Q}_1$ and $\overline{Q}_2$ can be estimated by analytical as well as by graphical methods.

Analytical Method Differentiation of Eq. 3.9 ($Q = x_1\overline{Q}_1 + x_2\overline{Q}_2$) yields

$$dQ = x_1 d\overline{Q}_1 + x_2 d\overline{Q}_2 + \overline{Q}_1 dx_1 + \overline{Q}_2 dx_2 \tag{3.14}$$

Combining Eqs.3.13 and 3.14, we get:

$$dQ = \overline{Q}_1 dx_1 + \overline{Q}_2 dx_2 \tag{3.15}$$

Multiplying Eq. 3.15 by $\frac{x_1}{dx_2}$ and putting $dx_1 = -dx_2$ (since $x_1 + x_2 = 1$) we get:

$$x_1 \frac{dQ}{dx_2} = x_1 \overline{Q}_1 \cdot \frac{dx_1}{dx_2} + x_1 \overline{Q}_2 = -x_1 \overline{Q}_1 + x_1 \overline{Q}_2$$

Adding this to Eq. 3.9 we get:

$$Q + x_1 \frac{dQ}{dx_2} = x_1 \overline{Q}_2 + x_2 \overline{Q}_2 = \overline{Q}_2$$

$$\therefore \overline{Q}_2 = Q + x_1 \frac{dQ}{dx_2} = Q + (1 - x_2) \frac{dQ}{dx_2} \tag{3.16}$$

$$\text{Similarly,} \quad \overline{Q}_1 = Q + (1 - x_1) \frac{dQ}{dx_1} \tag{3.17}$$

Graphical Method A graphical method, also known as the method of intercepts is illustrated in Fig. 3.4, wherein Q has been plotted as a function of composition x_2. In order to get the partial molar quantities $\overline{Q}_1$ and $\overline{Q}_2$ at a particular composition x_2 a

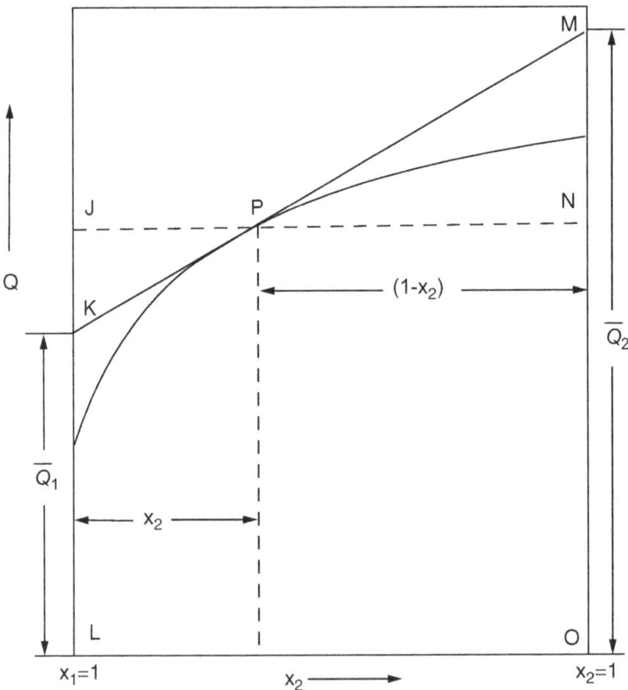

Fig. 3.4 Graphical method of estimating partial molar quantities from molar quantities

tangent KM, is drawn at point P on the curve (corresponding to the composition x_2). From the figure, it is demonstrated as how the ordinates (KL and MO) of this tangent at $x_2 = 0$ and $x_2 = 1$ are equal to $\overline{Q}_1$ and $\overline{Q}_2$, respectively. At point P, the value of Q is equal to the length of the line NO, and the slope of the tangent, as evident from the right angle triangle PMN is $\overline{MN}/(1-x_2)$. Equation 3.16, if presented in terms of the length will take the following form:

$$\overline{Q}_2 = Q + (1-x_2)\ \frac{dQ}{dx_2} = \overline{NO} + (1-x_2)\ \left[\frac{\overline{MN}}{(1-x_2)}\right] = \overline{NO} + \overline{MN} = \overline{MO}$$

Thus the ordinate of the tangent at $x_2 = 1$ is $\overline{Q}_2$.

Similarly, from the right angle triangle PKJ, slope $= \frac{-\overline{JK}}{\overline{JP}}$ and from Eq. 3.17, we can estimate:

$$\overline{Q}_1 = Q + (1-x_1)\ \frac{dQ}{dx_1} = \overline{JL} + (1-x_1)\ \left[\frac{-\overline{JK}}{x_2 = (1-x_1)}\right] = \overline{JL} - \overline{JK} = \overline{KL}$$

Hence, the ordinate of the tangent at $x_2 = 0$ is $\overline{Q}_1$.

3.4 Chemical Potential

The general equation for the free energy change of a system with temperature and pressure: $dG = VdP - SdT$, does not take into account any variation in free energy due to concentration changes. The total free energy of a system must vary when small amounts of constituents dn_A, dn_B, and so on are introduced into a homogeneous solution. Thus, the free energy of a system is dependent on four different variables i.e., $G = f\ (T, P, V, n_i)$, where n_i denotes the addition of n_i moles of a constituent i. The change in free energy can be expressed as total differential using three of the four variables. Any infinitesimal free energy change in terms of a change in temperature, pressure, and composition can be expressed as:

$$\begin{aligned} dG &= \left(\frac{\partial G}{\partial T}\right)_{P,\ n_i} dT + \left(\frac{\partial G}{\partial P}\right)_{T,\ n_i} dP + \sum \left(\frac{\partial G}{\partial n_i}\right)_{T,\ P} dn_i \\ &= -SdT + VdP + \sum \left(\frac{\partial G}{\partial n_i}\right)_{T,\ P} dn_i \end{aligned} \tag{3.18}$$

The coefficient $\left(\frac{\partial G}{\partial n_i}\right)_{T,P}$ is called the "chemical potential" and is denoted by μ_i.

$$\therefore dG = -SdT + VdP + \sum \mu_i dn_i \tag{3.19}$$

At constant pressure and temperature the first two terms are zero in Eq. 3.19 and if the system is at equilibrium, $dG = 0$, hence, $\sum \mu_i dn_i = 0$.

Physical Meaning of Chemical Potential Consider the change in free energy (dG) of a system produced by the addition of dn_A mole of component A at constant pressure and temperature. This change in free energy can be expressed as: $dG = \mu_A dn_A = \overline{G}_A dn_A$, where $\overline{G}_A$ is the partial molar free energy of component A in the solution. Chemical potential of either 1 g mole or 1 g atom of the substance dissolved in a solution of definite concentration is the partial molar free energy. Thus, $\overline{G}_A = \mu_A + \left(\frac{\partial G}{\partial n_A}\right)_{T,P}$.

Chemical Potential in Ideal Solutions In order to calculate the value of a chemical potential it is necessary to obtain a relation between μ and some measurable property. Consider the possibility of calculating the chemical potential of a single gas in an ideal gaseous mixture and then applying this result to other states of matter. It follows therefore that for a very small change of pressure at constant temperature, $dG = VdP$

Since $\overline{G}_A = \mu_A$ for 1 mol of A,
$d\mu = VdP$, and for 1 mol of perfect gas, $(V = RT/P)$

$$\therefore d\mu = RT\frac{dP}{P}$$

$$\text{on integration: } \mu = RT\ lnP + C$$

when $P = 1$ atm, $\mu = C = \mu^o$ (standard chemical potential)

$$\therefore \mu = \mu^o + RT\ lnP \tag{3.20}$$

Thus, the chemical potential of component A in a gaseous mixture exerting partial pressure p_A can be calculated from the equation:

$$\mu_A = \mu_A^o + RT\ ln\, p_A \tag{3.21}$$

$$\text{or } \Delta\mu_A = \mu_A - \mu_A^o = RT\ lnp_A \tag{3.21a}$$

$$\text{and } \Delta\mu_B = \mu_B - \mu_B^o = RT\ lnp_B \tag{3.21b}$$

Since $\Delta\mu$ (similar to the change in free energy of a system) is an extensive property, it depends on the quantity or mass of the system; therefore, the total free energy change associated with the formation of 1 g mol of the solution having x_A mole fraction of the constituent A and x_B mole fraction of the constituent B is the sum of the product of appropriate changes in chemical potential multiplied by the mole fractions of the components. Thus,

$$\Delta\mu_{sol} = \Delta G_{(1\ gmol\ sol)} = x_A \Delta\mu_A + x_B \Delta\mu_B$$

$$\therefore\ \Delta G_{(sol)} = RT(x_A ln\, p_A + x_B\, ln\, p_B) \tag{3.22}$$

These equations applicable to gases can also be used for liquid solutions. If the solution at a given temperature T is at equilibrium with its own vapors, the chemical potential of each component in the gaseous phase and in the liquid phase must be equal.

The partial pressure, p_A of the component A in a binary solution of A and B can be expressed in terms of its mole fraction x_A (from Raoult's law: $p_A = p_A^o.x_A$)

$$\therefore\ x_A = \frac{p_A}{p_A^o}\ \text{(ideal) and } a_A = \frac{p_A}{p_A^o}\ \text{(nonideal)}$$

Thus, from Eq. 3.21, we get:

$$\Delta\mu_A = \mu_A - \mu_A^o = \overline{G}_A - G_A^o = RT\ \ ln\frac{p_A}{p_A^o} = RTlnx_A \tag{3.23}$$

Where, G_A^o is the free energy of 1 g mol of component A in its standard state. Hence, from Eq. 3.23, for 1 g mol of a binary solution consisting x_A mole fraction of A and x_B mole fraction of B, we have:

$$\Delta G_{(sol)} = \Delta G^{M,id} = RT\,(x_A lnx_A + x_B lnx_B)\ \text{ for an ideal solution} \tag{3.24}$$

$$\Delta G_{(sol)} = \Delta G^{M} = RT\,(x_A lna_A + x_B lna_B)\ \text{ for a nonideal solution} \tag{3.25}$$

The free energy of 1 mol of component i in the solution designated as $\overline{G}_i$ is known as the partial molar free energy of the component, i. It represents simply the change in total free energy of the solution when 1 mol of component i is added to a large amount of the solution.

3.5 Fugacity

Lewis [7] pointed out that the use of Eq. 3.21 $(\mu_A = \mu_A^o + RT\ \ ln\, p_A)$ becomes difficult in the calculation of the chemical potential of the gas when it approaches zero pressure because the second term of the equation tends to minus infinite value. In order to avoid this problem he introduced a new function called fugacity. Thus fugacity was made equal to the vapor pressure of a perfect gas and hence, it may be regarded as a corrected pressure.

From the first and second laws of thermodynamics we know: $dG = VdP - S\ dT$ (Eq. 2.69). Hence at constant temperature, $dG = VdP$ and since for 1 mol of an ideal gas: $PV = RT$

$$\therefore dG = \frac{RT}{P}\ dP = RT\, dln\, P \tag{3.26}$$

If p_A and p_B respectively, denote the initial and final pressures of the system with corresponding free energies as G_{p_A} and G_{p_B}, integration of Eq. 3.26 between the limits p_A and p_B at constant temperature results:

$$\Delta G = G_{p_B} - G_{p_A} = RT\ ln\ \frac{P_B}{P_A} \tag{3.27}$$

On dividing both sides by *RT* and rearranging the terms we get:

$$ln\ p_B = \frac{G_{p_B}}{RT} + ln\ p_A - \frac{G_{p_A}}{RT} \tag{3.28}$$

On substituting fugacity Ø in place of pressures p_A and p_B, we get:

$$ln\ Ø_B = \frac{G_{p_B}}{RT} + ln\ Ø_A - \frac{G_{p_A}}{RT} \tag{3.29}$$

At low pressure $Ø = p$, hence the above equation is modified as:

$$ln\ Ø_B = \frac{G_{p_B}}{RT} + lim_{\,p_A \to 0}\left(ln\ Ø_A - \frac{G_{p_A}}{RT}\right) \tag{3.30}$$

Since the second term on the right hand of the equation is constant [2] at a constant temperature, on differentiation we get:

$$dln\ Ø_B = \frac{dG_{p_B}}{RT} \tag{3.31}$$

Equation 3.31 defines fugacity and for any general case it can be written as:

$$dlnØ = \frac{dG}{RT} \tag{3.32}$$

In case of a nonideal gas, the deviation from ideal behavior may be accounted by introducing a term fugacity Ø, for pressure. Hence, Eq. 3.26 can be modified as:

$$dG = RTdlnØ \tag{3.33}$$

Thus, activity (defined as $a_i = \frac{p_i}{p_i^o}$) can now be defined as the ratio of the fugacity of the substance in the state in which it happens to be to its fugacity in its standard state. That is, $a = (Ø/Ø^o)$. Substituting $Ø = a.\ Ø^o$ in Eq. 3.33 we get: $dG = RTdln(a.\ Ø^o)$. As fugacity in its standard state (Ø°) is constant at a particular temperature and composition, we can write:

$$dG = RTdlna \tag{3.34}$$

The standard state of a substance is commonly chosen as the pure liquid and solid at one atmospheric pressure and temperature under consideration. The activity of a pure substance in its standard state is unity. In metallurgical processes we are generally concerned with gases at high temperature and normal pressure of one atmosphere. Under these conditions the forces of inter molecular attraction between molecules becomes insignificant. Hence, the pressure exerted by gases are equal to fugacity that is, they are ideal and obey the ideal gas eq. ($PV = RT$). Contrary to this gases will not behave ideally at high pressure and low temperature. It would be interesting to know the magnitude of error involved in setting fugacity equal to pressure in case of nonideal gases. The extent of deviation in assuming ideal behavior for nonideal gases has been discussed in the following Section.

3.5.1 Fugacity of a Nonideal Gas

The behavior of nonideal gases exhibiting slight departure from ideality may be expressed by the following equation of state

$$V = \frac{RT}{P} - \alpha \tag{3.35}$$

The equation reduces to the ideal gas law when α is zero.

From Eqs. 3.26 ($dG = VdP = \frac{RT}{P}\, dP = RTdln\, P$) and Eq. 3.33 ($dG = RTdln\, Ø$) we get:

$$RTdlnØ = VdP \tag{3.36}$$

Substituting V from Eq. 3.35 in Eq. 3.36 yields

$$RTdlnØ = RTdlnP - \alpha dP \tag{3.37}$$

$$\text{or } dln\left[\frac{Ø}{P}\right] = -\frac{\alpha}{RT}dP \tag{3.38}$$

Integration of the above equation gives

$$ln\,\frac{Ø}{P} = -\frac{\alpha P}{RT} \tag{3.39}$$

$$\therefore \frac{Ø}{P} = e^{-\frac{\alpha P}{RT}} \tag{3.40}$$

$$\text{Or } \frac{\varnothing}{P} = 1 - \frac{\alpha P}{RT} \quad (e^{-x} = 1 - x, \ \textit{for small values of } x) \tag{3.41}$$

Substituting α from Eq. 3.35 in Eq. 3.41, we get:

$$\frac{\varnothing}{P} = 1 - \frac{P}{RT}\left[\frac{RT}{P} - V\right] = \frac{PV}{RT} = \frac{P}{\frac{RT}{V}} \tag{3.42}$$

$\frac{RT}{V}$ is referred to as the ideal pressure, that is, the pressure which would exist in the volume contained in 1 mol of ideal gas at temperature T. If P_i is the ideal pressure, the above equation can be expressed as:

$$\frac{\varnothing}{P} = \frac{P}{P_i} \tag{3.43}$$

From the above expression it is concluded that the actual pressure P is the geometric mean of the fugacity and the ideal pressure ($P = \sqrt{\varnothing P_i}$). However, it should be noted that the above considerations apply to a gas that exhibits only a moderate deviation from an ideal gas at the pressure and temperature under consideration.

3.6 The Free Energy Change Due to the Formation of a Solution

Since under standard conditions the free energy of a component i expressed by G_i^o and its partial molar free energy in the solution by $\overline{G}_i$, integration of Eq. 3.34 gives:

$$\int_{G_i^o}^{\overline{G}_i} dG = \int_{a_i^o}^{a_i} RTdlna \tag{3.44}$$

$$\text{or } \overline{G}_i - G_i^o = RTlna_i - RTlna_i^o = RTln\left(\frac{a_i}{a_i^o}\right) \tag{3.45}$$

since $a_i^o = 1$,

$$\overline{G}_i - G_i^o = RTlna_i = \Delta\overline{G}_i^M \tag{3.46}$$

The difference ($\overline{G}_i - G_i^o$) represents the partial molar free energy of mixing (on formation of the solution) of the component i and is denoted by $\Delta\overline{G}_i^M$. This is simply the free energy change [4] accompanying the dissolution of 1 mol of i in the solution. In a binary solution, A–B, containing x_A and x_B mole fractions of A and B, respectively, the integral molar free energy of mixing can be estimated as:

$$\Delta G^M = x_A \Delta \overline{G}_A^M + x_B \Delta \overline{G}_B^M = RT(x_A ln\, a_A + x_B\, ln\, a_B) \tag{3.47}$$

since for an ideal solution $a_i = x_i$

$$\Delta G^{Mid} = RT(x_A ln\, x_A + x_B\, ln\, x_B) \tag{3.48}$$

If at constant T and P, n_A number of mole of A and n_B of moles of B are mixed to form a binary solution, the free energy before mixing $= \left(n_A\ G_A^o + n_B G_B^o\right)$ and after mixing $= (n_A \overline{G}_A + n_B \overline{G}_B\ )$. The free energy change for the entire body of solution due to mixing, $\Delta G^{'M}$, referred as the integral free energy of mixing, is the difference between the two quantities, that is,

$$\Delta G^{'M} = \left(n_A \overline{G}_A + n_B \overline{G}_B\right) - \left(n_A\, G_A^o + n_B G_B^o\right) \tag{3.49}$$

$$= n_A \left(\overline{G}_A - G_A^o\right) + n_B \left(\overline{G}_B - G_B^o\right) \tag{3.50}$$

$$= n_A \Delta \overline{G}_A^M + n_B \Delta \overline{G}_B^M \tag{3.51}$$

$$= RT\, (n_A\, lna_A + n_B\, lna_B) \tag{3.52}$$

Dividing by (n_A + n_B),we get integral molar free energy of mixing for nonideal and ideal solutions as expressed by Eqs. 3.47 and 3.48, respectively. The integral molar entropy of mixing for an ideal solution is given as:

$$\Delta S^{Mid} = -R\, (x_A ln\, x_A + x_B\, ln\, x_B) \tag{3.53}$$

3.7 Equilibrium Constant

Consider a general chemical reaction taking place between different numbers of moles of reactants to form different numbers of moles of products at constant temperature and pressure.

$$lL + mM + \ldots = qQ + rR + \ldots \tag{3.54}$$

The change in free energy for the above reaction may be expressed as:

$$\Delta G = q\overline{G}_Q + r\overline{G}_R + \ldots - l\overline{G}_L - m\overline{G}_M - \ldots \tag{3.55}$$

when all the products and reactants are in their standard state:

$$\Delta G^o = qG_Q^o + rG_R^o + \ldots - lG_L^o - mG_M^o - \ldots \tag{3.56}$$

On subtracting Eq. 3.55 from Eq. 3.56, we get:

$$\Delta G^o - \Delta G = q\left(G_Q^o - \overline{G}_Q\right) + r\left(G_R^o - \overline{G}_R\right) + \ldots - l\left(G_L^o - \overline{G}_L\right) - m\left(G_M^o - \overline{G}_M\right) - \ldots$$

substituting, $\overline{G}_i - G_i^o = RTlna_i$ we get:

$$\Delta G^o - \Delta G = -qRTlna_Q - rRTlna_R - \ldots + lRTlna_L + mRTlna_M + \ldots$$

$$\Delta G = \Delta G^o + RTln\frac{a_Q^q.a_R^r}{a_L^l.a_M^m} \tag{3.57}$$

Equation 3.57 is known as van't Hoff isochore [2, 3] as discussed in Sect. 2.11.2. Thermodynamic equilibrium constant for the above chemical reaction is expressed as:

$$K = \frac{a_Q^q.a_R^r}{a_L^l.a_M^m} \tag{3.58}$$

Since, at equilibrium, $\Delta G = 0$

$$\therefore \Delta G^o = -RTlnK \tag{3.58a}$$

3.8 Gibbs–Duhem Integration

Thermodynamic equations for calculation of excess free energy and integral molar free energy of a solution need activity as well as activity coefficient of all the components present in the solution. However, experimental techniques, namely, chemical equilibria, vapor pressure and electrochemical can measure activity of only one component. For example,

1. Activity of mercury in mercury-thallium alloy is obtained by measuring the partial pressure of mercury over the alloy at a temperature, where, $p_{Hg} \gg p_{Tl}$, that is, at the working temperature partial pressure of thallium is negligible as compared to that of mercury. The same experiment cannot be used for measuring the partial pressure (hence activity) of thallium accurately.
2. Activity of carbon in Fe-C (or steel) can be determined by equilibrating CO-CO_2 or H_2-CH_4 gaseous mixture with steel at a particular temperature. However, we cannot obtain the activity of iron by this method. Similarly, a_S in Fe-S can be measured by using H_2-H_2S gaseous mixture but not a_{Fe}.

3. Concentration galvanic cell like *Cd(l)/LiCl – KCl*+$CdCl_2$/ *Cd – Pb(l)* determines the activity of cadmium only by measuring the open circuit emf of the cell, which is related to the partial molar free energy of mixing and activity as follows:

$$\Delta\overline{G}^{M}_{Cd} = RTlna_{Cd} = -nFE \tag{3.59}$$

We cannot use this cell: Pb (l)/LiCl-KCl + $PbCl_2$/Cd-Pb (l) to get a_{Pb} because the formation of more stable $CdCl_2$ will cause exchange reaction of the type: $PbCl_2$ (l) + Cd (l) = $CdCl_2$ (l) + Pb (l). Thus, cell emf will not be a true representative of the cell reaction but will represent a mixed potential.

Now the question is how to get the activity of the second component in a binary system. In order to calculate the activity of the second component we must couple the activity and atom/mole fractions of both the components with the aid of the Gibbs–Duhem equation [1–4, 6]:

$$\sum x_i d\overline{Q}_i = 0$$

where $\overline{Q}_i$ is any partial molar extensive property.

Since activity of a component is related to the partial molar free energy, we can write the Gibbs–Duhem equation for a binary system, A–B as under:

$$x_A d\Delta\overline{G}^{M}_{A} + x_B d\Delta\overline{G}^{M}_{B} = 0, \quad \text{since } \Delta\overline{G}^{M}_{i} = RTlna_i$$

$$\therefore \quad (x_A dlna_A + x_B dlna_B) = 0 \tag{3.60}$$

$$\text{or } dlna_A = -\frac{x_B}{x_A} dlna_B \tag{3.61}$$

If the variation of a_B with composition is known, the value of lna_A at the composition $x_A = x_A$ can be obtained by integration of the above equation.

$$\ln a_{A\ at\ x_A = x_A} = -\int_{na_B\ at\ x_A = 1}^{ln\,a_B\ at\ x_A = x_A} \left[\frac{x_B}{x_A}\right] dln\ a_B \tag{3.62}$$

This can be solved by graphical method of integration (Fig. 3.5). From the plot of $-\ ln\ a_B\ vs\ \frac{x_B}{x_A}$, the value of $\text{ln}a_A at\ x_A = x_A$ is obtained by estimating the shaded area under the curve.

The following points must be noted:

1. As $x_B \rightarrow 1$ (i.e. $x_A \rightarrow 0$), $a_B \rightarrow 1$, $\text{ln}a_B \rightarrow 0$ and $\frac{x_B}{x_A} \rightarrow \infty$. Thus the curve exhibits a tail to infinity as $x_B \rightarrow 1$ (y-axis).
2. As $x_B \rightarrow 0$ (i.e. $x_A \rightarrow 1$), $a_B \rightarrow 0$, $\text{ln}a_B \rightarrow -\infty$. Thus the curve exhibits a tail to minus infinity as $x_B \rightarrow 0$ (x-axis).

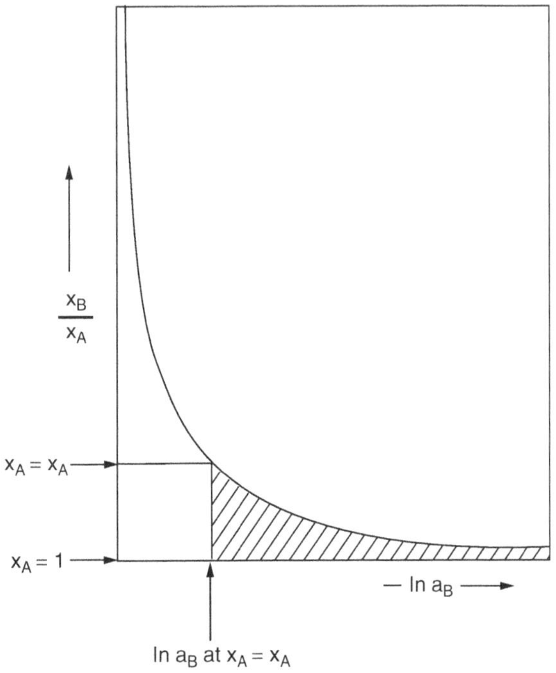

Fig. 3.5 A schematic illustration of the variation of lna_B with $\frac{x_B}{x_A}$ in a binary solution A–B

Since the calculation of $\ln a_A$ at any composition involves counting of squares under a curve which approaches minus infinity this introduces an uncertainty into the calculation. Thus the graphical method of integration lacks precision

3.8.1 An Improved Method of Integration

The tail to $-\infty$ as $x_B \rightarrow 0$ can be avoided by considering activity coefficient in the Gibbs–Duhem equation. In a binary solution, A–B we have:

$x_A + x_B = 1$, on differentiation we get:
$dx_A + dx_B = 0$, multiplying the first term by x_A/x_A and second by x_B/x_B, we get:

$$x_A \frac{dx_A}{x_A} + x_B \frac{dx_B}{x_B} = 0$$

$$\text{or } x_A d\ln x_A + x_B d ln x_B = 0 \tag{3.63}$$

Substracting Eq. 3.63 from Eq. 3.60, we get:

$$x_A d\ln\frac{a_A}{x_A} + x_B d\ \ ln\ \frac{a_B}{x_B} = 0 \quad \text{(by definition,}\ \ \gamma_i = a_i/x_i)$$

$$\therefore\ x_A dln\gamma_A + x_B dln\gamma_B = 0 \tag{3.64}$$

From Eq. 3.64, the activity coefficient of one component of a binary system can be estimated if that of the other is known over a range of composition. Thus on rearrangement of terms we can write as:

$$d\ln\gamma_A = -\frac{x_B}{x_A} d\ \ ln\,\gamma_B \tag{3.65}$$

Thus if the variation of γ_B is known over a range of composition, the value of ln γ_A at the composition $x_A = x_A$ can be obtained by integration of Eq. 3.65:

$$\ln\gamma_{A\ at\ x_A = x_A} = -\int_{ln\gamma_B\ at\ x_A = 1}^{ln\gamma_B\ at\ x_A = x_A} \left[\frac{x_B}{x_A}\right] dln\,\gamma_B \tag{3.66}$$

Figure 3.6 illustrates a typical variation of $ln\ \gamma_B$ with composition in a binary solution, A–B. The value of $\ln\gamma_A at\ x_A = x_A$ is given as the shaded area under the curve between the limits $ln\gamma_B\, at\, x_A = x_A$ and $ln\gamma_B\, at\, x_A = 1$. Following points must be noted:

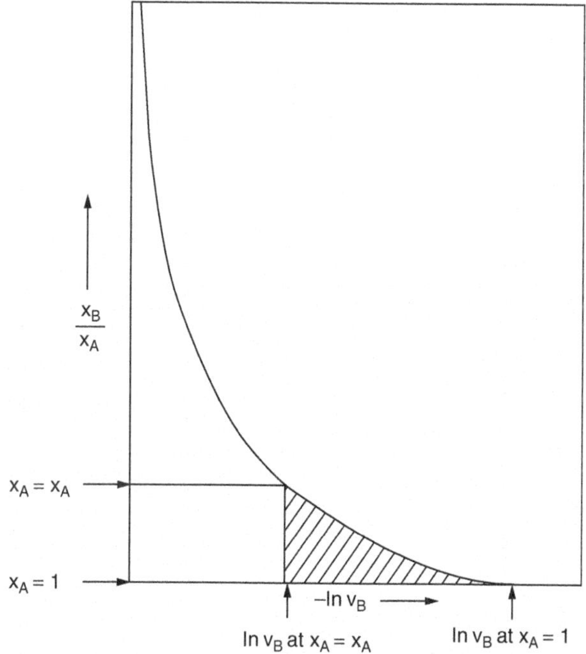

Fig. 3.6 A schematic illustration of the variation of $ln\gamma_B$ with $\frac{x_B}{x_A}$ in a binary solution A–B

1. As $x_B \rightarrow 1$ (*i.e.*, $x_A \rightarrow 0$), $a_B \rightarrow 1$, $\ln\gamma_B \rightarrow 0$ but $\frac{x_B}{x_A} \rightarrow \infty$, Thus the curve exhibits a tail to infinity as $x_A \rightarrow 0$ (y-axis).
2. As $x_B \rightarrow 0$, $x_A \rightarrow 1$, $a_B \rightarrow 0$, but γ_B is finite, hence $ln\ \gamma_B$ is finite (x-axis)

This method of integration also introduces an uncertainty on one axis.

3.8.2 α-Function in the Integration of the Gibbs–Duhem Integration

The problem of uncertainty in the integration of Eq. 3.66 has been solved with the aid of **α**-function which is defined as:

$$\alpha_i = \frac{\ln \gamma_i}{(1 - x_i)^2}$$

For a binary solution A–B,

$$\alpha_A = \frac{\ln \gamma_A}{(1 - x_A)^2} = \frac{\ln \gamma_A}{x_B^2} \quad \text{and} \quad \alpha_B = \frac{\ln \gamma_B}{x_A^2} \tag{3.67}$$

1. If $x_A \rightarrow 0$, $x_B \rightarrow 1$, $\gamma_B \rightarrow 0$. Using L. Hospitals rule, it can proved that α_B will always have a finite value.
2. If $x_B \rightarrow 0, x_A \rightarrow 1, \gamma_B \rightarrow \gamma_B^o$. Hence, α_B is finite at $x_B = 0$.

α_A would follow a similar behavior.

On differentiation of Eq. 3.67, we get

$$d \ln \gamma_A = 2\alpha_A x_B dx_B + x_B^2 d\alpha_A \tag{3.68}$$

$$d \ln \gamma_B = 2\alpha_B x_A dx_A + x_A^2 d\alpha_B \tag{3.69}$$

On substituting d ln γ_B in Eq. 3.65 ($d \ln \gamma_A = -\frac{x_B}{x_A} d\ ln\, \gamma_B$) we get:

$$\begin{aligned} d \ln \gamma_A &= -\frac{x_B}{x_A} 2\alpha_B x_A dx_A - \frac{x_B}{x_A} x_A^2 d\alpha_B \\ &= -2x_B \alpha_B dx_A - x_B x_A d\alpha_B \end{aligned} \tag{3.70}$$

On integration of Eq. 3.70 between $x_A = x_A\ to\ x_A = 1$,we get:

$$\ln \gamma_A = - \int_{x_A=1}^{x_A=x_A} 2x_B \alpha_B dx_A - \int_{\alpha_B \text{ at } x_A=1}^{\alpha_B \text{ at } x_A=x_A} x_B x_A d\alpha_B \tag{3.71}$$

Making use of the formula for integration by parts $\int d(xy) = \int y\,dx + \int x\,dy$, we can write:

$$\int d(x_B x_A \alpha_B) = \int x_B x_A d\alpha_B + \int \alpha_B d(x_B x_A)$$

$$\text{or} \int x_B x_A d\alpha_B = \int d(x_B x_A \alpha_B) - \int \alpha_B d(x_B x_A) \tag{3.72}$$

Substituting Eq. 3.72 into Eq. 3.71 and on integration from limit $x_A = 1$ to $x_A = x_A$, we get:

$$\begin{aligned}
\ln \gamma_A &= - \int 2x_B \alpha_B dx_A - \int d(x_B x_A \alpha_B) + \int \alpha_B d(x_B x_A) \\
&= - \int 2x_B \alpha_B dx_A - x_B x_A \alpha_B + \int \alpha_B x_B dx_A + \int \alpha_B x_A dx_B \\
&= - \int 2x_B \alpha_B dx_A - x_B x_A \alpha_B + \int \alpha_B x_B dx_A - \int \alpha_B x_A dx_A \\
&= - \; x_B x_A \alpha_B - \int (2x_B - x_B + x_A) \alpha_B dx_A \\
&= - \; x_B x_A \alpha_B - \int_{x_A=1}^{x_A=x_A} \alpha_B dx_A
\end{aligned} \tag{3.73}$$

Thus $\ln\gamma_A$ at $x_A = x_A$ is obtained as $-x_B x_A \alpha_B$ minus the area under the plot of $\alpha_B vs\, x_A$ from $x_A = x_A$ to $x_A = 1$ (Fig. 3.7). Since α_B is everywhere finite, this integration does not involve a tail to infinity. α_B may be negative or positive as per respective negative

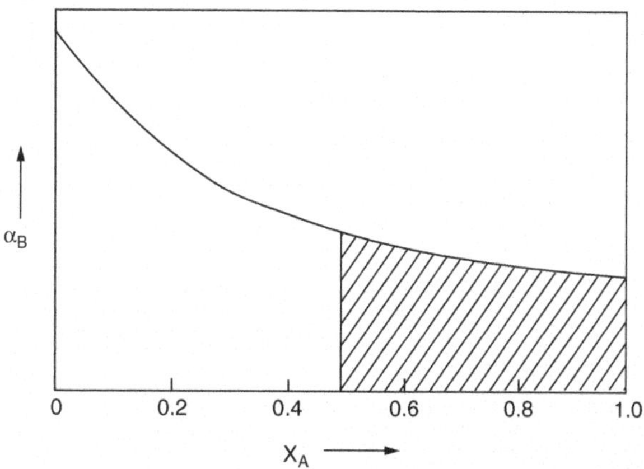

Fig. 3.7 Variation of α_B with composition in the binary system A–B

or positive deviation from Raoult's law in the system A–B. Accordingly, the integrand area under the curve from $x_A = x_A$ to $x_A = 1$ in the plot of α_B *vs* x_A,will be positive or negative.

3.9 Dilute Solutions

According to Henry's law, the partial pressure of a solute in a dilute solution is proportional to mole/atom fraction. If B is a solute, we can write: $p_B \propto x_B$.that is, $p_B = k.\ x_B$ dividing by p_B^o, we get:

$$\frac{p_B}{p_B^o} = \frac{k}{p_B^o}.x_B, \quad \text{that is,} \quad a_B = k' x_B \tag{3.74}$$

$$\text{Hence, } a_B \propto x_B$$

The constant, known as the activity coefficient of the solute B at infinite dilution, is equal to the slope of the curve at zero concentration of B, and is designated by γ_B^o (Fig. 3.8). Like Raoult's law, Henry's law is valid within a concentration range. The extent varies from one system to another, but it is valid only at low concentrations.

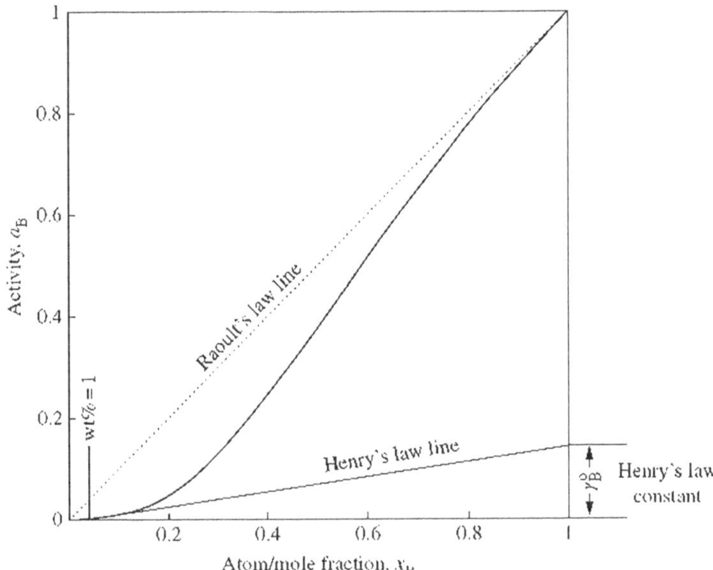

Fig. 3.8 Conversion of activities from one standard state to another (Reproduced from *Physical Chemistry of Metallurgical Processes* by M. Shamsuddin, [5] with the permission of The Minerals, Metals & Materials Society)

In concentrated solutions, the standard state has been defined as unit atmospheric pressure and unit activity for a pure substance at any temperature. In dilute solutions, relative standard states other than pure substances are being used. Henry's law offers two such standard states, called alternative standard states.

1. Infinitely dilute, atom/mole fraction standard state.
2. Infinitely dilute, wt% (w/o or %) standard state.

Solubility of Gases

It is important to note that the validity of Henry's law depends upon the proper choice of solute species. For example, nitrogen dissolves as different species in water and in liquid iron.

(a) In water, nitrogen dissolves molecularly as N_2:

$N_2(g) \leftrightarrows N_2$ (dissolved in water)

$K = \frac{a_{N_2}}{p_{N_2}}$ (as solubility of N_2 in water is low, according to Henry's law: Eq. 3.74) we can write:

$$a_{N_2} = k' x_{N_2}$$

$$\therefore \; K = \frac{k' x_{N_2}}{p_{N_2}}$$

$$\text{Hence } x_{N_2}\,(\text{solubility}) = \frac{K}{k'} \cdot p_{N_2} = k'' . p_{N_2} \tag{3.75}$$

Thus, the solubility of nitrogen in water is proportional to the partial pressure of nitrogen gas in equilibrium with water. Solubility can be expressed as mole fraction, cc per 100 g of water, or any other unit.

(b) In the second case under consideration, nitrogen dissolves atomically in solid or liquid metals:

$$N_2(g) \leftrightarrows 2\,[N] \;\; (\text{in Fe})$$

$$K = \frac{[a_N]^2}{p_{N_2}} = \frac{[k' x_N]^2}{p_{N_2}}$$

$$\therefore \; [x_N] \;\; (\text{solubility}) = \frac{\sqrt{K}}{k'} \; \sqrt{p_{N_2}} = k'' \sqrt{p_{N_2}}$$

Both the cases are in accordance with the experiment. Since all the common diatomic gases N_2, O_2, H_2, and so on dissolve atomically in metals, the general expression for solubility is given as:

$$s = k\sqrt{p_{N_2}} \tag{3.76}$$

This is known as Sievert's law and can be stated as—solubility of diatomic gases in metals is directly proportional to the square root of partial pressure of the gas in equilibrium with the metal.

3.9.1 Alternative Standard States

In dilute solutions two different standard states [8] are used.

1. *Infinitely dilute, atom fraction standard state*: The standard state is so defined that the Henrian activity approaches the atom fraction at infinite dilution.

$$a_B^H = x_B \ \ as \ \ x_B \to 0 \tag{3.77}$$

a_B^H denotes Henrian activity of B. In other words, in the concentration range where Henry 's law is obeyed, $a_B^H = x_B$, that is, Henrian activity = atom fraction. Beyond this concentration range, the activity can be related to the atom fraction by the expression:

$$a_B^H = f_B . x_B \tag{3.78}$$

where f_B is the Henrian activity coefficient of B, relative to the infinitely dilute atom fraction standard state.

The relation between the activity of B relative to the pure substance standard state and the activity of B relative to the infinitely dilute atom fraction standard state, is given by

$$\frac{activity\ of\ B\ relative\ to\ pure\ substance\ standard\ state}{activity\ of\ B\ relative\ to\ the\ infinitely\ dilute\ atom\ fraction\ standard\ state}$$
$$= \gamma_B^o \, (at \ \ constant \, x_B)$$

$$\left[a_i = \gamma_i . \ x_i, \ \ \gamma_i = \frac{a_i}{x_i} = \frac{a_i^R}{a_i^H} = \gamma_i^o \right]$$

where γ_B^o is the Raoultian activity coefficient of B at infinite dilution.

The free energy change accompanying the transfer of 1 mol of B from pure B as standard state to the infinitely dilute, atom fraction standard state may be calculated as follows:

B (pure substance standard state) $\to$ B (dilute, atom fraction standard state)

$$\Delta G^o = \Delta G_B^o(\text{pure substance standard state}) \\ - \Delta G_B^o(\text{infinitely dilute, atom fraction standard state})$$

$$= RTln\left[\frac{activity\ of\ B\ relative\ to\ pure\ substance\ standard\ state, a_B^R}{activity\ of\ B\ relative\ to\ the\ infinitely\ dilute\ atom\ fraction\ standard\ state, a_B^H}\right]$$

$$= RTln\gamma_B^o \tag{3.79}$$

2. *Infinitely dilute weight percent standard state*: This is the most widely used standard state in metallurgy and may be so defined that the Henrian activity approaches the wt% at infinite dilution, that is,

$$a_B^{wt\%} = wt\%\ B\ as\ wt\%\ B \to 0 \tag{3.80}$$

Assuming that the solution obeys Henry's law up to 1 wt% B, then Henrian activity, $a_B^{wt\%}$ is unity at this concentration and the standard state is 1 wt% solution. Deviation from the equality is measured in terms of Henrian activity coefficient, relative to infinitely dilute wt% standard state, that is,

$$a_B^{wt\%} = f_B.wt\%B \tag{3.81}$$

In both the cases, f_B has been used for Henrian activity coefficient of B.

The relation between the activity of B, a_B relative to the pure substance standard state and the activity of B relative to the infinitely dilute wt% standard state, is given by

$$\frac{activity\ of\ B\ relative\ to\ pure\ substance\ standard\ state, a_B^R}{activity\ of\ B\ relative\ to\ the\ infinitely\ dilute\ wt\%standard\ state, a_B^{wt\%}}$$

$$= \gamma_B^o \frac{x_B}{wt\%B}\ (at\ constant\ x_B)$$

$$\left[\text{In Henrian solution, activity } a_B^R = \gamma_B^o.x_B\right]$$

The free energy change accompanying the transfer of 1 mol of B from pure B as standard state to the infinitely dilute wt% standard state may be calculated as follows:

B (pure substance standard state) → B (dilute wt% standard state)

$$\Delta G^o = \Delta G_B^o(pure\ substance\ standard\ state) \\ - \Delta G_B^o(infinitely\ dilute\ wt\%standard\ state)$$

$$= RTln\left[\frac{activity\ of\ B\ relative\ to\ pure\ substance\ standard\ state,\ a_B^R}{activity\ of\ B\ relative\ to\ the\ infinitely\ dilute\ wt\%standard\ state,\ a_B^{wt\%}}\right]$$

$$= RTln\left[\gamma_B^o\ \frac{x_B}{wt\%B}\right] \tag{3.82}$$

$$= RTln\left[\gamma_B^o\ \frac{M_A}{100M_B}\right]$$

$$= RTln\gamma_B^o + RTln\ \left[\frac{M_A}{100M_B}\right] \tag{3.83}$$

where M_A and M_B refer to atomic weights of A and B, respectively.

The atom/mole fraction of B in solution of $A - B$ can be expressed as:

$$x_B = \frac{wt\%B/M_B}{(wt\%B/M_B) + (100 - wt\%B/M_A)} = \frac{wt\%B/M_B}{100/M_A}$$

(at infinitely dilute concentration of the solute B, $wt\ \%\ B$ is negligible as compared to 100) hence,

$$\frac{x_B}{wt\%\ B} = \frac{M_A}{100M_B} \tag{3.84}$$

$$\therefore\ wt\%B = \frac{x_{B.}100.M_B}{M_A} \tag{3.85}$$

3.9.2 Relation Between Different Standard States

According to the Raoult's law for ideal solution:

$a_B^R = x_B$. In Henrian solution: $a_B^R = \gamma_B^o.x_B$, where γ_B^o is the Raoultian activity coefficient at infinite dilution. This can also be expressed as:

$$\frac{a_B^R}{\gamma_B^o.x_B} = 1 \tag{3.86}$$

Henry's law gives: $a_B^H = f_B.x_B$ ($f_B = 1$ for Henrian solution)

$$\therefore\ a_B^H = x_B,\ \text{that is,}\ \frac{a_B^H}{x_B} = 1 \tag{3.87}$$

Similarly,

$$a_B^{1\ wt\%} = wt\%B, \ \text{that is,} \ \frac{a_B^{1\ wt\%}}{wt\%B} = 1 \tag{3.88}$$

From Eqs. 3.86, 3.87, and 3.88, we get:

$$\frac{a_B^R}{\gamma_B^o.x_B} = \frac{a_B^H}{x_B} = \frac{a_B^{1\ wt\%}}{wt\%B} \tag{3.89}$$

$$\text{Hence,} \quad \frac{a_B^R\,(pure)}{a_B\,(dilute\ wt\%)} = \frac{\gamma_B^o.x_B}{wt\%B} \tag{3.90}$$

For example in a *Fe-Si* system when $x_{Si} \rightarrow 0$

$$\frac{a_{Si}\,(pure)}{a_{Si}\,(dilute\ \ wt\%)} = \frac{\gamma_{Si}^o.\ x_{Si}}{wt\%\ \ Si} \qquad \text{since} \frac{x_{Si}}{wt\%\ \ Si} = \frac{M_{Fe}}{100M_{Si}}$$

$$\therefore \frac{a_{Si}\,(pure)}{a_{Si}\,(dilute\ \ wt\%)} = \frac{\gamma_{Si}^o.\ M_{Fe}}{100M_{Si}}$$

3.9.3 Activity and Interaction Coefficients in Multicomponent Systems

Although we have considered only binary solutions so far, in metallurgical processes, we rarely deal with binary solutions. Hence, we must take into account the interaction between various solutes in multicomponent systems. If the components of binary solutions interact with one another, those of multicomponent solutions will certainly interact in a more complex manner.

As an example, take a dilute solution of carbon in liquid iron. The Henrian activity coefficient of carbon in the binary solution, f_C^C, is raised by the addition of small amount of sulfur and lowered by the addition of small amount of chromium in the solution. Since in liquid iron, a number of solutes are present in dilute concentrations, each thermodynamic property of the system is influenced by the interaction of various solutes. Interaction causes marked changes in solute activities and the data for binary solutions do not apply to more complex systems.

Probable effects of one solute on the thermodynamic behavior of another solute can be expressed by the interaction parameter (ε) and interaction coefficient (e) [9], which are defined as:

$$\varepsilon_i^j = \left[\frac{\partial\, ln\,\gamma_i}{\partial x_j}\right]_{x_1 \rightarrow 1} \quad \text{and} \quad e_i^j = \left[\frac{\partial\, log f_i}{\partial \%j}\right]_{wt\%1 \rightarrow 100} \tag{3.91}$$

Thus, (ε) and (e) are related to the alternative standard states in dilute solutions. If the solute i interacts more strongly with the solute j than with the solvent 1 (in the system $1-i-j$), the activity coefficient of i, $\gamma_i/(f_i)$ decreases by the addition of j, so ε_i^j, (and also e_i^j) is a negative quantity. Conversely, if j interacts more strongly with the solvent as compared to the solute i, $\gamma_i/(f_i)$ will tend to increase and ε_i^j and also e_i^j will be positive (i.e., ε_i^j/e_i^j is positive, if j interacts more strongly with the solvent).

Therefore, if carbon is the solute considered, f_C is lowered by the addition of solutes that tend to form more stable carbide, for example, Cr, V or Nb, on the other hand, S and P raise the f_C. It is important to note that adequate representation of the equilibria encountered in steelmaking is possible only if the effects of the various solutes on the activity coefficient are taken into account. For instance, the presence of 2 $wt\ \%\ C$ in liquid iron increases the activity coefficient of silicon by a factor of 2.

The interaction coefficients result from the Taylor series expansion of the partial molar excess Gibbs free energy of mixing, $\overline{G}_i^{xs}$ of the solute component i, in the solvent, iron denoted as component 1. In atom fraction as the standard state we have:

$$\overline{G}_i^{xs} = RTln[\gamma_i]_{T,P,x_1,x_2,\ldots,x_n} \tag{3.92}$$

Expansion of Eq. 3.92 in case of infinitely dilute solution as reference state:

$$\begin{aligned} RTln\gamma_i &= RT\left[ln\,\gamma_i^o\right]_{T,\ P,\ x_1\to 1} + \sum_{j=2}^{m} RT\ \left[\frac{\partial\, ln\,\gamma_i}{\partial x_j} x_j\right]_{T,\ P,\ x_1\to 1} \\ &+ \sum_{j=2}^{m} \frac{1}{2} RT\ \left[\frac{\partial^2\, ln\,\gamma_i}{\partial x_j^2}\ x_j^2\right]_{T,\ P,\ x_1\to 1} \\ &+ \sum_{j=2}^{m}\sum_{k=2}^{m} RT\ \left[\frac{\partial^2\, ln\,\gamma_i}{\partial x_j \partial x_k} x_j x_k\right]_{T,\ P,\ x_1\to 1} + \ldots \end{aligned} \tag{3.93}$$

Accuracy of data does not permit to account for the third order terms, hence,

$$ln\,\gamma_i = ln\gamma_i^o + \sum_{j=2}^{m} \varepsilon_i^j x_j + \sum_{j=2}^{m} \rho_i^j x_j^2 + \sum_{j=2}^{m}\sum_{k=2}^{m} \rho_i^{j,k} x_j x_k \tag{3.94}$$

Equation 3.94 describes the thermodynamic properties of the ith component in an n-component system. By convention γ_i is the activity coefficient based on atom fraction as the composition coordinate, γ_i^o is the value at infinite dilution.

Partial derivative of γ is named as interaction parameter.

$\varepsilon_i^j \equiv$ interaction parameter of j on i in $1-i-j$ system simply tells the effect of j on the activity coefficient of i.

$$\varepsilon_i^j = \left[\frac{\partial\, ln\,\gamma_i}{\partial x_j}\right]_{x_1\to 1} \quad \text{in } i-i-j \text{ system (read as epsilon } j \text{ upon } i\text{)}$$

$$\varepsilon_i^i\,(\text{self interaction coefficient}) = \left[\frac{\partial\, ln\,\gamma_i}{\partial x_i}\right]_{x_1\to 1}$$

$$\varepsilon_2^3 = \left[\frac{\partial\, ln\,\gamma_2}{\partial x_3}\right]_{x_1\to 1} \quad (\text{epsilon 3 upon 2})$$

and ρ denotes second order interaction parameter

$$\rho_i^j = \frac{1}{2}\left[\frac{\partial^2 ln\gamma_i}{\partial x_j^2}\right]_{x_1\to 1}$$

$$\rho_i^{j,k} = \left[\frac{\partial^2\, ln\,\gamma_i}{\partial x_j \partial x_k}\right]_{x_1\to 1} \quad (\text{cross product})$$

In a simplified manner for the system 1–2–3–4, we can write:

$$ln\,\gamma_2 = ln\,\gamma_2^o + \varepsilon_2^2 x_2 + \varepsilon_2^3 x_3 + \varepsilon_2^4 x_4 + \ldots + \rho_2^2 x_2^2 + \rho_2^3 x_3^2 + \ldots + \rho_2^{2,3} x_2 x_3 + \ldots \tag{3.95}$$

In a system of 1–2–3 (2 and 3 are solutes and 1 is a solvent) at constant temperature and pressure it can be demonstrated that $\varepsilon_2^3 = \varepsilon_3^2$. This is known as the Wagner's Reciprocal Relationship.

In many instances it is convenient to use weight percent (wt%) as the composition coordinate. Hence, the activity coefficient, f_i in the zeoeth order form, $\log f_i^o$ disappears since the activity coefficient at infinite dilution is assigned the value of unity.

$$e_i^j = \left[\frac{\partial\, log f_i}{\partial \%j}\right]_{wt\%1\to 100} \quad \text{and } e_2^3 = \left[\frac{\partial\, log f_2}{\partial \%3}\right]_{wt\%1\to 100}$$

$$f_i = 1,\ f_i^o = 1,\ \log f_i^o = 0$$

Taylor series expansion:

$$\begin{aligned} log f_i = {} & \left[log f_i^o\right]_{\%1\to 100} + \sum\nolimits_{j=2}^m \left[\frac{\partial\, log f_i}{\partial(\%j)}\right]_{\%1\to 100} (\%j) \\ & + \sum\nolimits_{j=2}^m \frac{1}{2}\left[\frac{\partial^2\, log f_i}{\partial(\%j)^2}\right]_{\%1\to 100} (\%j)^2 \\ & + \sum\nolimits_{j=2}^m \sum\nolimits_{k=2}^m RT \left[\frac{\partial^2\, log f_i}{\partial(\%j)\partial(\%k)}\right]_{\%1\to 100} \partial(\%j)\partial(\%k) + \ldots \end{aligned} \tag{3.96}$$

or

$$\log f_i = \sum_{j=2}^{m} e_i^j(\%j) + \sum_{j=2}^{m} r_i^j(\%j)^2 + \sum_{j=2}^{m}\sum_{k=2}^{m} r_i^{(j,k)}\partial(\%j)\partial(\%k) + \ldots \tag{3.97}$$

where e and r denote first and second order interaction coefficients, respectively.

3.9.4 Relation Between ε and e

(x, pure 2 standard state) = 2 (wt% hypothetical standard state)

$$K = \frac{a_2^{(wt\%)}}{a_2} = \frac{f_2^{(wt\%)}.(wt\%2)}{x_2\gamma_2} \tag{3.98}$$

$$ln\,\gamma_2 = ln\left(\frac{wt\%\ 2}{x_2}\right) + lnf_2^{(wt\ \%)} - lnK \tag{3.99}$$

or

$$\left[\frac{ln\,\gamma_2}{\partial x_3}\right]_{x_1\to 1} = \left[\frac{\partial\ \ln(\%2/x_2)}{\partial x_3}\right]_{x_1\to 1} + \left[\frac{\partial\ lnf_2^{(wt\%)}}{\partial(wt\%3)}\right]_{\%1\to 100}\left[\frac{\partial(wt\%3)}{\partial x_3}\right]_{x_1\to 1} \tag{3.100}$$

We know that:

$$\left[\frac{\partial\ lnf_2^{(wt\%)}}{\partial(wt\%3)}\right]_{\%1\to 100} = \left[\frac{2.303\partial\ log f_2^{(wt\%)}}{\partial(wt\%3)}\right]_{\%1\to 100} = 2.303e_2^3 \tag{3.101}$$

From Eqs. 3.100 and 3.101 we can write:

$$\therefore\ \varepsilon_2^3 = \left[\frac{\partial\ \ln(\%2/x_2)}{\partial x_3}\right]_{x_1\to 1} + 2.303e_2^3\left[\frac{\partial(wt\%3)}{\partial x_3}\right]_{x_1\to 1} \tag{3.102}$$

If M_1, M_2, M_3 are, respectively, the atomic weights of the elements 1, 2, and 3, % 3 (*wt* % 3) [9] can be expressed as:

$$\%3 = \frac{x_3 M_3 100}{x_1 M_1 + x_2 M_2 + x_3 M_3} = \frac{x_3 M_3 100}{x_3(M_3 - M_1) + x_2(M_2 - M_1) + M_1} \tag{3.103}$$

$$\left[\frac{\partial \%3}{\partial x_3}\right]_{x_1 \to 1} = 100 \frac{M_3}{M_1} \tag{3.104}$$

$$\left[\frac{\partial \ln(\%2/x_2)}{\partial x_3}\right]_{x_1 \to 1} = \frac{M_1 - M_3}{M_1} \tag{3.105}$$

Substituting the above values in Eq. 3.102, we get:

$$\varepsilon_2^3 = 230.3 e_2^3 \frac{M_3}{M_1} + \frac{M_1 - M_3}{M_1} \tag{3.106}$$

or

$$\varepsilon_i^j = 230.3 \frac{M_j}{M_1} e_i^j + \frac{M_1 - M_j}{M_1}$$

In 1–*x*–*y*–*z* system, considering only first order interaction coefficients the above relations can be summarized as:

$$logf_x = \%x.e_x^x + \%y.e_x^y + \%z.e_x^z \tag{3.107}$$

$$ln\,\gamma_x = ln\gamma_x^o + \varepsilon_x^x x_x + \varepsilon_x^y x_y + \varepsilon_x^z x_z \tag{3.108}$$

where

$$\varepsilon_x^x = \left[\frac{\partial\, ln\,\gamma_x}{\partial x_x}\right]_{x_1 \to 1} \quad \text{and} \quad e_x^x = \left[\frac{\partial\, logf_x}{\partial \%x}\right]_{wt\%1 \to 100}$$

$$\varepsilon_x^y = \left[\frac{\partial\, ln\,\gamma_x}{\partial x_y}\right]_{x_1 \to 1} \quad \text{and} \quad e_x^y = \left[\frac{\partial\, logf_x}{\partial \%y}\right]_{wt\%1 \to 100}$$

3.9.5 Relation Between e_2^3 and e_3^2 in 1–2–3 System

In the system 1–2–3, if 1 is the solvent and 2 and 3 are the solutes (at infinitely dilute concentrations), according to Wagner's reciprocal relationship:

$$\varepsilon_2^3 = \varepsilon_3^2$$

And according to Eq. 3.106 ε and e are related as

$$\varepsilon_2^3 = 230.3 e_2^3 \frac{M_3}{M_1} + \frac{M_1 - M_3}{M_1}$$

If M_1 and M_3 are very close in atomic weights, we can approximate as

$$\varepsilon_2^3 = 230.3 e_2^3 \frac{M_3}{M_1} \tag{3.109}$$

and also

$$\varepsilon_3^2 = 230.3 e_3^2 \frac{M_2}{M_1} \tag{3.110}$$

$$\therefore\ e_2^3 = \frac{M_1}{230.3 M_3} \varepsilon_2^3 \qquad \text{(from Eq. 3.109)}$$

$$= \frac{M_1}{230.3 M_3} \varepsilon_3^2 \qquad \text{since } \varepsilon_2^3 = \varepsilon_3^2$$

$$= \frac{M_1}{230.3 M_3}\ 230.3\ e_3^2 \frac{M_2}{M_1} \qquad \text{(from Eq. 3.110)}$$

$$\therefore\ e_2^3 = e_3^2 \frac{M_2}{M_3}$$

Hence, for Fe-C-N system, we can write: $e_C^N = e_N^C \frac{M_C}{M_N}$

3.10 Regular Solutions

In Sects. 3.1 and 3.2 we have learnt about two classes of solutions: ideal and nonideal. The thermodynamic properties of ideal solutions are set by $a_i = x_i, \Delta\overline{H}_i^{M,id} = 0, \Delta\overline{V}_i^{M,id} = 0$, and $\Delta\overline{S}_i^{M,id} = -R\,lnx_i$. Hence, the molar free energy of mixing of the binary ideal solution, A–B having x_A mol of A and x_B mole of B is represented as $\Delta G^{M,id} = RT\ (x_A lnx_A + x_B lnx_B)$ and the molar entropy of mixing, $\Delta S^M = -R\ (x_A lnx_A + x_B lnx_B)$ whereas in the case of nonideal solutions we have, $a_i \neq x_i, \Delta\overline{H}_i^M \neq 0$, and $\Delta G^M = RT(x_A lna_A + x_B lna_B)$.

In the case of nonideal solutions, it is still possible to assume random mixing in certain cases but the enthalpy of mixing will no longer be zero because there will be heat changes due to changes in binding energy. This assumption of random mixing can only be made where there is a small deviation from ideal behavior, so that the enthalpy of mixing is quite small. Solutions of this type are called regular solution. Hildebrand defined a regular solution as one in which

$$\Delta\overline{H}_i^M \neq 0, \text{ and } \Delta\overline{S}_i^M = \Delta\overline{S}_i^{M,id} = -R \, ln x_i$$

For regular solutions the entropy of mixing is the same as for the ideal solution, hence,

$$\Delta H^M = \Delta G^M + T\Delta S^M = RT(x_A ln\, a_A + x_B\, ln\, a_B) - RT\, (x_A ln\, x_A + x_B\, ln\, x_B)$$
$$= RT(x_A ln\, \gamma_A + x_B\, ln\, \gamma_B) \quad (3.111)$$

where γ_A and γ_B are the activity coefficient of the components A and B, respectively.

$$\text{Since } \Delta H^M = x_A \Delta\overline{H}_A^M + x_B \Delta\overline{H}_B^M, \quad (3.112)$$

$$\Delta\overline{H}_A^M = RT \, ln\, \gamma_A \quad (3.113)$$

and

$$\Delta\overline{H}_B^M = RT \, ln\, \gamma_B \quad (3.114)$$

Thermodynamic properties of solutions are divided in two parts: ideal and excess [1–6]. Properties of regular solutions are discussed conveniently with the aid of excess part. Suppose Q denotes any extensive property of the solution, The molar property of the solution, Q is sum of ideal molar property Q^{id} and the excess molar property, Q^{xs}:

$$Q = Q^{id} + Q^{xs} \quad (3.115)$$

If x_A mol of A and x_B mol of B form the solution A–B, on subtracting the corresponding thermodynamic functions for the pure components A and B (i.e., $x_A Q_A^o + x_B Q_B^o$) from each side of Eq. 3.115, the change in the property on mixing can be expressed as:

$$\Delta Q^M = \Delta Q^{M,id} + Q^{xs}$$

Therefore, the integral molar free energy of mixing can be expressed as:

$$\Delta G^M = \Delta G^{M,id} + G^{xs} \quad (3.116)$$

$$\therefore\; G^{xs} = \Delta G^M - \Delta G^{M,id} \quad (3.117)$$

All the thermodynamic functions valid at constant composition, apply to Q as well as to Q^{xs}. Hence, we can relate G^{xs} and S^{xs} following the relation: $\left[\frac{\partial \Delta G}{\partial T}\right]_P = -\Delta S$,

$$\left[\frac{\partial G^{xs}}{\partial T}\right]_P = -S^{xs} \tag{3.118}$$

Substituting the values of $\Delta G^M$and$\Delta G^{M,\ id}$ from Eqs. 3.47 and 3.48 in Eq. 3.117, we get:

$$\begin{aligned} G^{xs} &= RT(x_A ln\, a_A + x_B\, ln\, a_B) - RT\ (x_A ln\, x_A + x_B\, ln\, x_B) \\ &= RT(x_A ln\, \gamma_A + x_B\, ln\, \gamma_B) \end{aligned} \tag{3.119}$$

$$= \Delta H^M = x_A \Delta \overline{H}_A^M + x_B \Delta \overline{H}_B^M \quad \text{(from 3.112)}$$

$$= x_A \overline{G}_A^{xs} + x_B \overline{G}_B^{xs} \tag{3.120}$$

Thus, thermodynamic discussions based on mathematical formalism led to the introduction of another class of solutions known as the "regular solution." Margules [10] for the first time in 1895, gave the concept of the following power series expression to represent the variation of activity coefficients of components in binary solutions at constant temperature.

$$\begin{aligned} ln\, \gamma_A &= \alpha_1 x_B + \frac{1}{2}\alpha_2\ x_B^2 + \frac{1}{3}\alpha_3\ x_B^3 + \cdots\ldots\ \ldots \\ ln\, \gamma_B &= \beta_1 x_A + \frac{1}{2}\beta_2\ x_A^2 + \frac{1}{3}\beta_3\ x_A^3 + \cdots\ldots\ \ldots \end{aligned} \tag{3.121}$$

With the aid of the Gibbs–Duhem equation (Eq. 3.64: $x_A dln\gamma_A + x_B dln\gamma_B = 0$) he showed that for these equations to hold over the entire composition range of composition, $\alpha_1 = \beta_1 = 0$. By comparison of the coefficients of the power series, Margules [10] further demonstrated that if $\alpha_2 = \beta_2$, the variation of the activity coefficients can be represented by the quadratic equations only. Hildebrand [11] proposed the following equations for a binary regular solution:

$$\begin{aligned} RT \ln \gamma_A &= \alpha' x_B^2 \\ RT \ln \gamma_B &= \alpha' x_A^2 \end{aligned} \tag{3.122}$$

If α for one component (say B) in a binary solution, A–B is independent of composition, use of Eq. 3.73 results in the following important equation:

$$\ln \gamma_A = -x_B x_A \alpha_B - \int_{x_A=1}^{x_A=x_A} \alpha_B dx_A \tag{3.73}$$

$$= -x_B x_A \alpha_B - \alpha_B(x_A - 1)$$
$$= -x_B x_A \alpha_B + \alpha_B x_B$$
$$= \alpha_B x_B(1 - x_A)$$
$$= \alpha_B x_B^2 \quad (3.123)$$

According to Eq. 3.67, $\ln \gamma_A = \alpha_A x_B^2$, hence for a binary regular solution:

$$\alpha_A = \alpha_B = \alpha \ \text{(a constant)} \quad (3.124)$$

Thus for a regular solution α is an inverse function of temperature (refer Eq. 3.122)

$$\alpha = \frac{\alpha'}{RT} \quad (3.125)$$

For a regular solution, $\ln \gamma_A = \alpha x_B^2$ and $\ln \gamma_B = \alpha x_A^2$, hence from Eq. 3.119, we get:

$$G^{xs} = RT\ \alpha x_A x_B \quad (3.126)$$

From Eqs. 3.125 and 3.126, we get:

$$G^{xs} = \alpha' x_A x_B \quad (3.127)$$

Thus G^{xs} for a regular solution is independent of temperature, and from Eq. 3.118 we have:

$\left[\frac{\partial G^{xs}}{\partial T}\right]_P = -S^{xs}$, and since S^{xs} is zero for a regular solution, G^{xs} and hence ΔH^M are independent of temperature. Therefore, from Eqs. 3.119, 3.120 and 3.126, we can relate:

$$\overline{G}_A^{xs} = RT_1 \ln \gamma_A(T_1) = RT_2\, ln\, \gamma_A(T_2) = \alpha' x_B^2$$

Thus, for a regular solution, we arrive on the following important relation, useful in converting activity values from one temperature to another temperature.

$$\frac{ln\, \gamma_A at\ the\ temperature\ T_2}{ln\, \gamma_A at\ the\ temperature\ T_1} = \frac{T_1}{T_2} \quad (3.128)$$

Linear variation of $\ln \gamma_A vs\ x_B^2$ (slope = α) at a given temperature indicates that the system A–B follows regular solution behavior. However, for strict adherence to the model, αT should be independent of T but it is not observed in many cases. In general we find $\ln \gamma_i$ varying linearly with x_B^2 but αT decreases with temperature [4].

3.11 Quasi-Chemical Model of Solutions

Quasi-Chemical [4, 12] model is useful in understanding the behavior of solutions formed by components having equi-molar volumes in the pure state. There is no change in volume on mixing of the pure components in formation of the solution. Furthermore the interatomic forces are considered only between nearest neighbor atoms (i.e., over short distances). Thus the energy of solution is estimated by summing the atom–atom bond energies.

Consider 1 mole of a crystal containing n_A number of atoms of A and n_B number of atoms of B. This solid solution contains three types of atomic bonds:

1. A–A bonds with E_{AA} energy for each bond,
2. B–B bonds with E_{BB} energy for each bond and
3. A–B bonds with E_{AB} energy for each bond.

The bond energies are negative quantities. When atoms are infinite distance apart the bond energy (E_{AA}/ E_{BB} /E_{AB}) is zero. If z is the coordination number of an atom in the crystal, each atom has z nearest neighbors. If there are P_{AA} number of A–A bonds, P_{BB} number of B–B bonds, and P_{AB} number of A–B bonds in the solution, the energy of the solution is obtained by the following summation.

$$E = P_{AA}.E_{AA} + P_{BB}.E_{BB} + P_{AB}.E_{AB} \tag{3.129}$$

Thus calculation of E needs estimation of the values of P_{AA}, P_{BB} and P_{AB}.

The number of A atoms × number of bonds per atom = number of A–B bonds + number of A–A bonds × 2.

(The factor 2 arises from the fact that each A–A bond involves 2 A atoms) Thus

$$n_A.\ z = P_{AB} + 2P_{AA}$$

$$\text{or } P_{AA} = \frac{n_A.z}{2} - \frac{P_{AB}}{2} \tag{3.130}$$

Similarly for B, n_B. z=P_{AB} + 2 P_{BB}

$$\text{or } P_{BB} = \frac{n_B.z}{2} - \frac{P_{AB}}{2} \tag{3.131}$$

On substitution of Eqs. 3.130 and 3.131 in Eq. 3.129, we get

$$E = \left[\frac{n_A.z}{2} - \frac{P_{AB}}{2}\right] E_{AA} + \left[\frac{n_B.z}{2} - \frac{P_{AB}}{2}\right] E_{BB} + P_{AB}E_{AB} \tag{3.132}$$

Let us now consider the energy of pure components before mixing, for n_A atoms in pure, A:

The number of A–A bonds × 2 = the number of atoms × number of bonds per atom = $n_A.\ z$

$$P_{AA} = \frac{1}{2} n_A.\ z$$

Similarly for n_B atoms in pure B, $P_{BB} = \frac{1}{2} n_B.z$

Thus, the energy change on formation of the solution:

$$\begin{aligned}\Delta E^M &= \text{(the energy of the solution)–(the energy of unmixed components)}\\ &= \frac{1}{2} n_A.\ z.\ E_{AA} + \frac{1}{2} n_B.\ z.\ E_{BB} + P_{AB}\ \left[E_{AB} - \frac{1}{2}\ (E_{AA} + E_{BB})\right]\\ &\quad - \left[\frac{1}{2} n_A.\ z.\ E_{AA} + \frac{1}{2} n_B.\ z.\ E_{BB}\right]\\ &= P_{AB}\ \left[E_{AB} - \frac{1}{2}\ (E_{AA} + E_{BB})\right]\end{aligned}$$

We know that for a mixing process: $\Delta H^M = \Delta E^M$ - P ΔV^M (by assumption, $\Delta V^M = 0$)

$$\therefore\ \Delta H^M = \Delta E^M = P_{AB}\ \left[E_{AB} - \frac{1}{2}\ (E_{AA} + E_{BB})\right] \tag{3.133}$$

From Eq. 3.133 it is clear that ΔH^M depends on P_{AB}, for given values of E_{AA}, E_{BB} and E_{AB}.

Since for the ideal solution, $\Delta H^M = 0$

Therefore, for an ideal solution:

$$E_{AB} = \frac{1}{2}\ (E_{AA} + E_{BB}) \tag{3.134}$$

In Sect. 3.2.1 we have noted that ideal mixing requires fulfilling the condition: $(A \leftrightarrow B) = (A \leftrightarrow A) = (B \leftrightarrow B)$ or $(A \leftrightarrow B) = \frac{1}{2}\ \{(A \leftrightarrow A) + (B \leftrightarrow B)\}$. According to Eq. 3.134, a necessary condition for ideal mixing is, $E_{AB} = \frac{1}{2}\ (E_{AA} + E_{BB})$, which is similar to that of $(A \leftrightarrow B) = \frac{1}{2}\ \{(A \leftrightarrow A) + (B \leftrightarrow B)\}$. On the other, if $E_{AB} > \frac{1}{2}\ (E_{AA} + E_{BB})$, ΔH^M, becomes negative which points to a negative deviation from Raoult's ideal behavior. ΔH^M is a positive quantity when $E_{AB} < \frac{1}{2}\ (E_{AA} + E_{BB})$, corresponding to positive deviations from ideality.

In case of random mixing of the n_A atoms of A with the n_B atoms of B, $\Delta H^M = 0$, the entropy of mixing is given as:

$$\Delta S^M = \Delta S^{M,id} = -R(x_A ln\, x_A + x_B\ ln\, x_B)$$

In solutions which show relatively small deviations from ideality, $\Delta H^M \leq RT$, approximately the same random distribution of atoms as occurs in an ideal solution, may be assumed. In such cases P_{AB} in Eq. 3.133 may be calculated as per the procedure outlined below:

Consider two neighboring lattice sites, labeled 1 and 2 in the mole of A–B crystal. The probability that site 1 is occupied by an A atom, is given as:

$$\frac{\textit{the number of A atoms in the crstal}}{\textit{the number of lattice sites in the crstal}} = \frac{n_A}{N_o} = x_A$$

Similarly the probability that site 2 is occupied by B atoms is x_B. In 1 mole of a crystal containing n_A number of atoms of A and n_B number of atoms of B, $x_A = \frac{n_A}{n_A + n_B} = \frac{n_A}{N_o}$, and $x_B = \frac{n_B}{N_o}$, where N_o is Avogadro number.

The probability that site 1 is occupied by an A atom and site 2 is simultaneously occupied by a B atom is thus $x_A.\ x_B$. Hence, the probability that a neighboring pair of sites contains an A–B pair in $2x_A.\ x_B$.

By a similar argument, the probability that the neighboring sites contain an A–A pair is x_A^2 and the neighbouring sites contain a B–B pair is x_B^2.

Since one mole the crystal contains ½ z No pair of lattice sites,

The number of A–B pairs = the number of pairs to sites × the probability of an A–B pair

$$\therefore\ P_{AB} = \frac{1}{2} z N_o \times\ 2x_A.\ x_B = z N_o x_A.\ x_B \tag{3.135}$$

Similarly, $P_{AA} = \frac{1}{2} z N_o \times x_A^2 = \frac{1}{2} z N_o\ x_A^2$

and $P_{BB} = \frac{1}{2} z N_o x_B^2$

Substituting Eq. 3.135 in Eq. 3.133, yields

$$\Delta H^M = z N_o x_A.\ x_B\ \left[E_{AB} - \frac{1}{2}\ (E_{AA} + E_{BB})\right]$$

If $z N_o\ \left[E_{AB} - \frac{1}{2}\ (E_{AA} + E_{BB})\right]$ known as the interaction energy parameter is designated as Ω,

$$\Delta H^M = \Omega\, x_A x_B \tag{3.136}$$

Equation 3.136 indicates that ΔH^M is a parabolic function of composition. As random mixing is assumed, the quasi-chemical model corresponds to the regular solution model i.e.

$$\Delta H^M = G^{xs} = \Omega\ x_A x_B = RT\alpha x_A x_B \tag{3.137}$$

$$\therefore \alpha = \frac{\Omega}{RT} \quad (3.138)$$

Let us apply this concept to estimate the value of partial molar heat of mixing:

$$\Delta\overline{H}_A^M = \Delta H^M + x_B \frac{\partial \Delta H^M}{\partial x_A}$$

Differentiation of Eq. 3.137 gives:

$$\frac{\partial \Delta H^M}{\partial x_A} = \Omega\ (x_B - x_A)$$

$$\text{Hence,}\quad \Delta\overline{H}_A^M = \Omega\ x_A x_B + x_B\ \Omega\ (x_B - x_A) = \Omega x_B^2 \quad (3.139a)$$

$$\text{Similarly,}\quad \Delta\overline{H}_B^M = \Omega x_A^2 \quad (3.139b)$$

Since the mixing is random, $\Delta\overline{S}_A^M = -R \ln x_A$ and $\Delta\overline{S}_B^M = -R \ln x_B$

$$\therefore\ \Delta\overline{G}_A^M = \Delta\overline{H}_A^M - T\Delta\overline{S}_A^M = \Omega x_B^2 + RTln x_A \quad (3.140)$$

$$\text{as}\ \Delta\overline{G}_A^M = RTln a_A = RTln\gamma_A + RTln x_A \quad (3.141)$$

On comparison of Eqs. 3.140 and 3.141 we get:

$$ln\,\gamma_A = \frac{\Omega}{RT} x_B^2 = \alpha\, x_B^2 \quad (3.142)$$

Thus to calculate ΔH^M one should know Ω and this can be obtained from γ. The value of γ thus depends on the value of Ω, which in turn depends on the relative value of the bond energies E_{AA}, E_{BB} and E_{AB}. If Ω is negative, $\gamma_A < 1$ and if positive, then $\gamma_A > 1$.

According to Henry's law $ln\gamma_A$ approaches constancy as x_B approaches unity. Thus as $x_B \rightarrow 1$, $ln\gamma_A \rightarrow \ln \gamma_A^o = \Omega / RT$ with this limiting value is being approached asymptotically. Similarly, by the relationship between Henry's Law and Raoult's Law, Raoult's Law is approached asymptotically by the component i as $x_A \rightarrow 1$.

The applicability of Quasi-Chemical model to actual solutions decreases as the magnitude of Ω increases i.e., if the magnitude of E_{AB} is significantly greater or less than the average of E_{AA} and E_{BB}, then random mixing of the A and B atoms cannot be assumed.

3.12 Darken's Formalism in Binary Metallic Solutions

The power series expressions (Eqs. 3.121) for composition dependence of activity coefficients introduced by Margules [10] have been mentioned in Sect. 3.10. Making use of the Gibbs–Duhem equation he further demonstrated that the first coefficient in the equations, α_1 and β_1 are zero, if these power series equations are considered to be valid for the entire range of composition in the system. If $\alpha_1 = \beta_1$ the system can be represented by the quadratic terms only. Porter [13] in 1921 showed that for quadratic approximation: $ln\, \gamma_1 = \alpha\, x_2^2$, and $ln\, \gamma_2 = \alpha\, x_1^2$ present adequate behavior of many non-electrolytes systems. It was also confirmed by Hildebrand [14] for other systems. On the basis of these observations Hildebrand proposed that systems following $ln\gamma_1 = \alpha\ (1 - x_1)^2$ be designated as Regular solutions. Since then considerable attention has been paid to the subject of the thermodynamics of binary solutions. The concept of regular solution as proposed and developed by Hildebrand and Scott [15, 16] had remarkable success in its application to binary system involving numerous organic substances, some halides, and a few elements (Br_2, I_2, P_4, S_8). But this principle achieved only a limited success in discussing the behavior of metallic solutions. It is apparent from the investigations of Hildebrand and co-workers that in large majority of the solutions they considered relatively weak intermolecular forces (van der Waals forces) and not the strong primary inter-atomic bond of metallic solutions. Darken [17] has pointed out the reason of failure. According to him the solubility parameter, defined by Hildebrand and Scott, was inadequate to characterize a metallic system. The entire approach of Hildebrand was designed primarily to explain positive deviation from Raoult's law (usually endothermic reactions) whereas metallic solutions are generally exothermic and sometimes very highly so, although metallic systems occasionally exhibit a miscibility gap. In the case of large negative deviation from Raoult's law Wagner [18] has pointed out that the concept of regular solution (zero excess entropy) involving random mixing is unrealistic.

Though Quasi-Chemical approach based on relative bond energies, developed by Herzefeld and Heitler [19] and Scatchard [20] received considerable popularity but failed to explain the behavior of metallic solutions. Wagner [18] has drawn attention about the limited applicability of all the approaches in a critical review based on electronic considerations. The need for cautions as pointed out by Wagner [18] in his review led Darken [17] to consider division of the entire concentration range (0 to unit atom fraction) into three regions: two terminal regions and a central region. The breadth of either terminal region varies, but usually covers a range of 0.15 to 0.5 atom fraction. The breath of the central region, though may not be symmetrically located, covers a range of similar extent. Darken has pointed out two apparent general features of binary metallic solutions which may be useful in understanding their nature and in the presentation of experimental data. (i) In each terminal region the second derivative of molar excess free energy with respect to atom fraction (called by Darken as excess stability) in substantially constant. (ii) In the central region the stability may vary strongly with composition, exhibiting a pronounced maximum, the maximum frequently occurs at compositions corresponding to small whole number ratio x_2/x_1 and in accord with the classical valence.

The Terminal Regions

In metallic solutions Raoult's law is strictly obeyed by the solvent designated by 1 in 1–2 system (i.e., when $x_1 \rightarrow 1$). In re-examining the available data at finite concentration, Darken [17] has pointed out that an expression of the type $ln\gamma_i = \alpha\,(1 - x_i)^2$ represents departure from Raoult's law at high solute concentration in systems: Sn-Au, Sn-Sb, Sn-Ag, and Fe-Si. The expression: $ln\gamma_i = \alpha\,(1 - x_i)^2$ may not be applicable to the entire binary system but may be applicable in certain parts of it. Activity coefficient of the solvent may be represented as $\ln\gamma_1 = \alpha\, x_2^2$, which holds up to a certain value of solute concentration, x_2, and for solute: $\ln\gamma_2 = \alpha\, x_1^2 + I$ (I is a constant). I is zero only if the equations apply over the entire composition range. It can be proved by Gibbs–Duhem equation:

$$d\ln\gamma_2 = -\frac{x_1}{x_2} d\ ln\gamma_1 \tag{3.143}$$

and we know that

$$\ln\gamma_1 = \alpha\, x_2^2 = \alpha(1 - x_1)^2 \tag{3.144}$$

On differentiation we get:

$$d\ln\gamma_1 = -2\,\alpha\,(1 - x_1)\,dx_1 \tag{3.145}$$

Substituting Eq. 3.145 in Eq. 3.143,

$$d\ln\gamma_2 = -\frac{x_1}{x_2}[-2\,\alpha\,(1 - x_1)\,dx_1]$$

$$\text{or } d\ln\gamma_2 = 2\alpha\, x_1\, dx_1$$

On integration: $\int dln\gamma_2 = \int 2\alpha\, x_1\, dx_1$

$$\ln\gamma_2 = \alpha\, x_1^2 + I \tag{3.146}$$

For evaluation of I, make use of a definite integral:

$$\int_{ln\gamma_2^o}^{ln\gamma_2} dln\gamma_2 = \int_{x_1=0}^{x_1} 2\alpha\, x_1\, dx_1 = \left[\alpha\, x_1^2\right]_1^{x_1} = \alpha\, x_1^2 - \alpha$$

$$\therefore\ ln\,\gamma_2 - ln\gamma_2^o = \alpha\,x_1^2 - \alpha$$

$$\text{or}\ \ ln\,\gamma_2 = \alpha\left(x_1^2 - 1\right) + ln\gamma_2^o = \alpha\,x_1^2 + \left(ln\gamma_2^o - \alpha\right) \tag{3.147}$$

On comparison of Eqs. 3.146 and 3.147, we can write:

$$I = ln\,\gamma_2^o - \alpha \tag{3.148}$$

Since $I = 0$ for a strictly regular solution, $\alpha = ln\gamma_2^o$ i.e. from Eq. 3.147, $ln\,\gamma_2 = \alpha\,x_1^2$.
Thus for a strictly regular solution:

$$ln\,\gamma_1^o = ln\,\gamma_2^o \tag{3.149}$$

If the solvent is designated by 1 and the solute by numeral 2, we can express Eq. 3.147 as:

$$ln\,\gamma_2 = ln\,\gamma_2^o + \alpha_{12}\left(x_1^2 - 1\right)$$

$$\text{or}\ \ ln\left(\gamma_2/\gamma_2^o\right) = \alpha_{12}\left(x_1^2 - 1\right) = \alpha_{12}\left[\left(1 - x_2\right)^2 - 1\right]$$

$$\text{or}\ \ ln\left(\gamma_2/\gamma_2^o\right) = \alpha_{12}\left(-2x_2 + x_2^2\right) \tag{2.150}$$

The above equation as proposed by Darken [17] is known as the quadratic formalism.

In the above equation Darken has adopted the convention for the subscripts to α that the solvent is named first and the solute second. He has applied the above formalism in analyzing the published thermodynamic data on liquid ferrous systems (Fe-X, where X stands for Al, Ag, Mn, Ni, and Si). The numerical values of $ln\gamma_X^o, ln\gamma_{Fe}^o, \alpha_{FeX}, and\ \alpha_{XFe}$ as reported by Darken [17] are presented in Table 3.1. The Table clearly demonstrate that Fe-Al and Fe-Mn conform to the regular solution behavior whereas Fe-Ni, Fe-Si, and Fe-Cu do not obey regular solution model.

Table 3.1 Parameters of activity coefficients of ferrous systems at 1600 °C (standard state is pure liquid)

System	Parameters			
	$\log\gamma_X^o$	α_X	$\log\gamma_{Fe}^o$	α_{Fe}
Fe-Al	−1.21	−1.21	−1.21	−1.21
Fe-Mn	0.1	0.1	0.1	0.1
Fe-Si	−2.75	−3.10	−1.64	−0.78
Fe-Ni	−0.18	−0.04	−0.40	−0.54
Fe-Cu	0.93	1.13	1.00	1.2

Adapted from L. S. Darken [17] with the permission of The Minerals, Metals & Materials Society

3.12.1 Stability Functions

Let us now consider an actual solution, A–B, whether gaseous, liquid or crystalline. If the solution is divided into two portions and then a small amount of component A is transferred from one portion to the other and a small amount of B in the opposite direction. The change in excess free energy in this transfer process would be zero if the solution is ideal. But a relatively large change in the free energy is expected in the case of a crystalline solution like sodium chloride, on transfer of a few sodium ions to chlorine sites and vice versa. In other words a large free energy change is expected in the vicinity of a known compound.

The free energy change for any process, by which a solution is broken down into two solutions differing only infinitesimally in composition, is found to be proportional to d^2G/dx_2^2 which is identical with d^2G/dx_1^2; where G represents the molar free energy of the system; the proportionality constant involves only arbitrary parameters and does not involve any property of the system; itself. Darken has named the second derivative of the molar free energy with mole fraction as "stability". The stability of an ideal solution is calculated as $\frac{RT}{x_1x_2}$. The excess stability defined as $d^2G^{xs}/dx_2^2 = d^2G^{xs}/dx_1^2$, is then equal to the stability minus ideal stability.

If the stability is anywhere negative, the solution is metastable and a miscibility gap occurs. The stability can be calculated by differentiating the general expression Eq. (3.17):

$$\overline{G}_2 = G + (1 - x_2)\frac{dG}{dx_2}$$

$$\text{Stability} = \frac{d^2G}{dx_2^2} = -2\frac{d\overline{G}_2}{d(1-x_2)^2} = -2RT\frac{d\ln a_2}{d(1-x_2)^2} = -4.606\ RT\frac{d\log a_2}{d(1-x_2)^2} \tag{3.151}$$

Similarly,

$$\text{Excess stability} = \frac{d^2G^{xs}}{dx_2^2} = -2\frac{d\overline{G}^{xs}}{d(1-x_2)^2} = -2RT\frac{d\ln\gamma_2}{d(1-x_2)^2} = -4.606\,RT\frac{d\log\gamma_2}{d(1-x_2)^2} \tag{3.152}$$

Thus the excess stability is readily calculated by multiplying the slope of a curve $\ln\gamma_i$ *vs* $(1 - x_i)^2$ by the factor 4.606 RT. The stability and excess stability are properties of the system rather than that of a component; alternatively, they may be regarded as joint properties of both the components. Since the slope $\frac{d\log a_2}{d(1-x_2)^2}$ is substantially constant in the terminal region, the excess stability is substantially constant.

Central Region The excess stability in the central region may be far from constant. A plot of the excess stability for the Mg-Bi system obtained by estimating the slope of log γ vs $(1\text{-}x_{Mg})^2$ plot shown in Fig. 3.9, is presented in Fig. 3.10. The pronounced peak is the most prominent feature of this figure. It is important to note that the peak occurs at $x_{Mg} = 0.6$ and $x_{Bi} = 0.4$, corresponding to the composition of the intermediate phase Mg_3Bi_2 and in accord with the classical valences. This peak is not just by a coincidence, it has been observed in H_2O-SO_3, Fe-O, CaO-SiO_2, FeO-SiO_2 etc. [17]. Although such peaks are not so abundant in metallic systems, they are by no means rare.

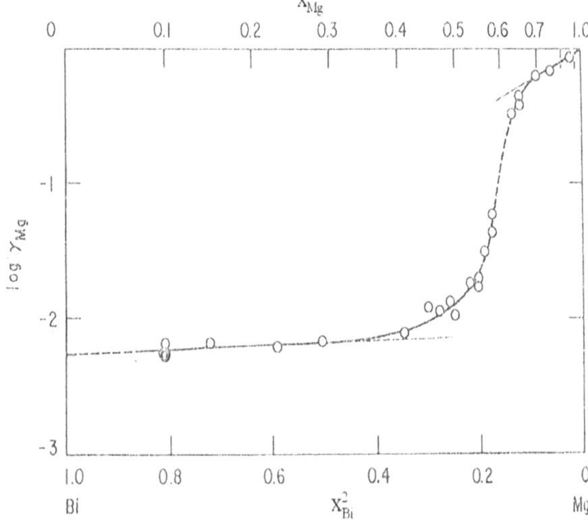

Fig. 3.9 Activity coefficient of magnesium in liquid Mg-Bi alloys at 700 °C (Reproduced from L. S. Darken [17] with the permission of The Minerals, Metals & Materials Society)

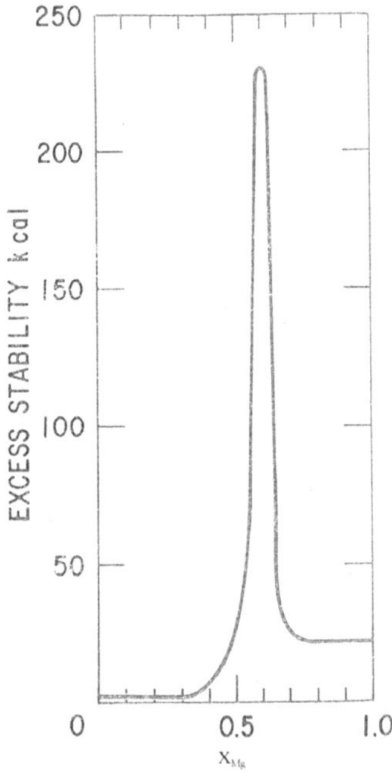

Fig. 3.10 Excess stability of liquid Mg-Bi alloys at 700 °C (Reproduced from L. S. Darken [17] with the permission of The Minerals, Metals & Materials Society)

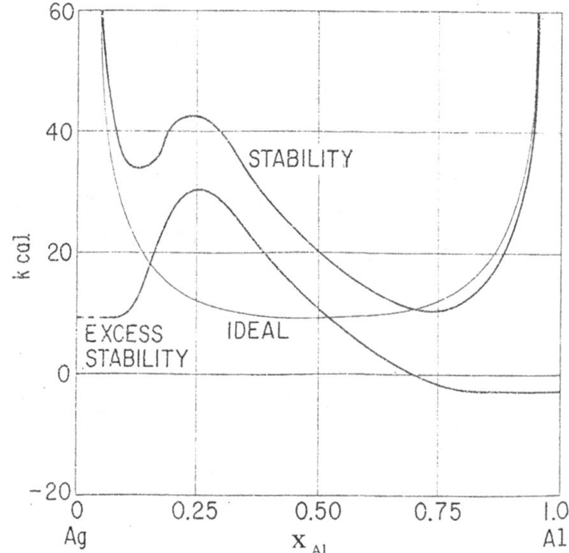

Fig. 3.11 Stability and excess stability of liquid Ag-Al alloys at 900 °C. (Reproduced from L. S. Darken [17] with the permission of The Minerals, Metals & Materials Society)

Stability peak in the liquid Ag-Al system at composition corresponding to the classical valences is shown in Fig. 3.11 whereas Fig. 3.12 depicts a similar peak in Cd-Sb system at the anticipated composition (i.e. $x_{Cd} = 0.6$, $x_{Sb} = 0.4$), (valence 2:3).

The Fe-Cu system is of particular interest since it exhibits a highly positive deviation from Raoult's Law that it is on the verge of immiscibility. As can be seen in Fig. 3.13 that a portion of the liquidus is almost flat and the addition of a small amount of any third component would give rise to two liquid layers. Thus the excess free energy is positive and the excess stability is negative over most of the composition range Even so, the excess stability shows a definite though small peak.

Contrary to the above observations the Fe-Si system presents a different type of behavior for the excess stability (Fig. 3.14). There is no apparent peak and the flat excess stability curves of two terminal regions are connected in the central region by a relatively simple transition. The similar trend is also noted in the figure for the Fe-Ni system.

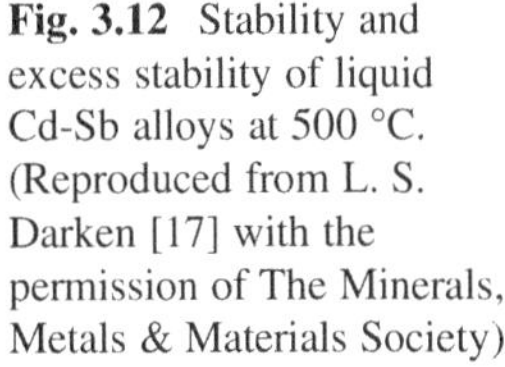

Fig. 3.12 Stability and excess stability of liquid Cd-Sb alloys at 500 °C. (Reproduced from L. S. Darken [17] with the permission of The Minerals, Metals & Materials Society)

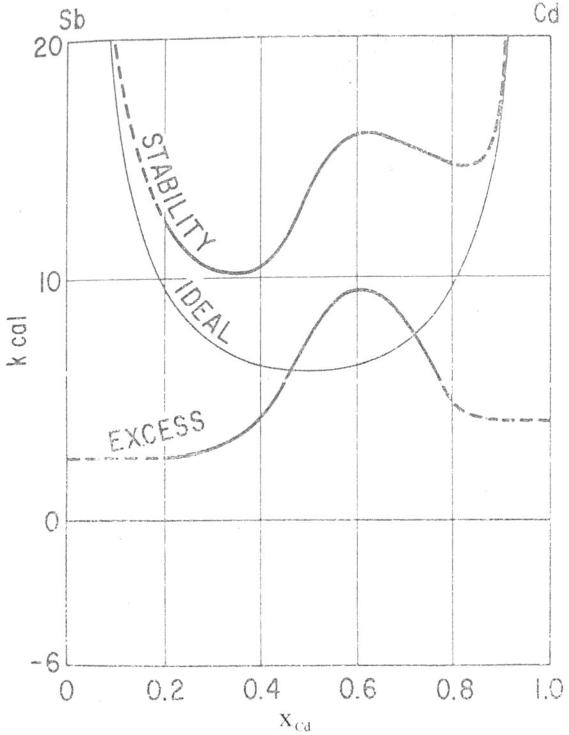

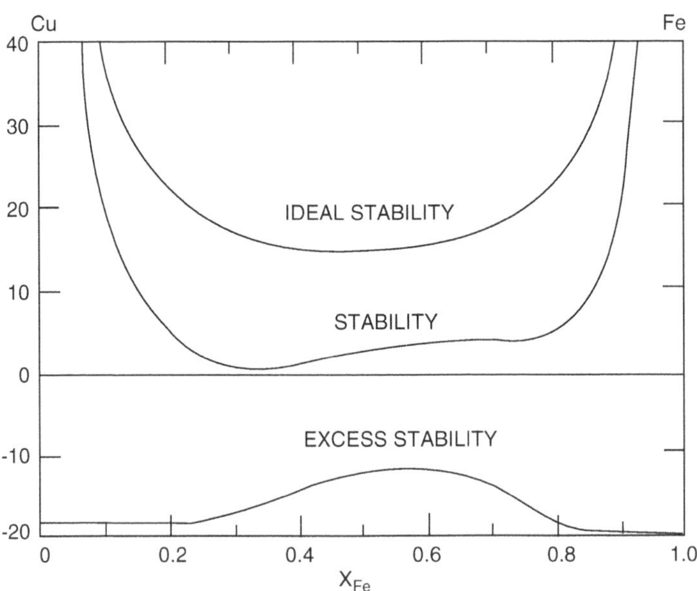

Fig. 3.13 Stability and excess stability of liquid Fe-Cu alloys at 1550 °C (Reproduced from L. S. Darken [76] with the permission of The Minerals, Metals & Materials Society)

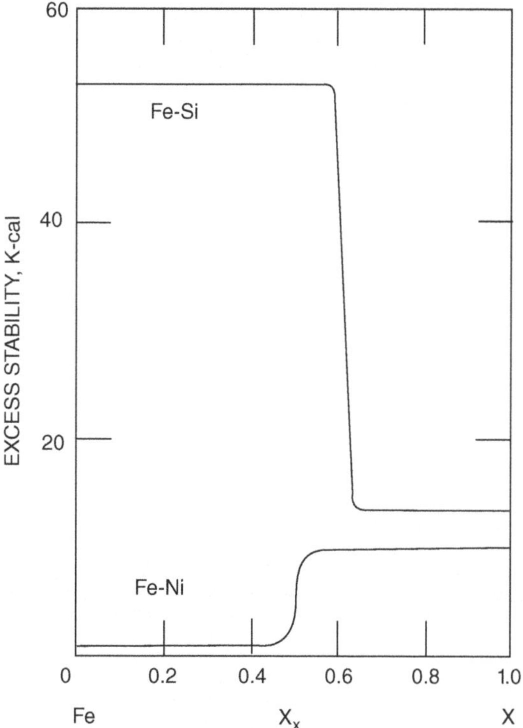

Fig. 3.14 Excess stability of liquid Fe-Si and Fe-Ni alloys at 1600 °C (Reproduced from L. S. Darken [17] with the permission of The Minerals, Metals & Materials Society)

The above phenomena are not restricted only to liquid binary metallic solutions. Similar variations in stability and excess stability functions have been observed in solid pseudo-binary systems CdTe-CdSe [21, 22]. As evident from Fig. 3.15, α_{CdTe} is not independent of composition, x_{CdTe}. This means system CdSe-CdTe does not follow a regular solution model. Further, the system does not satisfy Darken's quadratic formalism, because the data cannot be presented according to the equation:

$$ln\left(\gamma_{CdTe}/\gamma_{CdTe}^{o}\right) = \alpha_{CdSe}\left(-2x_{CdTe} + x_{CdTe}^{2}\right) \tag{3.153}$$

Hence, the thermodynamic behavior of CdSe-CdTe system has been analyzed in light of Darken's stability and excess stability functions [17] according to the expressions given below:

$$\text{Stability} = \frac{d^2G}{dx_{CdTe}^2} = -2\ RT\frac{d\,ln\ a_{CdTe}}{d(1-x_{CdTe})^2}$$

$$\text{Excess stability} = \frac{d^2G^{xs}}{dx_{CdTe}^2} = -2\ RT\frac{d\,ln\ \gamma_{CdTe}}{d(1-x_{CdTe})^2}$$

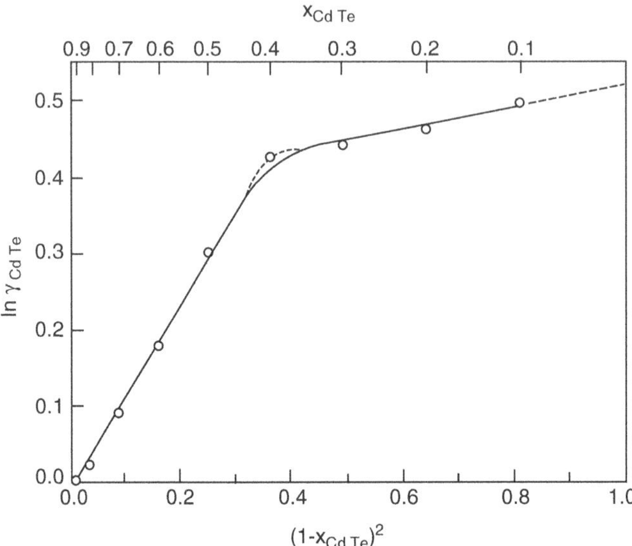

Fig. 3.15 $ln\gamma_{CdTe} vs\ (1 - x_{CdTe})^2$ plot at 840 K (Reproduced from M. Shamsuddin and A. Nasar [21] with the permission of The Minerals, Metals & Materials Society)

Thus, the values of stability and excess stability of the pseudobinary system CdSe-CdTe may be calculated by multiplying the slopes of $lna_{CdTe}\ vs\ (1 - x_{CdTe})^2$ and $ln\gamma_{CdTe} vs\ (1 - x_{CdTe})^2$ plots with $- 2\ RT$, respectively. The plot of $ln\gamma_{CdTe}\ vs\ (1 - x_{CdTe})^2$, shown in Fig. 3.15, is represented by two straight lines, intersecting at $x_{CdTe} = 0.4$ (on extrapolation). Thus, two values of excess stabilities of -1.96 and -17.46 kJ mol^{-1} at 840 K have been obtained in the two composition ranges: $x_{CdTe}=$ 0 to 0.35 and 0.45 to 1.0, respectively. The values of Darken's stability have been calculated by adding the values of excess stability to the ideal stability, defined as Ideal stability $= \frac{RT}{x_{CdTe}(1 - x_{CdTe})}$

The stability and excess stability of CdSe-CdTe system together with the ideal stability at 840 K are presented in Fig. 3.16. This figure shows pronounced changes in stability and excess stability parameters at composition, near $x_{CdTe} = 0.4$. The figure also demonstrates smooth variation of stability and constancy of excess stability in the two terminal regions: (a) $0 \leq x_{CdTe} \leq 0.35$ and (b) $0.45 \leq x_{CdTe} \leq 1.0$. The two terminal regions are connected by a transient or central region which extends from $x_{CdTe} = 0.35$ to $x_{CdTe} = 0.45$. The thermodynamic behavior seems to be relatively simple in the two terminal regions, whereas it is complicated in the central region. In this respect, behavior of CdSe-CdTe system in the temperature range 720 to 840 K has been found to be similar to those of Fe-Si and Fe-Ni [17] liquid solutions at 1873 K. According to Darken, [17] systems showing pronounced changes in excess stability vs composition plots instead of pronounced peaks have been termed as simple transient systems.

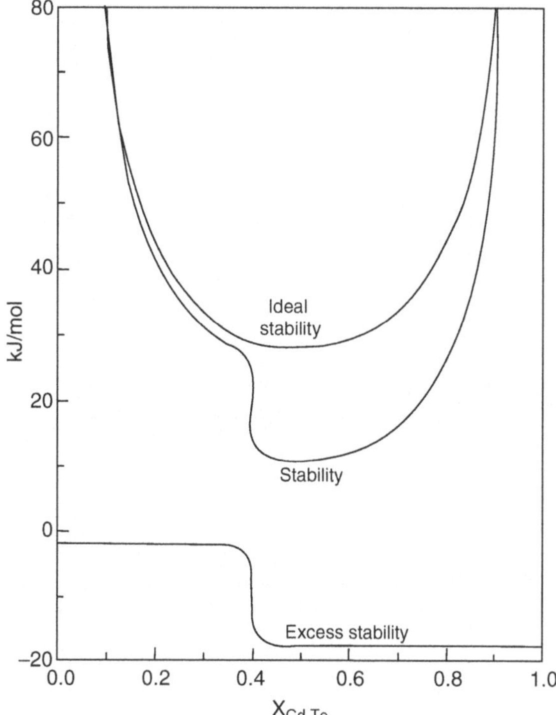

Fig. 3.16 Stability and excess stability functions of CdSe-CdTe system at 840 K (Reproduced from M. Shamsuddin and A. Nasar [21] with the permission of The Minerals, Metals & Materials Society)

Since Darken [17] introduced the concept of stability and excess stability parameters, there have been several attempts for improvement in the presentation of experimental data so that thermodynamic behavior of solutions for the entire range of composition can be discussed in a better manner. Schmid et al. [23] have extended Darken's concept of stability by introducing a new parameter termed as "relative stability" which combines the advantages of both stability and excess stability. The relative stability remains finite over the entire composition range and approaches unity at two ends. Chou [24] has defined a new term called "relative excess stability" which has the advantages of all three kinds of stability parameters and approaches zero at terminals.

Thus, use of relative stability and relative excess stability is advantageous over stability and excess stability, because the former have finite values in the terminal regions. Stability and excess stability of a system relative to ideal stability called relative stability and relative excess stability are expressed as according to the following expressions:

$$\text{Relative stability} = \frac{x_{CdTe}(1 - x_{CdTe})}{RT} \frac{d^2 \Delta G^M}{dx_{CdTe}^2} \tag{3.154}$$

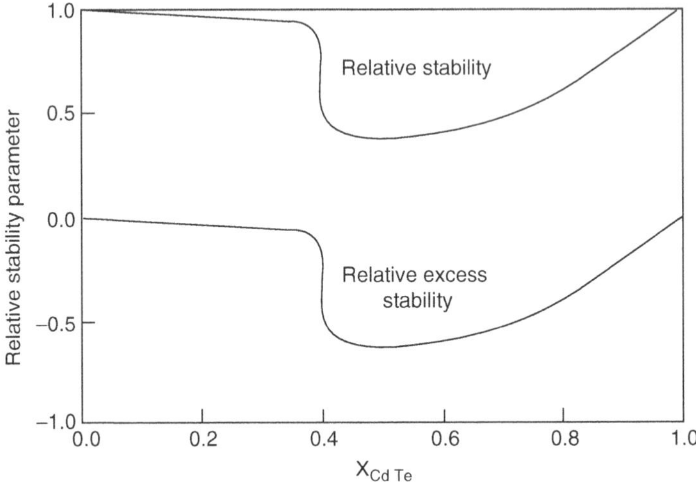

Fig. 3.17 Relative stability and relative excess stability functions of CdSe-CdTe system at 840 K (Reproduced from M. Shamsuddin and A. Nasar [21] with the permission of The Minerals, Metals & Materials Society)

$$\text{Relative excess stability} = \frac{x_{CdTe}(1 - x_{CdTe})}{RT} \frac{d^2 \Delta G^{xs}}{dx_{CdTe}^2} \tag{3.155}$$

The variation of relative stability and relative excess stability of CdSe-CdTe system at 840 K depicted in Fig. 3.17, also show pronounced changes near $x_{CdTe} = 0.4$. The pronounced changes in the values of the activity coefficient of CdTe and CdSe, stability, excess stability, relative stability, and relative excess stability in the system CdSe-CdTe near $x_{CdTe} = 0.4$ are in conformity with the structural changes in the phase diagram. The equilibrium phase diagram of CdSe-CdTe system [25] consists of two broad single-phase regions, one of solid solutions with hexagonal structure (wurtzite) of CdSe and the other of solid solutions with cubic structure (zinc blend) of CdTe, separated by a very narrow two-phase region (only about 3 mol pct CdTe), resulting from a eutectic reaction at 1364 K near 80 mol pct CdTe.

The pronounced changes in thermodynamic and as well as physical properties in the system, CdSe-CdTe (Fig. 3.18) near $x_{CdTe} = 0.4$ are in conformity with the structural changes in the phase diagram. The structural studies by a number of investigators [25–30] demonstrate that depending on composition, $CdTe_xSe_{1-x}$ alloys crystallize predominantly in the wurtzite ($x_{CdTe} = 0$ to 0.4) and zinc blend ($x_{CdTe} = 0.4$ to 1.0) structures. Tai et al. [31] have reported that a two-phase region in the system may exist in the composition range $0.4 \leq x_{CdTe} \leq 0.6$, depending on the time and temperature of annealing of samples. In this context, it is important to note that even with the long duration of experiments, to achieve reversible emf (varying from 24 to 72 hours), it is difficult to ascertain the exact position of the two-phase region in the CdSe-CdTe system. Thus, considering the simultaneous annealing of

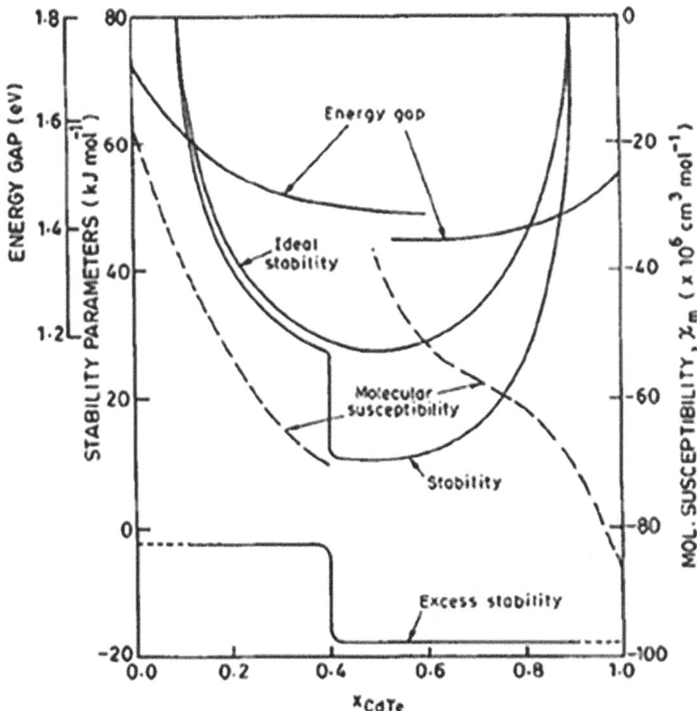

Fig. 3.18 Thermodynamic and physical properties of CdSe-CdTe system. (Reproduced from M. Shamsuddin and A. Nasar [22] with the permission of The Minerals, Metals & Materials Society)

samples while maintaining them in the temperature range of 720 to 840 K for sufficiently long duration to attain reversible emf, our results confirm that pronounced changes in partial molar quantities and Darken's stability and excess stability, relative stability as well as relative excess stability in the vicinity of x_{CdTe} = 0.4 at 840 K correspond to the two-phase region of the CdSe-CdTe phase diagram. A discontinuity in energy gap *vs* composition plot shown in Fig. 3.19 around this composition has been reported by Tai et al. [31]. A discontinuity has also been observed in the variation of magnetic susceptibility with composition near x_{CdTe} = 0.4. All these changes near x_{CdTe} = 0.4 are due to structural changes in the system.

The pseudobinary HgTe-CdTe system [32] forms solid solutions over the entire composition range. The continuous variation of activity with composition x_{CdTe} in HgTe-CdTe indicates that the system consists of a single phase field throughout the entire range of composition. A similar conclusion has been drawn by the continuous variation of Darken's stability and constancy of excess stability functions (Fig. 3.19). The continuous variation of electronic characteristics from semimetal to semiconductor with energy gap varying from −0.3 to 1.6 eV with increasing concentration of CdTe, reported by Long and Schmit, [33] has been presented in

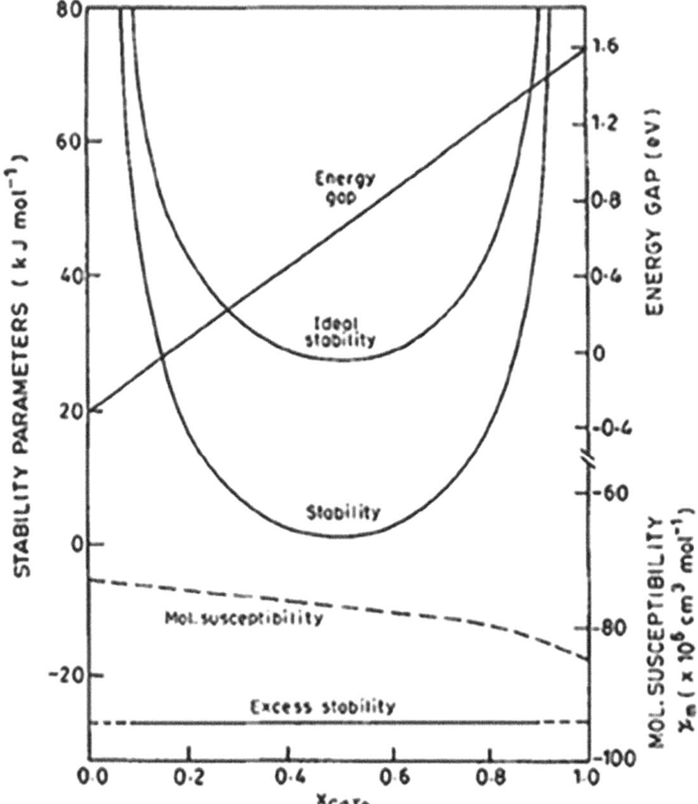

Fig. 3.19 Thermodynamic and physical properties of HgTe-CdTe system (Reproduced from M. Shamsuddin and A. Nasar [22] with the permission of The Minerals, Metals & Materials Society)

Fig. 3.19. As for the energy gap, continuous variation of magnetic susceptibility with composition in $Cd_xHg_{1-x}Te$ alloys has been observed at room temperature. Thus results on thermodynamic and physical properties are in conformity with the simple isomorphous phase diagram of the system.

3.13 Ionic Melts

Temkin [34] has made everlasting contributions in the development of ionic theory of fused salt mixtures and ionic solutions. Temkin's concept was extended by Herashymenko [35], Flood [36, 37] and Masson [38, 39] to explain the ionic behavior of slags.

3.13.1 Temkin Theory

This theory is based on the assumptions that components of fused salt mixture and slag are completely dissociated into ions. There is no interaction between ions of the same charge, hence there exists a state of complete randomness. There is mixing of cations (i^+) on cations sites and anions (j^-) on anion sites. Thus, fused salt mixture and slag consist of two separate sets of ideal solutions of cations and anions. Due to complete dissociation, all the ionic species in the salt mixture and slag are known. Based on this hypothesis the dissociation of component *ij*, present in the fused salt mixture can be expressed as:

$$ij = i^+ + j^- \tag{3.156}$$

$$K = \frac{a_{i^+} . a_{j^-}}{a_{ij}} \tag{3.157}$$

At equilibrium, $\Delta G^o = 0$, $\therefore K = 1$ ($\Delta G^o = -RT \ln K$)

Since cations and anions form two different sets of ideal solutions:

$$a_{i^+} = x_{i^+} \text{ and } a_{j^-} = x_{j^-} \tag{3.158}$$

$$\therefore K = \frac{x_{i^+} . x_{j^-}}{a_{ij}} = 1 \tag{3.159}$$

$$\text{or } a_{ij} = x_{i^+} . x_{j^-} \tag{3.160}$$

It is assumed that the above relation holds good at various molar fractions of the component *ij* present in the fused salt mixtures. Flood, Forland, and Grjotheim [36, 37] further extended Temkin's theory by considering equilibria between ions or compounds dissolved in the slag and elements dissolved in the metallic phase. It assumes ideal behavior for ions in the slag and nonideal for elements in metals. In general, Ca^{2+}, Mg^{2+}, Fe^{2+}, SiO_4^{4-}, PO_4^{3-}, $Al_2O_4^{2-}$, $Fe_2O_5^{4-}$, O^{2-} and so on are present in slag solutions. In order to calculate the activity of a basic oxide in a silicate slag, Masson [38, 39] has developed a theory to account more complex anions as compared to SiO_4^{4-}. Masson considers slags as complex solutions containing polymeric silicate anions. The character and quantity of basic oxide present in the slag control the degree of polymerization. Thus, in a highly basic slag, silica is present mainly as SiO_4^{4-} whereas less basic slag contains $\mathrm{SiO_4^{4-}}$, $\mathrm{Si_2O_7^{6-}}$, $\mathrm{Si_3O_{10}^{8-}}$ ... ... $\mathrm{Si_nO_{(3n+1)}^{2(n+1)-}}$ anions in equilibrium with each other.

Let us apply Temkin's principle to demonstrate that activities of salts AY_2 and BX in a salt mixture: $AY_2 - BX$, are given as $a_{AY_2} = x_A . x_Y^2$ and $a_{BX} = x_B . x_X$, where $x_{A^{2+}}, x_{Y^-}, x_{B^+}$ and x_{X^-} are ionic fractions and A^{2+} and B^+ are, respectively, divalent and univalent cations and X^- and Y^- are univalent anions.

According to Temkin's theory, fused salt mixtures of AY_2 and BX are solutions, which are completely dissociated into ions. There is no interaction between ions of the same charge and the state may be that of complete randomness. Hence, solution of salts may be assumed to be constituted of two ideal solutions; one of cations: A^{2+}, B^{+} and another of anions: X^{-}, Y^{-}.

The heat of mixing of an ideal solution is zero, that is, $\Delta H^{M,\,id} = 0$ and in the case of random mixing, entropy of mixing, $\Delta S^{M,\,id} = -R\sum x_i \ln x_i$.

Hence for the above system, we have:

$$\Delta S^{M,id} = -R\left(n_A ln\, x_A + n_B\, ln\, x_B + n_X\, ln\, x_X + n_Y\, ln\, x_Y\right) \tag{3.161}$$

$$\Delta G^{M,id} = \Delta H^{M,id} - T\Delta S^{M,id} = RT\left(n_A ln\, x_A + n_B\, ln\, x_B + n_X\, ln\, x_X + n_Y\, ln\, x_Y\right) \tag{3.162}$$

The mole fractions of anions and cations in the above system of the salt mixture, AY_2 *and* BX, comprising two ideal solutions can be expressed as:

$$x_A = \frac{n_A}{n_A + n_B}, x_B = \frac{n_B}{n_A + n_B}, x_X = \frac{n_X}{n_X + n_Y}, x_Y = \frac{n_Y}{n_X + n_Y}$$

$$\therefore \Delta G^M = RT\left[n_A ln\left(\frac{n_A}{n_A + n_B}\right) + n_B\, ln\left(\frac{n_B}{n_A + n_B}\right) + n_X\left(\frac{n_X}{n_X + n_Y}\right) + n_Y\, ln\left(\frac{n_Y}{n_X + n_Y}\right)\right] \tag{3.163}$$

The partial molar free energy of mixing of the salt AY_2 can be expressed as:

$$\overline{G}^M_{AY_2} = \left(\frac{\partial \Delta G^M}{\partial n_{AY_2}}\right) = RT ln a_{AY_2} \tag{3.164}$$

It may be noted that $dn_{AY_2} = \frac{1}{2}dn_Y = dn_A$

$$\therefore \left(\frac{\partial \Delta G^M}{\partial n_{AY_2}}\right)_{n_B,n_X} = \left(\frac{\partial \Delta G^M}{\partial n_A}\right)_{n_B,n_X,n_Y} + \left(\frac{\partial \Delta G^M}{\frac{1}{2}\partial n_Y}\right)_{n_B,n_X,n_A}$$

$$\text{or} \left(\frac{\partial \Delta G^M}{\partial n_{AY_2}}\right)_{n_B,n_X} = \left(\frac{\partial \Delta G^M}{\partial n_A}\right)_{n_B,n_X,n_Y} + 2\left(\frac{\partial \Delta G^M}{\partial n_Y}\right)_{n_B,n_X,n_A} \tag{3.165}$$

Differenting Eq.3.163 with respect to n_A at constant n_B, n_X, n_Y

$$\left(\frac{\partial \Delta G^M}{\partial n_A}\right)_{n_B,n_X,n_Y} = RT\left[ln\left(\frac{n_A}{n_A + n_B}\right) + n_A\frac{d}{dn_A}\left\{ln\left(\frac{n_A}{n_A + n_B}\right)\right\} + n_B\frac{d}{dn_A}\left\{ln\left(\frac{n_B}{n_A + n_B}\right)\right\}\right] \tag{3.166}$$

On differentiation of appropriate terms, we get:

$$\frac{d}{dn_A}\left[ln\left(\frac{n_A}{n_A+n_B}\right)\right]=\frac{n_B}{n_A}\frac{1}{(n_A+n_B)}$$

$$\frac{d}{dn_A}\left[ln\left(\frac{n_B}{n_A+n_B}\right)\right]=-\frac{1}{(n_A+n_B)}$$

Substituting these values in Eq. 3.166 we get:

$$\left(\frac{\partial \Delta G^M}{\partial n_A}\right)_{n_B,n_X,n_Y}=RT\left[ln\left(\frac{n_A}{n_A+n_B}\right)+n_A.\frac{n_B}{n_A}\frac{1}{(n_A+n_B)}-n_B.\frac{1}{(n_A+n_B)}\right]$$

$$=RT\ln\left(\frac{n_A}{n_A+n_B}\right)=RT\ ln\,x_A \tag{3.167}$$

Similarly,

$$\left(\frac{\partial \Delta G^M}{\partial n_Y}\right)_{n_A,n_B,n_X}$$

$$=RT\left[ln\left(\frac{n_Y}{n_X+n_Y}\right)+n_Y\frac{d}{dn_Y}\left\{ln\left(\frac{n_Y}{n_X+n_Y}\right)\right\}+n_X\frac{d}{dn_Y}\left\{ln\left(\frac{n_X}{n_X+n_Y}\right)\right\}\right] \tag{3.168}$$

$$\text{We Have: }\frac{d}{dn_Y}\left[ln\left(\frac{n_Y}{n_X+n_Y}\right)\right]=\frac{n_X}{n_Y}\frac{1}{(n_X+n_Y)}$$

$$\text{and }\frac{d}{dn_Y}\left[ln\left(\frac{n_X}{n_X+n_Y}\right)\right]=-\frac{1}{(n_X+n_Y)}$$

Substituting these values in Eq. 3.168 we get:

$$\left(\frac{\partial \Delta G^M}{\partial n_Y}\right)_{n_A,n_B,n_X}=RT\left[ln\left(\frac{n_Y}{n_X+n_Y}\right)+n_Y.\frac{n_X}{n_Y}\frac{1}{(n_X+n_Y)}-n_X.\frac{1}{(n_X+n_Y)}\right]$$

$$=RTln\left(\frac{n_Y}{n_X+n_Y}\right)=RT\ ln\,x_Y \tag{3.169}$$

From Eqs. 3.165, 3.167 and 3.169 we get:

$$\left(\frac{\partial \Delta G^M}{\partial n_{AY_2}}\right)_{n_B,n_X}=RT\ lnx_A+2\ RT\ lnx_Y=RT\ lnx_Ax_Y^2 \tag{3.170}$$

From Eqs. 3.164, and 3.170 we can write: $a_{AY_2}=x_{A^{2+}}.x_{Y^-}^2=x_A.x_Y^2$

Following the same procedure and noting that for the salt BX we can write:

$$\overline{G}_{BX}^{M} = \left(\frac{\partial \Delta G^M}{\partial n_{BX}}\right)_{n_A, n_Y} = RT ln a_{BX} \tag{3.171}$$

$$\text{and } dn_{BX} = dn_B = dn_X$$

$$\text{and } \left(\frac{\partial \Delta G^M}{\partial n_{BX}}\right)_{n_A, n_Y} = \left(\frac{\partial \Delta G^M}{\partial n_B}\right)_{n_A, n_Y, n_X} + \left(\frac{\partial \Delta G^M}{\partial n_X}\right)_{n_A, n_B, n_Y} \tag{3.172}$$

Differentiating Eq. 3.163 with respect to n_B at constant n_A, n_X, n_Y.

$$\left(\frac{\partial \Delta G^M}{\partial n_B}\right)_{n_B, n_X, n_Y} = RT\left[n_A \frac{d}{dn_B}\left\{ln\left(\frac{n_A}{n_A + n_B}\right)\right\} + ln\left(\frac{n_B}{n_A + n_B}\right) + n_B \frac{d}{dn_B}\left\{ln\left(\frac{n_B}{n_A + n_B}\right)\right\}\right] \tag{3.173}$$

On differentiation of appropriate terms, we get:

$$\frac{d}{dn_B}\left[ln\left(\frac{n_A}{n_A + n_B}\right)\right] = \frac{n_B}{n_A} \frac{1}{(n_A + n_B)}$$

$$\frac{d}{dn_B}\left[ln\left(\frac{n_B}{n_A + n_B}\right)\right] = -\frac{1}{(n_A + n_B)}$$

Subtituting these values in Eq. 3.173 we get:

$$\left(\frac{\partial \Delta G^M}{\partial n_B}\right)_{n_A, n_X, n_Y} = RT\left[ln\left(\frac{n_B}{n_A + n_B}\right) + n_A . \frac{n_B}{n_A} \frac{1}{(n_A + n_B)} - n_B . \frac{1}{(n_A + n_B)}\right]$$

$$= RT \ln\left(\frac{n_B}{n_A + n_B}\right) = RT\ ln\, x_B \tag{3.174}$$

Similarly,

$$\left(\frac{\partial \Delta G^M}{\partial n_X}\right)_{n_A, n_B, n_Y} = RT\ ln\, x_X \tag{3.175}$$

From Eqs. 3.172, 3.174 and 3.175, we can write:

$$\left(\frac{\partial \Delta G^M}{\partial n_{BX}}\right)_{n_A,n_Y} = RT\ \ln x_B + RT\ \ln x_X = RT\ \ln x_B x_X \tag{3.176}$$

Comparing Eqs. 3.171 and 3.176 we can write:

$$a_{BX} = x_B . x_X$$

References

1. G. N. Lewis, and M. Randall, Revised by Pitzer, K. S. and Brewer, L., in *Thermodynamics*, 2nd edn. (McGraw-Hill Co. Ltd., London, 1961)
2. J. Mackowiak, *Physical Chemistry for Metallurgists* (American Elsevier Publishing Co. Inc., New York, 1966)
3. R.H. Parker, *An Introduction to Chemical Metallurgy*, 2nd edn. (Pergamon, Oxford, 1978)
4. D.R. Gaskell, *Introduction to the Thermodynamics of Materials*, 4th edn. (Taylor & Francis, New York, 2003)
5. M. Shamsuddin, *Physical Chemistry of Metallurgical Processes*, 2nd edn. (TMS & Springer, Switzerland, 2021)
6. L.S. Darken, R.W. Gurry, *Physical Chemistry of Metals* (McGraw-Hill Co. Ltd, London, 1953)
7. G.N. Lewis, Proc. Amer. Acad. **37**, 40 (1901)
8. R.K. Dube, G.S. Upadhyay, *Problems in Metallurgical Thermodynamics and Kinetics* (Pergamon, Oxford, 1977)
9. C.H.P. Lupis, J.F. Elliott, Acta Met. **14**, 529 (1966)
10. M. Margules, Sitzber Akad. Wiss. Wein **104**, 1243 (1895)
11. J.H. Hildebrand, J. Amer. Chem. Soc. **51**, 66 (1929)
12. R.A. Swalin, *Thermodynamics of Solids* (John Wiley, New York, 1964)
13. A.W. Porter, Trans. Faraday Soc. **16**, 336 (1921)
14. J.H. Hildebrand, Proc. Nat. Acad. Sci. **13**, 267 (1927)
15. J.H. Hildebrand, R.L. Scott, *The Solubility of Non-Electrolytes* (Reinhold Publishing Corp, New York, 1950)
16. J.H. Hildebrand, R.L. Scott, *Regular Solutions, Prentice-Hall* (Englewood Cliff, New Jersey, 1962)
17. L.S. Darken, Trans. Met. Soc. AIME **239**, 80 (1967)
18. C. Wagner, *Thermodynamics of Alloys, Addison-Wesley* (Mass, Cambridge, 1952)
19. K.F. Herzfeld, W. Heitler, Zeit. Elektrochemie. **31**, 536 (1925)
20. G. Scatchard, Chem. Rev. **8**, 321 (1931)
21. M. Shamsuddin, A. Nasar, Metall. Trans. B **23B**, 467 (1992)
22. M. Shamsuddin, A. Nasar, Metall. Trans. B **26B**, 569 (1995)
23. R. Schmid, Y. Chuang, Y.A. Chang, Calphad **9**, 383 (1985)
24. K.C. Chou, Bet. Bunsen-Ges. Phys. Chem. **93**, 569 (1989).
25. A.J. Strauss, J. Steininger, J. Electrochem. Soc. **117**, 1420 (1970)
26. T.H. Weng, J. Electrochem. Soc. **117**, 725 (1970)
27. C. Levy-Clement, R. Triboulet, J. Rioux, A. Etcheberry, S. Licht, R. Tenne, J. Appl. Phys. **58**, 4703 (1985)
28. G. Hodes, J. Manassen, D. Cohen, J. Amer. Chem. Soc. **102**, 5962 (1980)
29. J. Litwin, Phys. Status Solidi **5**, 551 (1964)
30. A.D. Stuckes, G. Farreil, J. Phys. Chem. Solids **25**, 477 (1964)
31. H. Tai, S. Nakashima, S. Hori, Phys. Status Solidi **30**, K115 (1975)

32. J. Steininger, J. Appl. Phys. **41**, 2713 (1970)
33. D. Long, J.L. Schmit, in *Semiconductor and Semimetals*, ed. by R.K. Willardson, A.C. Beer, vol. 5, (Academic Press, New York, 1970)
34. M. Temkin, Acta Physicochimica, URSS **20**, 411 (1945)
35. P. Herashymenko, G. E. Speight, Part I-Derivation of fundamental relationships. J. Iron Steel. Inst. **166**, 169 (1950), Part II-Application to the basic open hearth process. J. Iron & Steel. Inst., **166**, 269 (1950)
36. H. Flood, K. Grjotheim, J. Iron Steel Inst. **171**, 64 (1952)
37. H. Flood, T. Forland, K. Grjotheim, Physical Chemistry of Melts. Inst. Min. Met. (1953), London
38. C.R. Masson, Proc. Roy. Soc. **A287**, 201 (1965)., London
39. S.G. Whiteway, I.B. Smith, C.R. Masson, Can. J. Chem. **48**, 1456 (1970)

Chapter 4
Calorimetry

Calorimetry is defined as the science of measurement of energy changes in the form of heat; the apparatus in which this change is measured is called a calorimeter. The heat or enthalpy of formation of compound can be directly measured by calorimetry. The enthalpy effect accompanying any physical or chemical process or reaction is the most direct and sensible compared to any other effect and hence can be precisely determined.

Historically, calorimetry was the first to be employed among the thermodynamic measurement techniques by M. Berthelot [1] in France and J. Thomsen [2] in Denmark in the nineteenth century. They determined the heats of formation of several compounds as guided by the perception that heat of formation was the measure of driving force for the formation of compounds. Some of their values are still not superseded. Although it was later recognized that the true driving force is the free energy of formation, heat of formation is still a major contributor to the measure of stability for reactions involving condensed ordered phases (e.g. most intermetallic compounds) where entropy of formation is small.

Since the heat effects are directly measured by the calorimetric technique the corresponding result can be reported with an estimated error which can be relied upon. The quantities obtained directly from equilibrium techniques viz. chemical equilibria, electrochemical and vapor pressure methods are the free energies from which corresponding heats can be calculated only when the temperature coefficients of the free energy are known. A small error in the temperature coefficient may result in a large error in the heat effect.

Calorimetric techniques measure the enthalpy or heat of a reaction, changes in heat content of a substance associated with changes in temperatures, changes of state such as melting and boiling, changes of chemical combination such as reactions, or alloying, or dissolution. The changes of physical structure like heat of order-disorder transformation, and energy stored in cold working can also be measured by alloy solution calorimetry by the alternate dissolution of ordered and disordered and annealed and cold worked samples, respectively. The successful and versatile use

M. Shamsuddin, *Thermodynamic Measurement Techniques*, The Minerals, Metals & Materials Series, https://doi.org/10.1007/978-3-031-47118-6_4

of the calorimetric technique in thermodynamic measurements has encouraged the development of various types of calorimeters.

4.1 Classification of Calorimeters

Calorimeters have been classified into five main groups viz. isothermal, adiabatic, isoperibol, constant temperature gradient, and variable temperature gradient on the basis of the simplified heat flux equation:

$$q = k\,(T_s - T_c) \tag{4.1}$$

where q, T_s and T_c are the heat flux, the temperature of the surrounding and the calorimeter, respectively, and k is the overall heat transfer coefficient. A number of book chapters and reviews [3–7] have appeared on calorimetry highlighting its importance, classification, experimentation and instrumentation.

4.1.1 Isothermal Calorimeter

When T_s and T_c are always equal and remain constant during the course of investigation the calorimeter is named isothermal. The heat effect produces a phase change of a substance at the transformation temperature. The measurement is usually associated with a change in volume [8, 9]. The heat effect is calculated from the latent heat of transformation and the amount of phase changed.

The Bunsen Ice Calorimeter is considered to be a true representative of the isothermal group of calorimeters. The reaction container is surrounded by a mixture of ice and water. The proportion of which is normally kept constant by an outer vessel, also maintained at 0 °C. Any heat transfer due to reaction in the inner vessel causes melting of ice or freezing. Thus heat of reaction is translated into latent heat of fusion of ice at 0 °C, and the quantity of heat is measured by the change in volume of the ice-water mixture. The apparatus is made throughout of glass. The main bulb, enclosing the reaction tube, is filled with distilled water, while its lower portion and the capillary tube contains mercury. The calorimeter is placed in a beaker containing ice, and the whole apparatus is stored in a refrigerator when not in operation. The coating of ice required around the reaction tube is produced by successive addition of small amounts of solid carbon dioxide to a quantity of alcohol contained in tube. The ice layer should initially be 10–15 mm thick. The heat effect is determined by the change in weight of mercury in the cup (1 cal = 15.49 mg Hg.). Ice calorimeter permits the study of reactions which are slow or in which the heat exchange is small.

4.1.2 Adiabatic Calorimeter

When T_s and T_c are equal but not constant, the calorimeter is designated as an adiabatic calorimeter [10]. The temperature of the surrounding, T_s, is equalized with that of the calorimeter. The heat effect is calculated from the temperature change of the calorimeter and its water equivalent. Such calorimeters can be used for the determination of heat capacities as well as heats of reaction, but because of the requirement, $T_s = T_c$, its application is limited to relatively slow reactions. Adiabatic calorimeter designed by Dench [11] operative in the temperature range: 500–1400 °C for the determination of heat capacities and heat of transition as well as heat of reaction is shown in Fig. 4.1.

The specimen A, prepared by compacting a mixture of the powder reactants into cylindrical form (19 mm dia × 32 mm height) is placed in the adiabatic enclosure B,

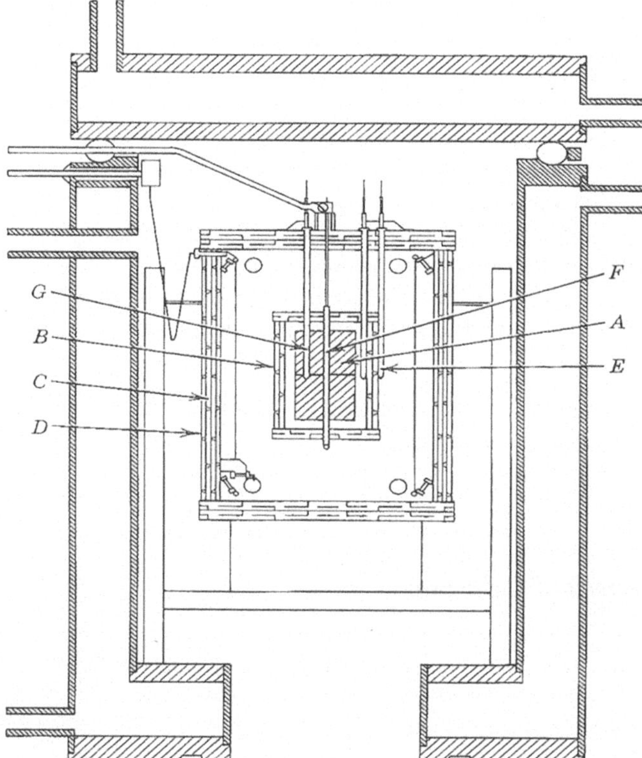

Fig. 4.1 Adiabetic reaction calorimeter. (Reproduced from O. Kubaschewski, and C.B. Alcock, Metallurgical Thermochemistry [3] 1967, p 132 (Originally from W. A. Dench [11] Trans. Faraday Soc. **59** (1963) 1279)) A – specimen; B – adiabetic enclosure; C – furnace; D – vacuum chamber; E – differential thermocouples; F – tungsten resistance heater; G – specimen thermocouple

consisting of three concentric cylindrical radiation shields made of tantalum sheet. The vacuum furnace C, made of tantalum winding surrounds the enclosure. The furnace is enclosed in a water-cooled brass vacuum chamber D. Three Pt-Pt/Rh differential thermocouples E, in alumina sheaths are placed around the middle of the adiabatic enclosure, each having one junction against the outside of the enclosure walls. Any difference between the inside and outside of the enclosure is indicated by a deflection in a galvanometer to which the differential thermocouples are connected in series. Heat may be supplied to the specimen via a tungsten resistor heating coil in an alumina sheath F, placed in a hole down the axis of the specimen. The energy supplied to the heater is measured by means of a precision watt-hour meter. A Pt/Pt-Rh thermocouple G is placed in a second hole of the specimen through an alumina sheath to measure the specimen temperature.

For measurement of heat of formation the following procedure is adopted:

(i) First the highest temperature i.e. the "safe temperature" at which the powder mixture may be held without any appreciable reaction is determined.
(ii) The specimen is then heated adiabatically from the safe temperature to the reaction temperature using the internal tungsten heater.
(iii) It is then switched off and temperature of the sample is measured as a function of time.
(iv) The reaction begins during heating and after switching off, the temperature of the specimen will continue to rise, if the reaction is exothermic, or will begin to fall if it is endothermic.
(v) The temperature will gradually decrease as the reaction progresses. A constant temperature is attained on completion of the reaction.
(vi) In separate experiments the changes in heat contents of the empty calorimeter and the reactants are determined from the "safe temperature" to the reaction temperature.

The heat of formation/reaction is obtained by subtracting the change in heat content of the reactants and the empty calorimeter from the measured input energy to the tungsten heater.

4.1.3 Isoperibol Calorimeter

A calorimeter where T_s remains constant and T_c changes during the course of a reaction, is known as isoperibol type. The heat effect produces a temperature change in the calorimeter which gradually returns to equilibrium with its surroundings by exchange of heat. Such calorimeters can be employed for measurement of a number of thermodynamic parameters; hence they have received the maximum attention with regard to the advancement of components. Isoperibol calorimeters are available in different designs to suit different investigations, such as Bomb, Solution and Drop types. In recent years micro-calorimeters have drawn maximum attention of the thermodynamic investigators. In this section principle and construction of some calorimeters will be briefly discussed.

4.1.3.1 Bomb Calorimeter

The bomb calorimeter refers to an apparatus in which combustion is carried out with a gas (oxygen, nitrogen, fluorine or chlorine [5, 12, 13] as one of the reactants depending upon the reactivity of the material under investigation. In principle the calorimeter compares the quantity of heat liberated from the combustion of the sample with a known input of electrical energy. In general bomb calorimeter is calibrated with standard benzoic acid whose heat of combustion is well established. As the combustion reaction is carried out at constant volume this technique measures the change in internal energy (ΔU). The enthalpy change (ΔH) can be estimated by the expression:

$$\Delta H = \Delta U + \Delta n\, RT \tag{4.2}$$

where Δn is the change in number of moles on formation of the combustion products at temperature T K and R is the gas constant.

In order to measure heat of combustion, a known weight of sample is placed in the heavy-walled bomb made of corrosion resistant alloy. The bomb is filled with oxygen gas at 25 atm pressure and then immersed in a calorimeter containing known amount of water and fitted with a stirrer, a thermocouple and an electric heating coil. The reaction (combustion) is initiated by passing electric current in the sample through the coil. The combustion being exothermic, temperature of the water in calorimeter rises. This is measured by a platinum thermometer sensitive to 10^{-4} K. Subsequently, separate experiment is conducted to measure the heat change (supplied through the coil in the form of electrical energy) to get the same temperature rise. Bomb calorimetry is useful in estimation of enthalpy/heat of formation of compounds which cannot be formed directly from its constituent elements. For example, methane cannot be synthesized by direct combustion of carbon and hydrogen but its enthalpy of formation can be calculated from the values of heat of combustion of methane, carbon and hydrogen determined in a bomb calorimeter.

4.1.3.2 Solution Calorimeters

The technique of measurement of heats of formation or mixing in metallic system by direct reaction between the constituent elements is limited to liquid alloys and in the case of solids to systems with high exothermicity since otherwise the solid constituents will take a much longer time to form a homogeneous solid product. Hence the indirect method where the homogeneous alloy is brought into a state in which the enthalpy with respect to the constituents in their pure standard states can be determined, is more promising. The liquid metal solution calorimeter was developed on this principle.

Since the kinetics of most of the solid-state reactions or processes is slow, involving a small quantity of net heat changes and very high activation barriers, these heats of reactions cannot be measured with precision by direct calorimetry. In such cases the heat effect can be measured by following an indirect route, for example, rapid dissolution of the substance under investigation in a suitable solvent. The solvent is so selected that both the compound as well as the mechanical mixture of the constituent elements dissolve in it fairly rapidly with a small heat effect. Calorimeters using aqueous solvents have been employed for determination of heats of formation of oxide systems [14]. However, such solvents are discarded in studying metallurgical reactions due to the limitations of temperature of operation and evolution of large volumes of gases with acidic solvents. These problems have been overcome by the use of liquid metals and alloys [15, 16] as solvents for alloys and intermetallic compounds and fused salts [17] for oxides, aluminates and silicates.

The process of dissolution in a liquid metal or alloy is rapid, so that the end state of the reaction is not only well defined and reproducible, but is also attained quickly. By alternatively dissolving a material from the standard state (A^o) and from any other state (A) in a suitable solvent, and then applying the Hess's Law of constant heat summation the difference in the enthalpy between the two states, $H_A - H_{A^o} = \Delta H\ (T)$, is obtained as the difference between the heat effects of the two dissolution processes as shown below:

$$A_T^o \rightarrow \underline{A},\ \ (T', N_A),\ \ \Delta H_1 \tag{4.3}$$

$$A_T \rightarrow \underline{A}\ \ (T', N_A),\ \ \Delta H_2 \tag{4.4}$$

$$A_T^o \rightarrow A_T, \Delta H_T = \Delta H_1 - \Delta H_2 \tag{4.5}$$

In the above equations, T' and N_A refer to the temperature and composition, respectively of the solvent after the dissolution, T refers to the temperature from which the materials A^o and A are added to the calorimeter, and ΔH values refer to the heats of the corresponding dissolution reactions. $\underline{A}$ refers to the end state of the substance (metal, alloy or oxide) added in the calorimeter.

Bever and Ticknor [15] for the first time realized the advantages of liquid metal solution calorimetry and fabricated the first calorimeter in 1951 with tin as solvent. Since then a number of designs and modifications [16–44] have been introduced to increase its degree of accuracy. A liquid metal solution calorimeter of any design has following essential components, namely, the dissolution chamber and its solvent, the vacuum system, the thermostat and the devices for the temperature measurement and control. The thermostat may consist either of a vessel containing 50–60 liters of low melting nitrate salt [19–22] or oil [23] or a highly conducting massive metal block, for example, silver [25], nickel plated copper [26–28], aluminum [24, 29, 30], nickel [31] and special steel [32]. Oriani and Murphy [33] designed a differential tin solution calorimeter to further reduce the heat effect due to dissolution. From time to time attempts have been made to increase the temperature [34, 35] of operation of the calorimeter by changing the design of the thermostat, container material and the mode of heating. The calorimeter developed by Vieth and Pool [36] using liquid uranium as the solvent and boron nitride as the isothermal block is attractive and can be operated up to 1570 K. Gupta and Jena [37] have designed and operated a metal solution calorimeter consisting of several radiation shields as the constant temperature environment. Shamsuddin and Misra [38] have designed solution calorimeters which can be operated in different temperature ranges. The essential features of a number of liquid metal solution calorimeters operated by different investigators have been summarized in Table 4.1.

Table 4.1 Salient features of liquid metal/alloy solution calorimeters

Investigators [References]	Thermostat, medium and temperature range (K)	Solvent bath, composition and stirring techniques	Calibration, temperature measuring technique	General areas of applications
1. Bever and coworkers [19, 20] (1952, 1962)	Eutectic mixture of Na, Li and K nitrates, 515–725	Pb, Sn, Bi and Bi-Sn alloy, helical glass stirrer	Chemical, two iron-constantan thermocouples	Heats of formation, mixing and solution, stored energy, annealing, energy of order-disorder transformation, excess energy of splat cooled phases.
2. Kleppa [24] (1955)	Aluminum block, 525–775	Sn, graphite rod	Electrical, chromel-alumel thermocouples	Heats of formation, of solid solutions and intermetallic compounds.
3. Wittig and Huber [18] (1956)	Metal block up to 600	Zn and low melting alloys, rod with globular end	Electrical, differential thermocouples of Ni-CrNi	Heats of mixing and formation.
4. Hultgren and coworkers [26] (1957)	Nickel plated copper block 525–575	Sn, impeller or plunger of molybdenum	Chemical, copper-constantan thermocouples	Heats of formation heat capacity and energy of order-disorder transformation.
5. Oriani and Murphy [33] (1958)	Twin calorimeter	Sn (about 700 g), silica rod	Electrical, chromel-constantan thermocouples	Heats of formation and solution and energy of order- disorder transformation.
6. Kleppa and coworkers [25] (1966)	Twin calorimeter, silver block up to 900	Sn (about 700 g), reciprocating stirrers	Electrical, Pt/Pt-10% Rh thermocouples	Heats of solution and formation of transition metals.

(continued)

Table 4.1 (continued)

Investigators [References]	Thermostat, medium and temperature range (K)	Solvent bath, composition and stirring techniques	Calibration, temperature measuring technique	General areas of applications
7. Morris and Pratt [27] (1965)	Nickel plated copper block, 525–875	Sn(240 g), molybdenum	Chemical, iron-constantan thermocouples	Heat of formation and heat capacity
8.Leach and coworkers [21] (1966)	Mixture of Na and K nitrates (55/45 ~mole ratio), 515–775	Sn, 6 mm dia, stainless steel tube with peddle blades	Pure Sn or W, chromel-alumel thermocouples	Heats of formation and solution, energy changes accompanying solid-state reaction
9. Pool [29] (1965)	Aluminum block (vacuum or argon), 525–775	Sn (about 75 g), tantalum stirrer	Chemical, thermopile	Heats of formation and solution, enthalpy interaction coefficient
10. Elliott and coworkers [34, 35] (1964, 1967)	Induction heating, argon, 1275–1875	Cu, Ni, (Fe ~100 g atom), automatic stirring	Chemical, Pt/Pt-10% Rh thermocouples	Partial atomic heat of solution, interaction coefficient, heat of fusion
11. Mathieu et al. [31] (1966)	Nickel block up to 1475	Sn, graphite vessel	Chemical, (Al_2O_3) thermopile (8 thermocouples)	Heats of solution
12. Misra et al. [23] (1968)	Mobil oil, 360–400	Wood' s alloy (Bi-50, Pb-27, Sn-13 & Cd – 10 wt %), helical glass stirrer	Chemical, copper or tungsten, 2 copper-constantan thermocouples	Heats of formation, mixing and solution, energy of order-disorder transformation, excess energy of splat cooled metastable phases
13. Predel and Mohs [30] (1969)	Aluminum block, about 400	Hg (80 g), glass stirrer	Chemical, gallium	Heat of mixing
14. Predel and Ruge [32] (1972)	Special high temperature steel block, 720–970	Sn (30 g), quartz stirrer, argon atmosphere	Chemical, pure Sn, chrromel-alumal thermocouples	Enthalpies of formation and mixing
15. Blair and Downie [28] (1970)	Copper block, 570–670	Sn, molybdenum impeller	Chemical, differential temperature measurement by micrograph recorder	Heat of solution
16. Boom et al. [22] (1973)	Eutectic mixture of liquid salts	Sn bath in silica Dewar flask	Sn and Bi, chromel-constantan thermocouples connected to strip chart recorder	Heats of formation/solution

(continued)

Table 4.1 (continued)

Investigators [References]	Thermostat, medium and temperature range (K)	Solvent bath, composition and stirring techniques	Calibration, temperature measuring technique	General areas of applications
17. Vieth and Pool [36] (1972)	Boron nitride block, 770–1570	U, (200 g)	Thermopile consisting of 120 thermocouple junctions of W/W-25 Re	Heat of solution
18. Gupta and Jena [37] (1973)	Copper radiation shields, 470–520	Bi-Sn alloy, stainless steel stirrer	Chemical, W, chromel-alumel thermocouples	Partial and integral enthalpy of formation, heat capacity
19. Shamsuddin and Misra [41, 42] (1972)	(i) Heat treatment salt, 500–800	Bi bath in glass Dewar flask	Chemical, Bi, two chromel-alumel thermocouples	Heat of formation
	(ii) Stainless steel block, 500–1200	Sn bath in silica tube	-do-	Heat of formation
20. Kleppa [40] (1960)	Calvet-type twin microcalorimeter with aluminum block, 500–700	Low melting metal	Electrical, 96 + 96 Pt-Pt 13 Rh thermocouples (thermopile)	Thermochemistry of low-melting molten salts, heat of mixing of low melting alloys
21. Kelppa [44] (1960–1970)	Calvet- type twin microcalorimeter with nickel block, 1000–1100	Molten metals or oxide melts	Electrical, Pt-Pt 13 Rh thermopiles	Heat of formation of intermetallics, minerals and complex oxides, heat of mixing of alloys
22. Kleppa [44] (1960–1970)	Calvet- type twin microcalorimeter, inconel block, 1300–1400	Molten metals or oxide melts	Electrical, Pt-Pt 13 Rh thermocouples	Heat of formation of intermetallics, minerals and complex oxides and heat of mixing
23. Kleppa [44] (1980–2000)	Calvet-type twin rcaction calorimeter, alumina discs, 1400–1500	Liquid copper	Electrical, Pt-Pt 13 Rh thermocouples	Heat of formation of borides of Mn, Fe, Co and Ni and LaB_6

These calorimeters can be employed for determination of a number of thermodynamic parameters, viz. heat of formation, solution or mixing, energy of order-disorder transformation, enthalpy interaction coefficient, heats of precipitation from liquid alloys, heat capacity of metallic melts, excess energy of metastable states and stored energy of deformation.

The liquid metal/alloy solution calorimeter [42] shown in Fig. 4.2, had a long necked Dewar flask as a calorimeter vessel. This could hold 500 g of molten bismuth or tin as a solvent under a vacuum of 0.05 micron. The Dewar flask was immersed in a thermostat containing crescent TS-150, heat treatment salt bath maintained at

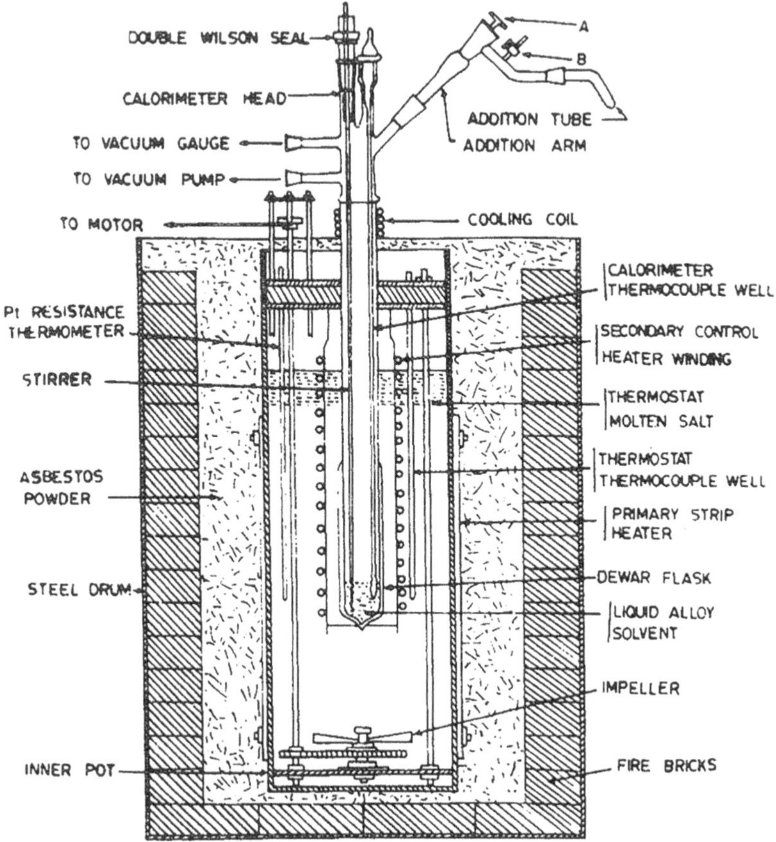

Fig. 4.2 Liquid metal/alloy solution calorimeter. (Reproduced from M. Shamsuddin and S. Misra [42] with the permission of The Minerals, Metals & Materials Society)

500 K by means of a set of primary heaters mounted around the thermostat container and a secondary control heater (in the form of a spiral surrounding the Dewar flask). The thermostat and the molten solvent were stirred vigorously and their temperatures were controlled within ±0.05 K and ±0.02 K, respectively, by means of voltage stabilization of primary heating circuit in conjunction with an electronic relay in the control circuit. This calorimeter has been used to determine heat of formation of SnTe [38] and PbTe [42].

PbS and PbSe do not dissolve in any of the melts of tin, lead, bismuth or in any binary or ternary alloys of these metals, up to a temperature of 800 K. An increase of temperature beyond 800 K is not possible due to the evaporation of the salt and softening of glass under high vacuum in the long duration of experimentation. This led to the development a high temperature solution calorimeter [41] with tin solvent capable of operating up to 1100 K. The high temperature calorimeter shown in

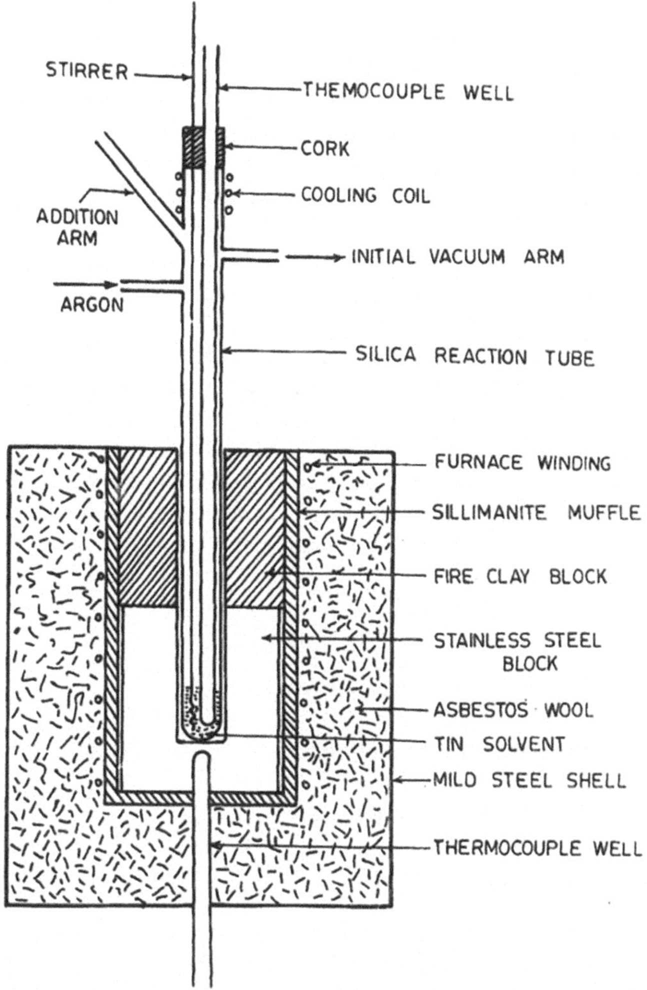

Fig. 4.3 High temperature liquid metal/alloy solution calorimeter. (Reproduced from Shamsuddin [41] Trans. Ind. Inst. Metals, **28** (1975) 303 with the permission of The Indian Institute of Metals)

Fig. 4.3 consisted of an A_1 grade Kanthal wound vertical muffle furnace closed at the bottom. A cylindrical stainless steel block of 87 mm dia. and 100 mm height is located centrally in the muffle furnace. The block serves the purpose of the isothermal reference surrounding and was provided with a 40 mm dia. cavity in which 38 mm diameter silica tube closed at the bottom was placed. This tube serving as the reaction compartment of the calorimeter could hold 200 g of tin solvent.

There were five sets of thermocouples, two of which were embedded in the solvent with a silica thermocouple well; the other three sets were inserted from the bottom in a cylindrical hole of 12 mm in the block. One of the latter was differentially connected to one of the solvent thermocouples, the second and third were employed to measure the temperature of the block and control its temperature. There were arrangements at the top of the calorimeter reaction tube for evacuation, introduction of argon and charging of the sample under investigation. The tube top was cooled by copper coils carrying water. The temperature of the reference block was controlled by electronic controller. During the course of investigation the temperature of the block and the differential temperature between the block and the solvent were measured by a potentiometer or by digital multi-meter.

Both the calorimeters are operated in an identical manner. Alternate samples of the compound and mechanical mixture corresponding to the stoichiometric composition are added from the reference temperature, T' (273 K or 298 K) to the solvent temperature, T (500–1100 K, depending upon the dissolution characteristics of the compound in a solvent). At least four samples, each, of the compound and mechanical mixtures of the same composition are added in a typical run. The difference in heat effects on dissolution of the two types of addition yields heat of formation of the compound. For example, to get the heat of formation of the compound AB, alternatively the compound AB and the mechanical mixture, (A + B) are added. The difference in two types of dissolution yields the heat of formation (ΔH) in the following manner:

$$\mathrm{A + B \rightarrow \underline{AB}; \Delta H_1} \tag{4.6}$$

$$\mathrm{AB \rightarrow \underline{AB}; \Delta H_2} \tag{4.7}$$

Hence, for the compound formation reaction:

$$\mathrm{A + B \rightarrow AB; \Delta H = \Delta H_1 - \Delta H_2} \tag{4.8}$$

This has been demonstrated in Fig. 4.4.

Microcalorimeter The sensitivity of the above two types of calorimeters decreases with the increase of operating temperature. Calvet and Prat [39, 43] suggested the most satisfactory approach to sort out this problem in their books. According to them at high temperature calorimetry should be based on the integrated heat flux principle. This provided the physical basis for the Calvet type microcalorimeters. Presently it is possible to construct a calorimeter of such sensitivity and accuracy that one can study closely the thermal effects which accompany slow phenomena. Kleppa [40, 44] very successfully designed and operated a number of microcalorimeters for thermodynamic measurements at different temperatures using various constructional materials for reference surrounding, namely, aluminum, nickel, inconel and alumina blocks. A brief description of the Calvet-type twin microcalorimeters designed by Kleppa is given below.

The microcalorimeter shown in Fig. 4.5 had (96 + 96) chromel-alumel thermocouples in its thermopile and its twin construction to a large extent eliminated most problems associated with precise temperature control. The calorimeter built from

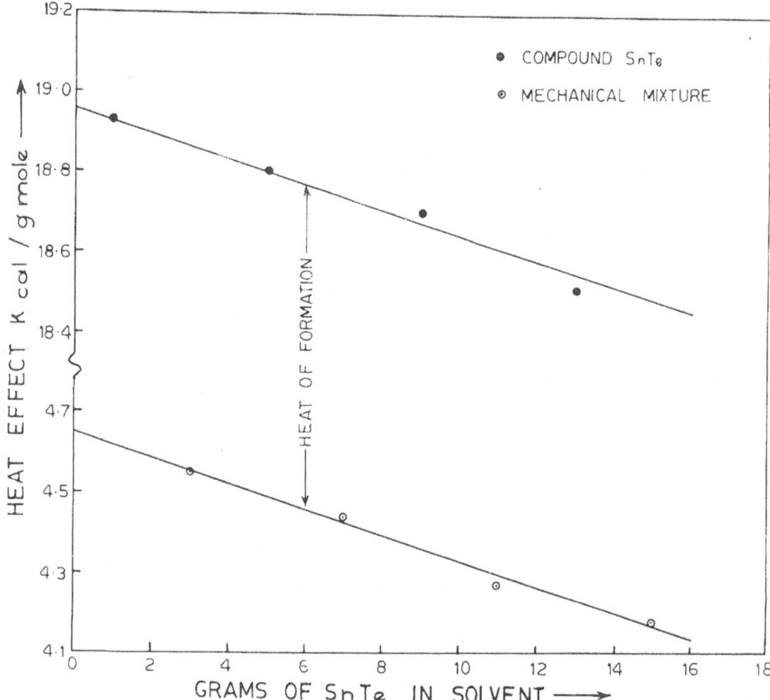

Fig. 4.4 Heat effects associated with the addition of samples of the compound, SnTe and the mechanical mixture Sn + Te corresponding to the composition of SnTe in liquid bismuth. (Reproduced from Shamsuddin [38], Ph.D. Thesis: Thermodynamics of Group IV Chalcogenides, Banaras Hindu University, 1974)

aluminum block was limited to temperature below 500 °C. It was used in Kleppa's laboratory for more than 20 years for innumerable investigations on many different kinds of materials. As a modification its original chromel-alumel thermopile was replaced by another thermopile based on Pt/Pt-13 Rh wire. Although it reduced the sensitivity of the calorimeter, it increased the life of the thermopile from about one year to more than 20 years.

In order to increase the operating temperature to 800 °C and 1100 °C Kleppa built other Calvet-type twin calorimeters using nickel and inconel reference blocks, respectively. In collaboration with Kelppa, Darby [25] developed a twin calorimeter based on pure silver block. The most ambitious project of Kleppa [44] led to the construction of a twin calorimeter shown in Fig. 4.6. It was constructed from 2.5 cm. thick refractory alumina discs and was heated in a very large furnace by three large separate heating elements made of Pt-40 Rh wire and was maintained nearly at 1100 °C for a number of years. This calorimeter allowed precision solution and reaction calorimetry to entirely new class of materials for example, dissolution of a number of transition metals [45] in liquid copper and of the liquid alloys formed

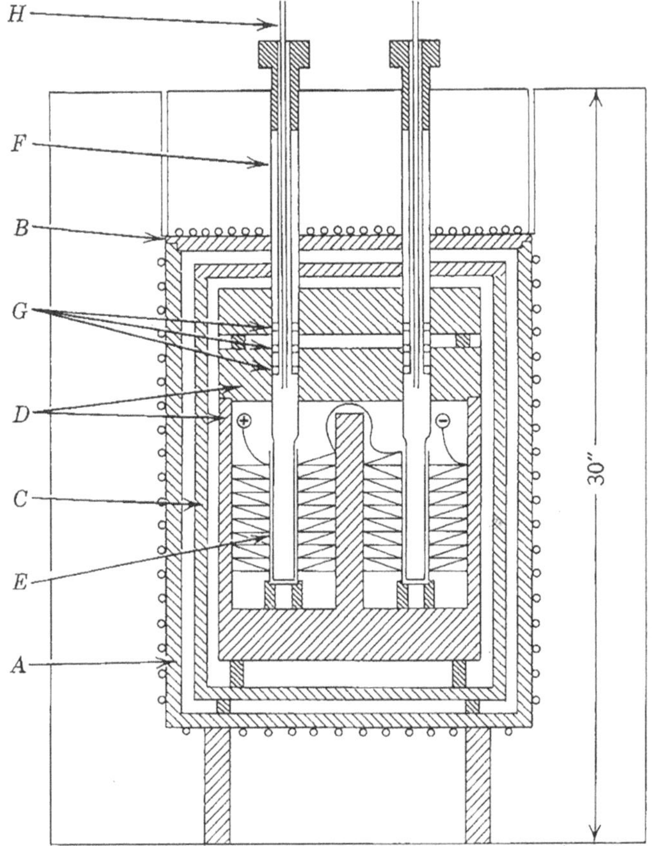

Fig. 4.5 Schematic diagram of Calvet microcalorimeter. (Reproduced with permission from O. J. Kleppa [40], J. Phys. Chem. 64 (1960) 1937. Copyright 1937 American Chemical Society) A – main heater; B – top heater; C – heavy shield; D – aluminum jacket; E – calorimeter; F – protection tube; G – radiation shields; H – manipulation tube; ⊕ – Θ – thermopile

among the noble metals copper, silver and gold and to determine heats of formation of the borides [46–48] of manganese, iron, cobalt and nickel.

Direct Synthesis Calorimeter In the 1980s Kleppa constructed a high temperature direct synthesis calorimeter shown in Fig. 4.7 for determination of heat of formation of congruently melting compounds of copper [49, 50] with Ti, Zr, Hf, Sc, Ya and La and alloys of neodymium [51], dispronium [52] and samarium [53] with Ni, Ru, Rh, Pd, Ir and Pt, borides, germanides and aluminides of 3d, 4d and 5d transition metals [54, 55]. In these investigations two sets of colorimetric experiments [44]: (1) reaction experiments, in which mixture of the two elements was dropped from room temperature into the high temperature calorimeter to react chemically in the liquid

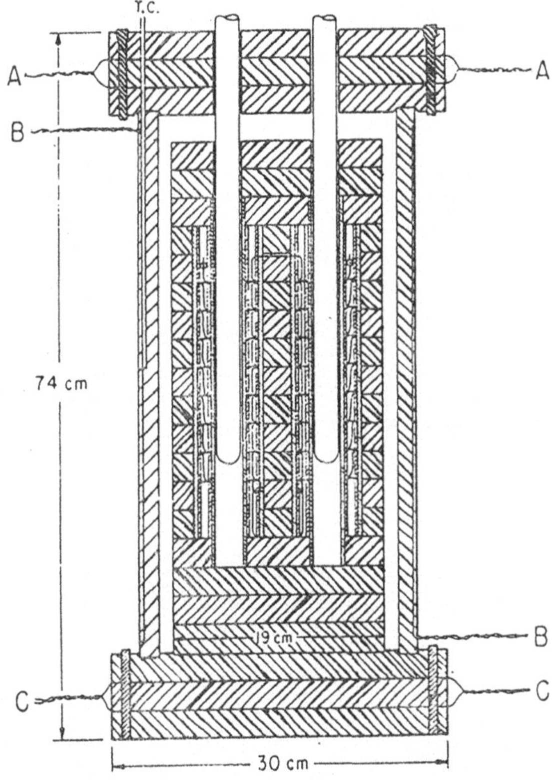

Fig. 4.6 High temperature (up to 1400 K) Calvet microcalorimeter. (Reproduced from O. J. Kleppa [44] J. Phase Equilibria, **15** (1994) 244) A – top heater; B – main heater; C – bottom heater; TC – thermocouple/thermopile

state, and (2) heat content experiments, in which the solidified congruently melting compounds or the alloy were dropped under similar conditions of step (1). The difference between the two heat effects gives the desired heat of formation of the solid compound/alloy at room temperature.

4.1.4 Constant Temperature Gradient Calorimeter

This type of calorimeter utilizes the principle of thermal analysis involving temperature-time or inverse rate curves during heating or cooling. The technique of thermal analysis has been extensively popular for studies of phase transformations and phase diagrams. The quantitative use of this technique for calorimetry has not been duly appreciated owing to the complications introduced by the variables T_S, T_c and time, t in the dynamic method.

An improved method of quantitative thermal analysis was devised by imposing the constancy of (T_S-T_C). Thus the technique is called constant temperature gradient calorimetry, although the term constant heat flow calorimetry is also in use

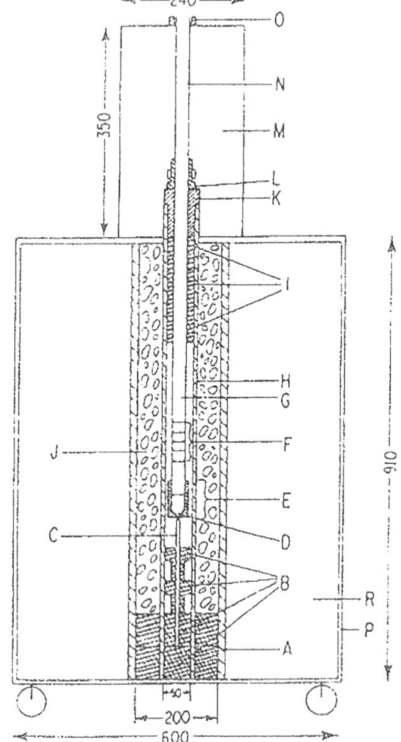

Fig. 4.7 High temperature calorimeter for 1400 K and above. (Reproduced from O. J. Kleppa [44] J. Phase Equilibria, **15** (1994) 244) A – furnace shell; B & I – insulation brick; C – alumina supporting tube; D – alumina tube; E – reference section; F – colorimeter working section; G – calorimeter tube; H – alumina furnace core; J – alumina beads; K – lavite support; L – sleeve; M – ceramic wool; N – alumina extension tube; O – bakelite clamp; P – water-cooled jacket; R – refractory insulation

[3, 6]. The constant temperature gradient calorimeter has been successfully utilized for determination of heat capacity and heat and temperature of transformations. The details of its design, construction operation, control of temperature and its gradient has been discussed by the author in a review paper [6], only a brief account will be given here.

A Constant Temperature Gradient Calorimeter shown in Fig. 4.8 has the following essential components, namely, the calorimeter proper with radiation shields, the vacuum system, the temperature gradient control unit, the temperature measuring device and the continuous time counting unit. The figure shows the general feature of the entire assembly. There is provision for inserting purified argon in the system. The specimen container consists of a high-density graphite crucible with a lid supported at the bottom by a four-bore alumina tube. The crucible is surrounded in succession by a thin wall cylindrical jacket of stainless steel (inner reference shield), the furnace with heater coils strung vertically and horizontally, a stainless steel radiation shield. The entire assembly is enclosed in a water cooled brass bell jar and evacuated by a vacuum pumping system and filled with high purity argon. This type of calorimeter has not been popular due to difficulty of operating condition by maintaining a constancy of (T_s-T_c).

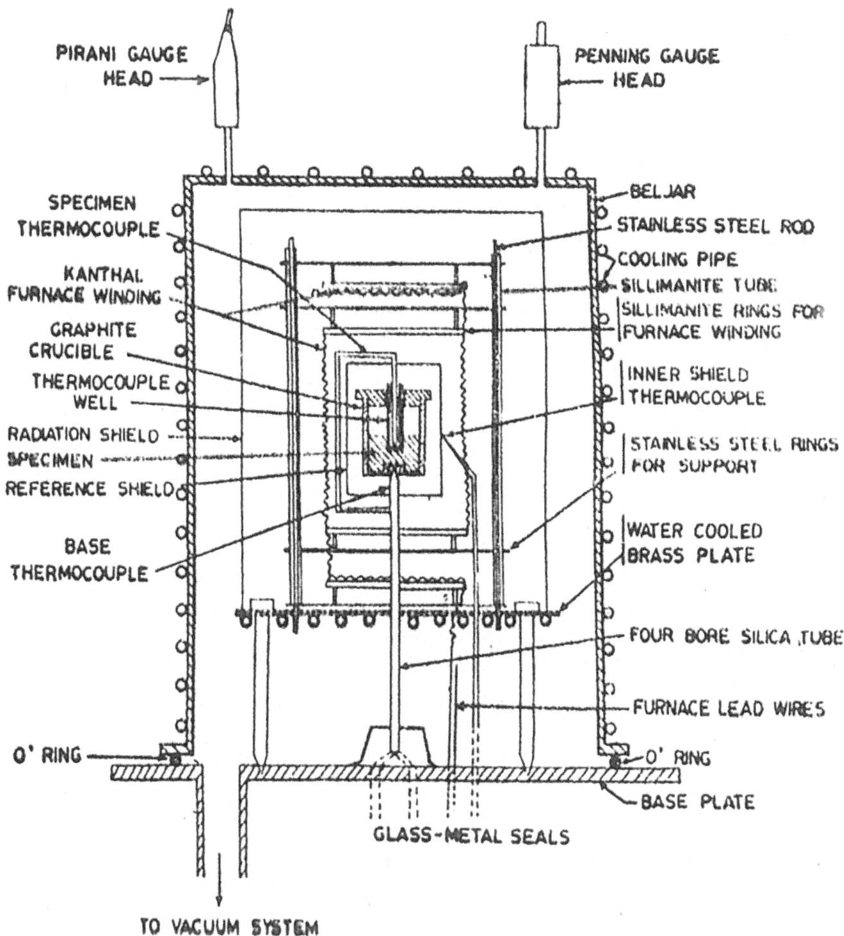

Fig. 4.8 Schematic diagram of constant temperature gradient calorimeter. (Reproduced from Shamsuddin and Misra [6] Trans. Ind. Inst. Met., **27** (1974) 282 with the permission of The Indian Institute of Metals)

4.1.5 Variable Temperature Gradient Calorimeter

When T_s and T_c change independently and hence (T_s-T_c) is a variable, the technique is known as variable temperature gradient calorimetry. The principle of Variable Temperature Gradient (VTG) calorimetry is based on differential thermal analysis. Transformations in metals and alloys during heating and cooling give rise to arrests in their plots between differential temperature (between the sample and its surroundings) and time. The duration of arrests accompanying a transformation can be translated to the corresponding heat effect on the basis of heat transfer considerations. Since, this technique employs the differential temperature versus time plot, the apparatus is also known as differential thermal analysis (DTA) calorimeter.

A DTA calorimeter is usually employed to determine the heat of any 1st order phase transformation during which the temperature of the sample remains constant and that of the surrounding block changes. The calorimeter can also be used to determine the heat of formation for which reaction, the temperatures of both the sample and the block change. In either case the differential temperature, ΔT between the sample and the surrounding metallic block deviates from its normal value. ΔT decreases after the completion of the exothermic reaction and increases after the completion of endothermic reaction due to the rapid heat transfer between the calorimeter and the surrounding. Hence, in the post-transformation period, the ΔT value approaches rapidly the normal background value. Thus, the transformation period in each case is confined to the ΔT vs t peak. According to Newton's law of heat transfer:

$$\frac{\delta Q}{\delta t} = k.\Delta T \tag{4.9}$$

where, δQ is the heat transferred due to excess differential temperature ΔT, above the background or normal value in time δt, and k is the heat transfer coefficient.

The total excess heat transferred due to the phase transformation or heat of formation can be obtained by integrating the above equation:

$$Q = \int_{t_1}^{t_2} k\Delta T.\delta t \approx k \int_{t_1}^{t_2} \Delta T.\delta t = ku \tag{4.10}$$

The limits of integration indicate the times of the beginning and end of transformation and u, the area of the peak above the background. The heat of formation of a compound may be obtained by comparing its peak area in the differential thermal analysis (DTA) curve with the area corresponding to a known heat of transformation of a calibrating element transforming in the same temperature range.

Differential thermal analysis calorimetry measures directly the heats of formation, fusion and transformation by employing a simple apparatus. The use of the thermal analysis calorimeter is however limited because the data obtained include the following sources of errors: (a) temperature dependence of calibration, (b) shape of transformation peak in ΔT vs time curves, (c) dependence of the calibration on the amount of transformation, (d) uncertainty in the measurement of time, (e) uncertainty in the measurement of peak area, (f) variation of the rate of heat transfer with the sample size. For estimating the accuracy of measurement by this technique some of the data have to be independently determined by another technique; such as liquid metal solution calorimetry.

A number of DTA calorimeters and analyzers designed and developed on this principle are under operation in different research laboratories for carrying out successful investigations on different materials: metals, alloys, intermetallic compounds, salts and ceramics. The design of construction changes with the type of material to be investigated and the operating conditions.

The main advantages of DTA calorimetry are the simple construction and operation and the direct sensing of heat effects. The technique is capable of measuring the rise or fall of temperature during the formation of a compound. The degree of accuracy depends on the resolution of peak, hence this technique is not very useful where the evolution or absorption of heat is small or reaction is slow. Its application for heat capacity measurements is also difficult. In recent years there is a popular trend to use Differential Scanning Calorimeters (DSC) which are commercially available with the trade names: Perkin Elmer, SETARAM etc.

A DTA calorimeter can be employed to determine a number of thermodynamic quantities, for example, heats or enthalpies of mixing [56], heat and temperature of transformation [57], heat of solidification [58], stored energy of cold working [59], heat capacity [60]. The use of DTA technique in establishing the phase boundaries [61–65] is well known. Nasar and Shamsuddin have determined Heats of formation of highly exothermic compounds, namely, CdS [66], CdSe [67], CdTe [68], ZnTe [69], ZnSe [70] and PbS [71] from their constituent elements using this calorimeter.

4.1.5.1 Differential Thermal Analysis Calorimeter

The calorimeter [72], shown schematically in Fig. 4.9 is fabricated out of indigenous materials. It consists of a Kanthal (A_1) wound vertical furnace with a cylindrical sillimanite muffle closed at the bottom. A cylindrical copper or stainless steel block of 75 mm dia is placed centrally in the furnace muffle. The block is insulated from the furnace muffle by means of a sillimanite tube. A central cavity of 30 mm dia contains a 19 mm I.D. silica well which is insulated from the block by another sillimanite tube. A sample container of graphite (12 mm I.D. and 60 mm long) is placed in the silica well. There are three sets of chromel-alumel thermocouples in the calorimeter, one of which was embedded in the sample. The other two sets are inserted in a cylindrical hole of 10 mm dia in the block, one of which is used for measuring the temperature of the block and the other is differentially connected to the sample thermocouple. The top of the block is insulated by a fireclay disc having holes identical to that of the copper or stainless steel block.

In order to conduct experiments on elements/compounds having appreciable vapor pressure at moderate to high temperatures the calorimeter is provided with the arrangement to insert inert gas (argon) from the top of the silica well. Before inserting argon the calorimeter reaction tube is evacuated through a three way stop cock by means of a rotary pump. The third limb of the stop cock is connected to a mercury manometer, which indicates the desired pressure of argon in the silica well. A slight negative pressure of high purity argon atmosphere is maintained in the calorimeter to take care of the expansion of the gas due to heating.

The furnace is heated at a reproducible and controlled rate from room temperature to the desired temperature with the aid of a temperature programmer. The temperature of the block is recorded by means of a calibrated temperature indicator or a

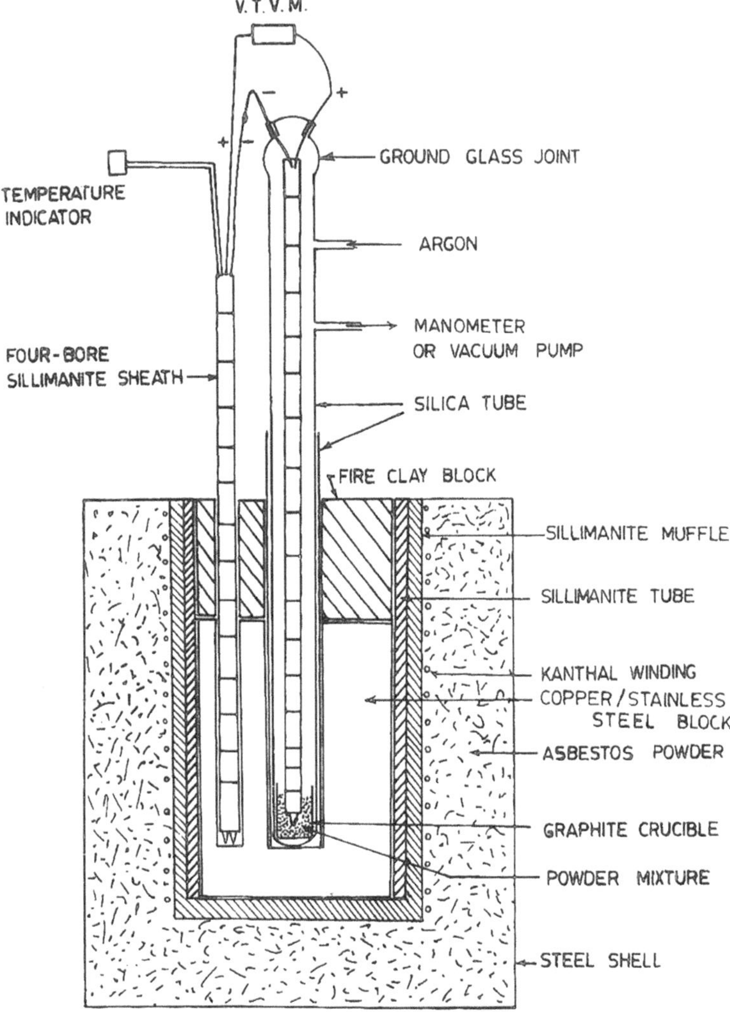

Fig. 4.9 Schematic diagram of differential thermal analysis calorimeter. (Reproduced from M. Shamsuddin and S. Misra, [72] J. Thermal Analysis, **7** (1975) 310)

potentiometer. The differential emf is measured by a D.C. Microvoltmeter with an input impedance of 1 mega-ohm. 0.5–1.5 cm^3 of a thorough mixture of fresh powders (−150 mesh) of the constituent elements in the stoichiometric proportion to form the compound is taken in the graphite crucible with a chromel-alumel thermocouple embedded in it. After inserting the crucible in the calorimeter furnace, the silica well is evacuated and purged with purified argon repeatedly for three times. The furnace reference block is heated at the rate of 2.5–3°/min. The temperature of the reference block and differential temperature ΔT between the sample and the

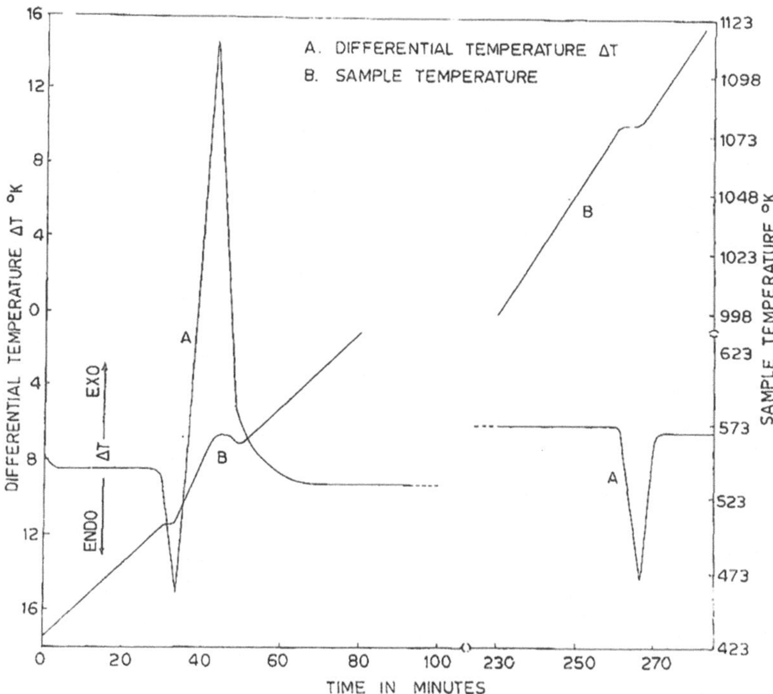

Fig. 4.10 Differential temperature, and sample temperature vs. time, t plot of the DTA calorimetric run with the powder mixture of tin and tellurium. (Reproduced from M. Shamsuddin and S. Misra [72], J. Thermal Analysis, **7** (1975) 311)

reference block are recorded in intervals of 30 s. The calorimeter is calibrated chemically by using high purity elements (Se, Te, Pb, Sn, Ge, Sb and Cu) and pure inorganic salts (KBr, KCl and NaCl) of known melting temperatures and heats of fusion. As the characteristic of the block may change with time and use, it is only proper to calibrate the apparatus for each investigation separately.

A typical DTA calorimetric run with the powder mixture of tin and tellurium [72] is shown in Fig. 4.10. 3.225 g of a thorough mixture of fresh powders (−150 mesh) of tin and tellurium in stoichiometric proportion is heated at the rate of 3.5 °C /min in the graphite crucible under the cover of purified argon. The rate of heating was controlled by a programmer. The temperature of the block and the differential temperature, ΔT between the sample and the reference surrounding (copper block) are noted at intervals of 30 s. As noticed in Fig. 4.10, the ΔT value show an endothermic peak (downward) corresponding to the melting point of tin (505 K) and thereafter begins to increase rapidly reaching an exothermic peak of 14.6 °C corresponding to the sample temperature of 565 K. After this the ΔT starts decreasing and returns to the normal background value.

The upward peak is due to the exothermic formation of tin telluride by the reaction between liquid tin and solid tellurium. Since the reaction between liquid tin and solid tellurium is accomplished between 505 and 565 K the calorimeter has to be calibrated by studying the fusion of pure tin, bismuth and lead (melting points of 505, 544 and 600 K, respectively) under conditions identical to those of the runs with the mixture of tine and tellurium. In the same DTA calorimeter, even after the SnTe compound formation reaction was complete, the product was heated beyond the melting point of tin telluride. A sharp endothermic (downward) peak was observed during the fusion of SnTe. To estimate the heat of fusion of the compound the calorimeter was calibrated by studying the fusion of sodium chloride which has nearly the same melting point as that of tin telluride.

4.2 Recent Trends of Automation in Calorimetric Measurements

The successful and versatile use of the calorimetric technique in thermodynamic measurements has encouraged the development of various types of calorimeters. In recent years a few commercial establishments have developed calorimeters that can measure different heat changes such as the enthalpy or heat of a reaction, changes in heat content of a substance associated with changes in temperatures, changes of state, changes of chemical combination such as reactions or alloying, dissolution or changes of physical structure, annealing, recovery and grain growth etc. in the same unit by change of modules. The MHTC L96 Evo system, developed by Setaram Instrumentation, KEP Technologies, has been designed for high-temperature, calorimetric and thermo gravimetric measurements. It consists of a high-temperature furnace with the provision of working under control atmosphere. A multipurpose software package piloting one or several modules depending on the use for a drop-type or dynamic-type calorimetric measurements, differential thermal analysis, differential scanning calorimeter, and differential thermo gravimetric analysis are provided. In drop calorimetric technique, an automatic sample dropping device has been used to drop the sample after a fixed interval of time. Maximum of twenty three samples can be loaded in the automatic dropping device. The balance used for thermo gravimetric analysis is attached with the unit which allows up to a 100 g sample with suspension wires and crucible. This system can be operated under vacuum as well as in controlled gas atmosphere.

The central unit, integrated in the structure of the L96 Evo device, is composed of several electronic modules: (i) CPU module (ii) command module (electro-valves, furnace, etc.) (iii) two temperature acquisition modules (iv) TMA signal measurement module (v) TG signal measurement module (vi) DTA/DSC signal low level amplification module and (vii) a gas panel control module. L96 Evo TG/DSC/DTA/TMA has CPU module for programming temperature control for the thermal analyzer. This module controls the heating power applied to the L96 Evo furnace. In addition there are separate modules for Temperature acquisition, TMA acquisition, DTA/DSC amplification and gas panel. These modules execute the conditioning and the acquisition of two temperature signals (furnace temperature and sample temperature) via two thermocouples. For a good temperature scanning and control, the heating actions need to be set. There are four heating actions. The heating actions offer a scanning temperature possibility on all of the scanning range with varying scanning rate.

Before starting the experiment, a sample scanning cycle needs to be prepared. In order to do this, the software offers a sequence-gathering table which carries out heating and cooling operations, as well as controls the periods of constant temperature. The gas flow rate can be programmed during heating and cooling period.

The sample can be a liquid or a solid (powder, granule, film, etc.). In both cases, the sample represents the product to be analyzed and must be used carefully. Alumina crucibles are used for metallic samples whereas platinum crucibles are preferred for minerals under investigation. If the sample is organic, either crucible can be used but the platinum crucible is preferred.

Depending on the temperature under study, for each transducer, Drop, HFDSC or DTA/DSC, different maximum working temperatures are managed with the aid of different types of thermocouples. However, while performing the experiments, a safety temperature has to be set at the selected rod's maximum operating temperature so as to avoid any damage to the transducer, specially, the Platinel rod. The safety temperature has to be set at a lower value keeping in mind the decomposition temperature of the sample under investigation and the heat bearing capacity of crucibles (e.g. aluminum). Temperature sensing device varies according to the requirement, for example, chromel-alumel from ambient to 1200 °C, Pt/Pt- 10% Rh from ambient to 1500 °C, and Pt- 6% Rh/Pt- 6% Rh from 200 to 1600 °C.

References

1. M. Berthelot, (as quoted in Metallurgical Thermochemistry by O. Kubaschewski, and C.B. Alcock, [3])
2. J. Thomsen, Thermochemistry (London, 1908)
3. O. Kubaschewski, C.B. Alcock, *Metallurgical Thermochemistry*, 4th edn. (Pergamon, Oxford, 1967)
4. J.S.L. Leach, in *Physicochemical Measurements in Metals Research*, Vol. IV, Part I, ed. by R.A. Rapp (Interscience, New York, 1970), p. 197
5. E. Huber, Jr., C.E. Holley, in *Physicochemical Measurements in Metals Research*, Vol. IV, Part I, ed. by R.A. Rapp (Interscience, New York, 1970), p. 243
6. M. Shamsuddin, S. Misra, Trans. Ind. Inst. Met. **27**, 281 (1974)
7. M. Shamsuddin, High Temp. Sci. **28**, 255 (1990)
8. R. Hultgren, P. Newcombe, R.L. Orr, L. Warner, Natl. Phys. Lab. Gt. Brit., Proc. Symp. **9** (1959)., Paper III, p. 2
9. W. Oelsen, W. Tebbe, O. Oelsen, Arch. Eisenhuettenw. **22**, 689 (1956)
10. O. Kubaschewski, H. Villa, Z. Elektrochem. **53**, 32 (1949)
11. W.A. Dench, Trans. Faraday Soc. **59**, 1279 (1963)
12. E. Greenberg, J.L. Settle, H.M. Feder, W.N. Hobbard, J. Phys. Chem. **65**, 1168 (1961)
13. P. Gross, C. Hayman, Trans. Faraday Soc. **60**, 45 (1964)

14. K.K. Prasad, H.S. Ray, K.P. Abraham, *Chemical and Metallurgical Thermodynamics* (New Age International Publishers, New Delhi, 2007)
15. M.B. Bever, L.B. Ticknor, J. Appl. Phys. **22**, 1297 (1951)
16. S. Kravitz, J.S.L. Leach, Acta. Met. **14**, 1485 (1966)
17. O.J. Kleppa, T. Yokokawa, J. Am. Chem. Soc. **86**, 2749 (1964)
18. F.E. Wittig, F. Huber, Z. Elektrochem. **60**, 1181 (1956)
19. L.B. Ticknor, M.B. Bever, Trans. Met. Soc. AIME **194**, 941 (1952)
20. B.W. Howlett, J.S.L. Leach, L.B. Ticknor, M.B. Bever, Rev. Sci. Instr. **33**, 619 (1962)
21. A.K. Jena, J.S.L. Leach, J. Sci. Instr. **43**, 206 (1966)
22. R. Boom, P.C.M. Vendel, F.R. de Boer, Acta Met. **21**, 807 (1973)
23. S. Misra, H.P. Singh, P.V. Nayak, Ind. J. Tech. **6**, 254 (1968)
24. O.J. Kleppa, J. Am. Chem. Soc. **77**, 175 (1955)
25. J.B. Darby Jr., R. Kleb, O.J. Kleppa, Rev. Sci. Instr. **37**, 164 (1966)
26. R.L. Orr, A. Goldberg, R. Hultgren, Rev. Sci. Instr. **28**, 767 (1957)
27. A.W.H. Morris, J.N. Pratt, Br. J. Appl. Phys. **16**, 517–525 (1965)
28. G.R. Blair, D.B. Downie, Met. Sci. J. **4**, 1 (1970)
29. M.J. Pool, Trans. Met. Soc. AIME **233**, 1711 (1965)
30. B. Predel, R. Mohs, J. Less Common Metals **18**, 267 (1969)
31. J.C. Mathieu, F. Durand, E. Bonnier, *Thermodynamics*, vol I (IAEA, Vienna, 1966)
32. B. Predel, H. Ruge, Met. Sci. Engg. **9**, 141 (1972)
33. R.A. Oriani, W.K. Murphy, J. Phys. Chem. **62**, 327 (1958)
34. M.G. Benz, J.F. Elliott, Trans. Met. Soc. AIME **230**, 706 (1964)
35. R. Dokken, J.F. Elliott, Trans. Met. Soc. AIME **233**, 1351 (1965)
36. D.L. Vieth, M.J. Pool, *Symposium 'Thermodynamik der Legierungen'* (Munster, 1972)
37. B.K. Gupta, A.K. Jena, Trans. Ind. Inst. Met. **26**, 44 (1973)
38. Shamsuddin, Ph.D. Thesis, *Thermodynamic Investigations of Group IV Chalcogenides* (Banaras Hindu University, 1974)
39. E. Calvet, H. Prat, *Microcalorimetrie* (Masson et. Cie, Paris, 1956)
40. O.J. Kleppa, J. Phys. Chem. **64**, 1937 (1960)
41. Shamsuddin, Trans. Ind. Inst. Met. **28**, 302 (1975)
42. M. Shamsuddin, S. Misra, in *Chemical Metallurgy – A Tribute to Carl Wagner*, ed. by N.A. Gockcen, (TMS – AIME, Warrendale, 1981), p. 241
43. E. Calvet, H. Prat, *Recent Progress in Microcalorimeters* (Mc Millan, New York, 1963)
44. O.J. Kleppa, J. Phase Equilibria **15**, 240 (1994)
45. O.J. Kleppa, J. Non-cryst. Solids **61**, 101 (1984)
46. O.J. Kleppa, S. Sato, J. Chem. Thermodyn. **14**, 133 (1982)
47. L. Topor, O.J. Kleppa, J. Chem. Thermodyn. **16**, 993 (1984)
48. S. Sato, O.J. Kleppa, Met. Trans. **13B**, 251 (1982)
49. O.J. Kleppa, S. Watanabe, Met. Trans. **13B**, 391 (1982)
50. S. Watanabe, O.J. Kleppa, Met. Trans. **15B**(15), 357 (1984)
51. Q. Guo, O.J. Kleppa, Met. Mater. Trans. **26B**, 275 (1995)
52. Q. Guo, O.J. Kleppa, Met. Mater. Trans. **27B**, 417 (1996)
53. Q. Guo, O.J. Kleppa, Met. Mater. Trans. **29B**, 815 (1998)
54. S.V. Meschel, O.J. Kleppa, in *Metallic Alloys: Experimental and Theoretical Perspectives*, ed. by J.S. Faulkner, (Kluwe Academic Publishers, Dordrecht, 1994)
55. S.V. Meschel, O.J. Kleppa, J. Alloys Compd. **205**, 165 (1994)
56. B. Predel, H. Sandig, Z. Metallk. **60**, 126 (1969)
57. B. Predel, W. Schwermann, J. Inst. Metals **99**(169), 209 (1971)
58. T. Heumann, B. Predel, Z. Metallk. **57**, 50 (1966)
59. J.L. White, K. Koyama, Rev. Sci. Instr. **34**, 1104 (1963)
60. T.K. Engel, K.C. Jordan, G.W. Otto, D.M. Scott, Rev. Sci. Instr. **35**, 875 (1964)
61. B. Predel, Z. Metallk. **56**, 791 (1965)
62. B. Predel, W. Schwermann, Z. Metallk. **61**, 58 (1970)

63. K.L. Komarek, K. Wessely, Monatshefte fuer Chemie **103**(896), 923 (1972)
64. D.E. Grimes, Trans. Met. Soc. AIME **233**, 1442 (1965)
65. V.P. Zlomanov, W.B. White, R. Roy, Met. Trans. **2B**, 121 (1971)
66. A. Nasar, M. Shamsuddin, Thermochim. Acta **197**, 3 (1992)
67. A. Nasar, M. Shamsuddin, J. Less Common Metals **158**, 131 (1990)
68. M. Shamsuddin, A. Nasar, High Temp. Sci. **28**, 245 (1990)
69. A. Nasar, M. Shamsuddin, J. Less Common Metals **161**, 193 (1990)
70. A. Nasar, M. Shamsuddin, Z. Metallk. **81**, 244 (1990)
71. Shamsuddin, S. Misra, Curr. Sci. **42**, 119 (1973)
72. Shamsuddin, S. Misra, J. Therm. Anal. **7**, 309 (1975)

Chapter 5
Chemical Equilibria

The standard free energy of formation, ΔG^o and the equilibrium constant, K are related by the expression, $\Delta G^o = -$ RT ln K. Hence the measurement of equilibrium constant under various conditions would give the thermodynamic data such as free energy, enthalpy and entropy and partial quantities. However, slow attainment of chemical equilibrium in many systems, especially where all the phases are present in the solid state limits the applicability of this technique. It is a common experience of many investigators about this technique that some of the solid-state reactions take months to attain equilibrium. The systems involving gases and vapors can be most conveniently studied. These can be classified into two groups:

(a) Vapor in equilibrium with metals or alloys in solid or liquid state: If all the components in an alloy have comparable vapor pressures, vapors in equilibrium will consist of vapors in proportion to vapor pressure of the components.
(b) Formation of a gas molecule as a result of a chemical reaction: For example, metal sulfide and oxide may react to form sulfur dioxide according to the reaction:

$$\mathrm{MS} + 2\ \mathrm{MO} = 3\mathrm{M} + \mathrm{SO_2} \tag{5.1}$$

In such a case, measurement of p_{SO_2} gives the equilibrium constant of the reaction $\left(K = p_{SO_2}\right)$ provided all other constituents (e.g. MS, MO, and M) are condensed phases. Most of the books consider the measurement under category (a) separately as vapor pressure measurements. However, by employing these methods one can also study the reaction of type (b). For this reason it is considered appropriate to include vapor pressure measurement under this heading but details have been discussed in the following Chap. 6. Solid or liquid/gas equilibria can thus be divided into two categories:

(i) Measurement of vapor pressure.
(ii) Other methods to study solid or liquid/gas equilibria.

M. Shamsuddin, *Thermodynamic Measurement Techniques*, The Minerals, Metals & Materials Series, https://doi.org/10.1007/978-3-031-47118-6_5

5.1 Measurement of Vapor Pressure

The following thermodynamic quantities can be estimated from the measurement of vapor/gas pressures provided complete thermodynamic equilibrium is ensured.

(i) Using Clausius–Clapeyron equation, one can calculate heat of vaporization if the vapor pressure is known at different temperatures.
(ii) If, in a multicomponent system vapor pressure of one component is much lower than the others the vapor pressure over the alloy will give the activity of the component in the alloy, provided vapor pressure of pure element is known (activity $= p_i/p_i^o$).
(iii) As mentioned earlier in a reaction between sulfide and oxide the pressure of sulfur dioxide is related to the free energy of formation of sulfide and oxide by the relation:

$$-RTlnp_{SO_2} = (2\Delta G_{MO} + \Delta G_{MS}) \tag{5.2}$$

If free energy of formation of either oxide or sulfide is known one can calculate the other.

(iv) In a reaction like:

$$MC = M + C \tag{5.3}$$

or

$$MO = M + \tfrac{1}{2}O_2 \tag{5.4}$$

If the pressure of M in Eq. (5.3) is far more than MC or C or if the dissociation pressure, p_{O_2} is large compared to vapor pressure of MO or M (Eq. 5.4) the measurement of vapor/gas pressure would give the free energy of formation of MC ($\Delta G^o = -RTlnp_M$) or MO ($\Delta G^o = -RTlnp_{O_2}^{1/2}$) respectively. Details about techniques for vapor measurement will be discussed in Chap. 6.

5.2 Solid-Liquid-Gas Equilibrium

This technique is most convenient for the equilibria involving a gas. It has been successfully employed to study different types of heterogeneous equilibria, namely, solid-gas, molten metal-gas, molten salt-gas, molten metal-salt-gas, solid-liquid and liquid-liquid. For example, slag-metal is the most common and important reaction in pyrometallurgy. The free energy of formation of Ag_2S, PbS and FeS has been determined by using H_2-H_2S gas mixtures. Activity of oxygen, carbon, sulfur and nitrogen in steel has been determined by equilibriating steel or iron samples with H_2-H_2O, H_2-CH_4, H_2-H_2S, and H_2-NH_3 gas mixture, respectively by studying the following equilibria:

$$H_2(g) + [O]_{(in\ Fe)} = H_2O\ (g) \qquad K = \frac{p_{H_2O}}{p_{H_2}\cdot[a_O]} \tag{5.5}$$

$$2H_2(g) + [C]_{(in\ Fe)} = CH_4\ (g) \qquad K = \frac{p_{CH_4}}{p_{H_2}^2\cdot[a_C]} \tag{5.6}$$

$$H_2(g) + [S]_{(in\ Fe)} = H_2S\ (g) \qquad K = \frac{p_{H_2S}}{p_{H_2}\cdot[a_S]} \tag{5.7}$$

$$\frac{3}{2}H_2(g) + [N]_{(in\ Fe)} = NH_3\ (g) \qquad K = \frac{p_{NH_3}}{p_{H_2}^{3/2}\cdot[a_N]} \tag{5.8}$$

Experimental method to obtain various gaseous mixtures will be discussed in the next section. Suitable amounts of iron/steel samples heated in appropriate gaseous mixture and equilibrium concentration is determined by either quenching or by taking the samples of both the phases. Several experimental precautions suggested by Richardson and Alcock [1] are listed below:

1. Crucible material should react as little as possible with the system being investigated.
2. Equilibrium is established primarily at the phase boundary between reacting phases. In order to remove the concentration difference and to enhance equilibrium mechanical stirring is usually employed.
3. If a subsequent chemical analysis is to be made the entire system should be cooled quickly.
4. It is always desirable to determine all constituents in the phases, although in principle it is necessary to analyze only the constituents of one phase.
5. Equilibrium should be approached from opposite sides in different experiments and result be compared for consistency.

5.3 Preparation of Gas Mixtures

The studies on reaction equilibria between gases and condensed phases require the exposure of the condensed phase to gas mixture with accurately controlled chemical potentials. The required chemical potentials can be established by using appropriate gaseous mixtures. In such studies two techniques are employed. In the first case the condensed phases are brought to equilibrium with the gas mixture of known chemical potential and the condensed phases are chemically analyzed after quenching and the chemical potentials are correlated with the composition. Alternatively, the gaseous mixtures are equilibrated with the condensed phase of the desired

Table 5.1 Gaseous mixtures used for establishing chemical potentials

Element	Gaseous mixture
1. Carbon	CO-CO_2 and CH_4-H_2
2. Chlorine	HCl-H_2
3. Nitrogen	N_2-inert gas, NH_3-N_2
4. Oxygen	O_2-inert gas, H_2-H_2O, H_2-CO_2 and CO-CO_2
5. Silicon	H_2-H_2O-SiO
6. Sulfur	H_2-H_2S, H_2-S_2, SO_2-O_2 and SO_2-CO-CO_2
7. Sulfur/Oxygen	SO_2-O_2, S_2-SO_2, SO_2-H_2-CO_2, H_2-H_2S-H_2O and SO_2-CO-CO_2

composition and the chemical potentials are determined by analysis of the gas mixture. There are only few elements whose chemical potentials can be controlled in practice at elevated temperatures by mixing the appropriate gases at room temperature. The important ones are oxygen, carbon, sulfur, nitrogen, silicon and chlorine. The choice of the gas mixtures for imposing the given potential depends on the potential required, the potentials to be avoided and also on the practical difficulties involved in mixing and analyzing the gases.

The gaseous mixtures which have been used for establishing the chemical potentials of the elements mentioned above are listed in Table 5.1.

The gases mixed in known proportions at room temperature interact with each other at the elevated temperature giving rise to a variety of thermodynamically stable gaseous species. From available thermodynamic data pertaining to each species, it will be possible to identify those species which are formed in appreciable amounts under the prevailing conditions. Thus there will be several independent equilibria, each associated with an equilibrium constant corresponding to the gaseous species formed. The equilibrium constant is known from the standard free energy change of the reaction. Considering a closed system, for every component or element an expression for its conservation can be written. For a system at equilibrium the number of such independent equilibria is equal to the number of species appearing in the chemical equation minus the number of components. In addition, an equation may be formulated representing the one possible arbitrary restriction of either fixed pressure or fixed volume on the system in addition to those of constancy of temperature and of masses. Thus two types of equations can be written: (1) those expressing the state of equilibrium and (2) those expressing the arbitrary restrictions including those of conservation of masses. It will be possible to solve these simultaneous equations if the number of unknowns equals the number of equations. The unknowns are the masses of various gaseous species formed at equilibrium. On solving the equation the composition of the gas mixture and the partial pressure of each gaseous species at equilibrium at the elevated temperature can be known. The reverse process of calculating the ratio in which the gases are to be mixed at room temperature in order to get a desired equilibrium chemical potential of a given species at high temperature is much simpler.

The method of estimating the ratio in which gases are to be mixed at room temperature in order to get the required potential at the high temperature can be worked out. In these calculations, it is assumed that the gases behave ideally. This is justified in view of the high temperatures and ordinary pressures involved. The calculations also assume that the gases reach equilibrium instantaneously, because at high temperatures reactions between gases occur at high speed.

A brief account on gaseous mixtures used for controlling the chemical potentials of some of the common elements is given below:

1. **Controlled Carbon Potentials:** The gas mixtures commonly used for establishing carbon potential are CO-CO_2 [2, 3] and H_2-CH_4 [4, 5]. Both the gas mixtures can give rise to carbon deposition from the gas phase depending upon temperature and pressure. In the case of CO-CO_2 mixture, at low temperatures carbon can be deposited by the reaction:

$$2\,CO \rightarrow CO_2 + C \tag{5.9}$$

In the case of H_2-CH_4 mixture, with increasing temperature carbon deposition occurs by the reaction:

$$CH_4 \rightarrow 2H_2 + C \tag{5.10}$$

2. **Controlled Chlorine Potential:** Required chlorine potential is obtained by mixing H_2 and HCl gases [6, 7].
3. **Controlled Nitrogen Potentials:** To a limited extent, partial pressures of nitrogen can be controlled by diluting pure nitrogen with purified hydrogen or oxygen free dry helium or argon. But the nitrogen activities are usually established by using NH_3-H_2 [8, 9] mixtures. However, ammonia is likely to be cracked into nitrogen and hydrogen above 600 °C and below it in the presence of catalysts. So the extent of cracking has to be determined by analysis.
4. **Controlled Oxygen Potentials:** By mixing pure oxygen gas with purified inert gases like argon, helium or nitrogen, oxygen partial pressures in the range 1×10^{-3} atm can be obtained. Still lower values can be obtained by the dissociation of pure carbon dioxide at high temperatures. Low oxygen pressures are generally obtained by using a mixture of two gases like H_2-H_2O [10, 11] or H_2-CO_2 [12, 13] or CO-CO_2 [14]. However H_2-CO_2 mixtures have been preferred due to its ease of preparation in laboratories.
5. **Controlled Silicon Potential:** Reduction of silica by dry hydrogen gives silicon monoxide gas which is useful in preparing gas mixtures of known silicon potential. The following two reactions are considered:

$$SiO_2\,(s) + H_2\,(g) = SiO\,(g) + H_2O\,(g) \tag{5.11}$$

$$SiO\,(g) + H_2\,(g) = Si\,(g) + H_2O\,(g) \tag{5.12}$$

The equilibrium constant, *K* for reaction (5.12), is given as:

$$K = \frac{p_{H_2O}.p_{Si}}{p_{SiO}.p_{H_2}} \tag{5.13}$$

Since the vapor pressure of silicon is negligibly small at equilibrium,

$$p_{SiO} = p_{H_2O} \tag{5.14}$$

$$\therefore p_{Si} = K.p_{H_2} \tag{5.15}$$

Thus, we find that the partial pressure of silicon in equilibrium with the gas phase is directly proportional to the partial pressure of hydrogen. Hence, the activity of silicon a_{Si} in the condensed phase in equilibrium with the gas phase is given as

$$a_{Si} = K.p_{H_2} \tag{5.16}$$

This method was used by Bowles et al. [15] to measure activity of silicon in liquid silicon-metal system at 1560 °C.

6. **Controlled Sulfur Potentials:** Sulfur potentials can be established by mixing hydrogen and hydrogen sulfide gases available in tanks [16]. H_2-H_2S gas mixtures can also be prepared by bubbling dry hydrogen gas through molten sulfur kept at a constant temperature. The hydrogen gas saturated with sulfur vapor [17] is passed through a column of alumina chips maintained at 500 °C to convert sulfur into hydrogen sulfide. For high sulfur potentials, special techniques have to be employed to avoid sulfur deposition in cooler parts of the reaction tube.
7. **Controlled Oxygen and Sulfur Potentials:** Simultaneous sulfur and oxygen potentials are obtained by an appropriate mixture of sulfur and oxygen-bearing gaseous species. However, the most appropriate ternary gas combinations are: H_2-CO_2-SO_2 [18], CO-CO_2-SO_2 [19] and H_2-H_2S-H_2O [20]. The compositions of the ingoing gas and the partial pressures of oxygen and sulfur obtained by calculation, for the gas mixtures of H_2-CO_2-SO_2 and CO-CO_2-SO_2 at iron and steelmaking temperatures are available in literature [19].
8. **Controlled Phosphorus Potentials:** The phosphorus potential cannot be established at high temperatures by means of PH_3-H_2 mixtures [16] due to the non-stable nature of PH_3. However, low potential can be obtained by combining a controlled oxygen potential obtained from H_2-H_2O with a P_2O_5 potential obtained from two solid phases such as CaO and $4CaO.P_2O_5$. At high temperature the tetra calcium phosphate is reduced by dry hydrogen according to the reaction [21]:

$$4\,CaO.P_2O_5\ (s) + 5H_2\ (g) = 4CaO\ (s) + P_2\ (g) + 5\ H_2O\ (g). \tag{5.17}$$

By stoichiometry partial pressure of P_2 is one-fifth of H_2O. Hence by measuring the water vapor content of the exit gas, the equilibrating partial pressure of P_2 can be estimated.

5.4 Experimental Methods for Preparation of Gas Mixtures

The multicomponent gas mixtures are prepared by mixing the individual gases in the calculated proportions with the help of capillary flow meters maintained at constant pressure head with the help of 'blow offs' or bleeders. The pressure drop across the capillary is measured by a manometer and the capillary tube is calibrated by measuring the flow rate of the gas with a soap bubble flow meter which can be constructed in the laboratory or with a commercially available wet gas meter. For experiments involving fast flow rates it is more convenient to use a gas rotameter. The liquids used in the 'blow offs' and manometers must be inert towards the gases used and also must have negligible small vapor pressures. Dibutyl phthalate is commonly used, though vacuum pump oil, liquid paraffin etc. are also in use. Details can be obtained from the review papers of Richardson and Alcock [1] and Schwerdtfeger and Turkdogan [16].

5.5 Equilibration Techniques

The gas mixture of the desired composition is next allowed to react with the system under study until equilibrium is attained. The available methods can be divided into two main groups:

1. The flowing gas system with open circuit experiments and
2. The circulating gas system with closed circuit experiments.

5.5.1 *The Flowing Gas Method*

In this case the equilibrium is maintained by passing a continuous stream of the gas mixture over the sample kept in the furnace and the gases are run to waste. It is sometimes impossible to establish the chemical potential required solely by mixing the appropriate gases. In such cases the use of a condensed phase may be required. Here the gas is passed over a particular condensed phase to saturate it with a desired vapor species and the gas mixture of controlled chemical potential so obtained, is then allowed to react with the experimental sample contained in the same reaction vessel. For example, by passing dry hydrogen over silica maintained in a furnace, a gas phase of known silicon potential is obtained and this gas is next passed over the

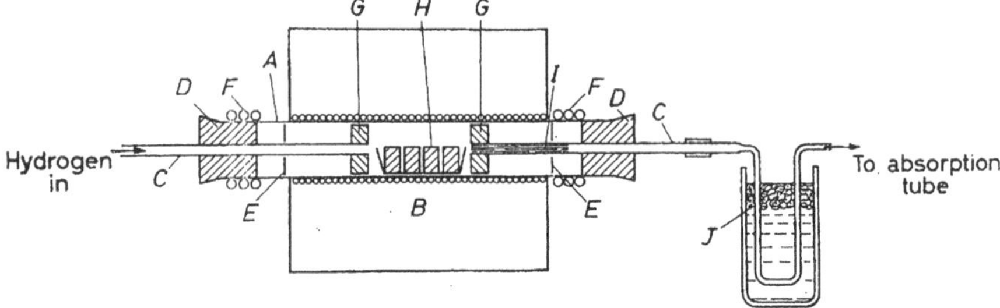

Fig. 5.1 Schematic arrangement of experimental set-up for measuring reduction equilibria involving phosphates. (Reproduced from J. B. Bookey [21] J. Iron and Steel Institute, **172** (1952) 61). A – recrystallized alumina tube, B – platinum wound furnace, C – alumina tubes for gas inlet and outlet, D – rubber stoppers, E – metal radiation shields, F – water cooling traps, G – alumina plugs, H – phosphate samples, I – thermocouple tubing used to decrease the cross-section of the exit tube, J – ice trap

silicon alloy where silicon activity is being studied [15]. The technique has been successfully used for studying many gas-solid equilibria.

In certain gas mixtures simultaneously, the chemical potentials of phosphorus and oxygen may be controlled by reducing a phosphate of known composition. The schematic diagram of the gas flow system employed by Bookey [21] for studying the equilibrium [4 $CaO.P_2O_5$ (s) + $5H_2$ (g) = 4CaO (s) + P_2 (g) + 5 H_2O (g)] in the temperature range 1250–1500 °C is shown in Fig. 5.1. From the stoichiometry of the reaction, $p_{H_2} = \left(\frac{1}{5}\right) p_{H_2O,}$ the equilibrium partial pressure of phosphorus can be estimated by measuring the water vapor present in the exit gas.

The phosphate sample, H consisting of a mixture of the appropriate solid reactants and products is taken in a phosphided molybdenum boat. A recrystallized alumina tube, A holding the boat is placed in a platinum wound tubular furnace. Hydrogen gas is allowed to flow through another tube, C having rubber stoppers, D at both ends serving as inlet and outlet for the gas. In order to protect the stoppers, both ends of the tube, C were provided with water cooling tubes, F and metallic radiation shields, E. The outgoing gas is first passed through an ice U-tube, J to condense phosphorus vapor (without condensing water vapor) from the gas. In the next step the resulting gas is passed through anhydrone-filled weighing bottles for absorption of water vapor. The desired partial pressure of phosphorus in the reaction zone can be managed by temperature variation, dilution of the dried hydrogen with an inert gas or by introducing water vapor to the inlet hydrogen.

According to Bookey [21] the equilibrium was attained rapidly and the results remained unaffected even with four fold increase in the gas flow rates.

5.5.2 Circulating Gas Method

In this method, gases are circulated continuously through the apparatus by means of a suitable pump until equilibrium is established. The method will not work if the circulating gases deposit solids or liquids on cooling. For example, CO rich mixtures of CO-CO_2 can deposit carbon on cooling. The circulating method is not suitable at temperatures above 1600 °C when the common refractories used become porous and gaseous impurities can enter the system and upset the chemical potentials established..

In these closed circuit experiments, the reacting species are transferred from one condensed phase at temperature T_1 to another condensed phase at temperature T_2 via a suitable circulating gas. Thus sulfur potential can be established by passing hydrogen through a metal-metal sulfide mixture kept at a constant temperature in a control furnace. The sulfur potential ultimately built up depends on p_{H_2S} in the system and this in turn depends on the temperature of the control furnace. Systems like Ag-Ag_2S, Co-Co_9S_8, Cu-Cu_2S and Fe-FeS are suitable for control of sulfur potentials. A high CO_2/CO ratio in CO_2-CO gas mixture can be prepared by passing CO_2 through the appropriate metal-metal oxide mixture (e.g., Ni-NiO). The desired partial pressure of CH_4 can be obtained by allowing H_2 to react with heated graphite.

Turkdogan et al. [22] have studied the solubility of sulfur in iron alloys by this method. The closed circuit experimental set-up used by them is shown Fig. 5.2. H_2-H_2S gas mixture is circulated in the closed system by a pump. The Fe-FeS mixture is placed at a known temperature in one furnace (whose temperature can be varied). Metallic iron is placed in another furnace at a fixed temperature. The partial pressure of sulfur in the gas phase i.e., the p_{H_2O}/p_{H_2} ratio is controlled by varying the temperature of the sulfide furnace.

In all cases, the establishment of true equilibrium is necessary for the success of the experiment. At high temperature gases react and come to equilibrium instantaneously, whereas much longer time is required for attaining equilibrium when condensed phases are involved. This is because, transport steps are very often rate controlling in reactions involving condensed phases and diffusivities of various species in liquids are extremely small and still smaller in solids. Moreover, in order to assist the approach to equilibrium, the area of contact between the condensed phase and the gases should be made as large as possible. In case of solids, this can be achieved by using finely divided particles. In case of liquid samples small beads or drops are used. This helps the attainment of equilibrium by diffusion in a comparatively shorter period in the absence of stirring or transport by convection.

It should be pointed out that the aim of these experiments is attainment of true equilibrium and it must be understood that once equilibrium is attained, the equilibrium composition is independent of time and so it can be established from both ends.

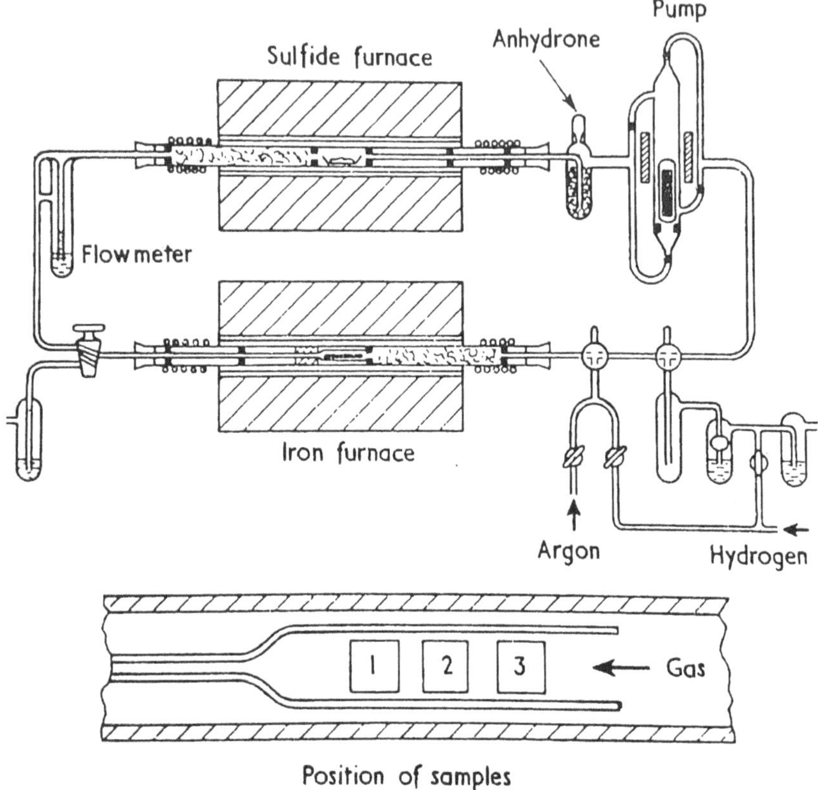

Fig. 5.2 Sulfide furnace for study of solubility of sulfur in iron alloys. (Reproduced from E. T. Turkdogan, et al. [22] J. Iron and Steel Institute, **180** (1955) 349)

For example, suppose an iron-carbon alloy is being equilibrated with a gas phase of a given carbon potential obtained by mixing hydrogen and methane. If initially the carbon activity in the alloy is less than that of the gas, carbon will be picked up from the gas phase into the alloy until the equilibrium composition is reached. If initially the carbon activity in the alloy is more, carbon will be lost from the alloy until the same equilibrium composition is reached. In an experiment of this type it is advisable to carry out two experiments, one with an alloy of higher carbon content and then making sure that the same equilibrium composition is reached in both cases. This ensures the absence of any interfering factor which prevents the attainment of true equilibrium. The other factors to be considered in these equilibrium studies are quenching techniques and sampling technique, containers and refractories to be used and gas analysis.

5.6 Quenching and Sampling Techniques

In the study of reaction equilibria, it is necessary to know the composition of the condensed phase under equilibrium conditions at elevated temperature. This is done by cooling the sample at a sufficiently fast rate to prevent further reaction occurring during cooling before subjecting it to chemical analysis. Different quenching techniques are used to freeze the equilibrium in this way. One method frequently used is to pull the reacted sample held in a container suddenly to the cool end of the reaction tube. Sometimes an inert gas like helium is injected on to the sample at the cold end and this is known as helium quench. If a vertical furnace assembly is used, the reacted sample is quenched by dropping it suddenly into a suitable chilled liquid medium. If a large amount of melt is equilibrated, a portion can be solidified quickly by suction into a chilled mold. Samples of the reacting liquid can be taken this way by suction. Taylor and Chipman [23] used a copper sampling tube connected to an aspirator bulb by a steel tube.

5.7 Choice of Containers and Refractory Materials

The containers like crucibles or boats for holding the specimen must have good mechanical properties, high melting point, low vapor pressure and should not react with the condensed phases or atmosphere used. This is also true of the refractory materials used as furnace tubes and reaction tubes.

The melting points of the refractory materials give an idea about the temperatures at which they can be used. Tubes can be used at temperatures below 300 °C of the melting point and crucibles as containers some 50 °C below. The next step is to determine whether the material is thermodynamically stable under the conditions of the experiment. There are several compilations of thermodynamic data [24, 25, 26] which may be used. Phase Diagram for Ceramists [27] is also useful in deciding the choice of the container material. For holding liquid metals, the refractory oxide materials, namely, silica, alumina, zirconia, magnesia, beryllia, calcia, thoria and mullite are used as container materials. The oxygen potential in the atmosphere should be high enough to keep the dissociation of the oxide to a minimum and low enough to prevent the formation of new phases by the reaction of the oxide, the metal and oxygen. The displacement reaction between the metal and the oxide refractory also should not occur. One must also consider the possibility of other side reactions occurring during the course of experimentation. If the carbon, nitrogen and sulfur potentials are sufficiently high, the oxide crucible will be attacked. On the other hand carbide, nitride and sulfide will be formed, if the oxygen potential is low. The molten oxides, halides, silicates are held in noble metal crucible made of platinum, palladium, iridium and rhodium. These also become unsuitable under extreme conditions. At low oxygen pressures both silica and manganese oxide can get reduced and form low melting Pt-Si or Pt-Mn alloys. Iridium crucibles are used for holding liquid lead

[28]. If the oxygen potential is sufficiently low, crucibles made of iron, molybdenum or tungsten can be used for holding molten oxides and salts. Liquid metals and slags can be held in graphite containers in reducing atmospheres.

There are however certain limitations, because of the possibility of carbon pick up and the formation of carbides. Sometimes the sample equilibrated has a tendency to creep out and it is difficult to hold it in any container. In such cases small amounts of the liquid can be held in a spiral of metal wire wound to form a conical cup.

5.8 Gas Analysis

An analysis of the gas composition is important in the measurement of equilibrium between gases and condensed phases. A variety of techniques [16] are used in the analysis of gases involved in equilibrium studies. For example, infra-red gas analyzers have been employed for continuous measurement of CO in CO_2, and CH_4 and H_2O in H_2. Thermal conductivity methods have been used for H_2O in H_2 mixtures and the radio-sulfur technique for continuous analysis of H_2S present in H_2.

5.9 Experimental Techniques

A few typical experimental techniques used in investigations of metals, slags and salts in equilibrium with gases are discussed below.

5.9.1 Equilibrium Between Gases and Solids

By passing CO-CO_2 or H_2-CH_4 gas mixture of accurately controlled carbon potential over iron carbon alloys, the activity of the carbon in the alloys can be determined. Smith [3] used the flowing gas technique with both CO-CO_2 and H_2-CH_4 gas mixtures to determine the dependence of carbon activity on carbon content in pure Fe-C alloys. Later Zupp and Stevenson [29] determined the activity of carbon in iron carbon-vanadium alloys by equilibrating Fe-C and Fe-C-V alloys in a sealed silica capsule containing hydrogen gas. At equilibrium, the chemical potential of carbon is the same in both alloys and from the known activity in the binary that in the ternary can be calculated. The composition of the ternary is determined by chemical analysis after quenching. The activity of sulfur in solid iron was determined by Turkdogan et al. [22] by circulating H_2-H_2S through a second furnace containing a mixture of Fe-FeS at an appropriate temperature.

5.9.2 Equilibrium Between Molten Metals and Gases

In a typical study Morris and Williams [30] measured the effect of silicon on the activity of sulfur dissolved in molten iron at 1615 °C. H_2-H_2S gas mixture was bubbled through 50 g samples of Fe-S alloy kept in the furnace. The melt was sampled from time to time until the sulfur content was constant and equilibrium was ensured. The bubbling ensured vigorous stirring and enabled the attainment of equilibrium at a faster rate.

The activity of dissolved oxygen in liquid copper has been measured by equilibration with CO-CO_2 gas mixture. Diaz and Richardson [31] have used CO_2-CO gas mixtures for controlling the oxygen content of liquid copper and the activity was measured by a solid electrolyte galvanic cell technique.

5.9.3 Equilibrium Between Molten Slags or Salts and Gases

The determination of sulfide capacity of slags [20] by slag-gas equilibration is a typical example. The molten slags were exposed to a gas mixture with oxygen and sulfur potentials prevailing in iron and steel making furnaces, prepared by mixing H_2, CO_2, SO_2 and N_2. After equilibration, the samples were pulled to the cold end of the reaction tube where the slag immediately solidified owing to its small size and weight.

5.9.4 Equilibrium Between Metal-Slag-Gas

An example of equilibria involving metal, slag and gas is the measurement of manganese oxide activity in slags. A thin foil of platinum is embedded in a slag containing MnO kept in a small platinum cup which can hold about half a gram of the slag. The slag is then exposed at high temperature to a gas mixture of accurately controlled low oxygen potential prepared by mixing H_2 and CO_2 in the appropriate proportions. During equilibration, the manganese reduced from the oxide formed an alloy with platinum.

$$(\mathrm{MnO})\ (\mathrm{l}) = \mathrm{Mn}-\mathrm{Pt}\ (\mathrm{s}) + \tfrac{1}{2}\,\mathrm{O_2}\ (\mathrm{g}) \tag{5.18}$$

$$K = \frac{a_{Mn\ (in\ Pt)}}{a_{MnO}} \cdot p_{O_2}^{1/2} \tag{5.19}$$

After equilibration, the sample is quenched and the manganese oxide content of the slag and the manganese content of the platinum alloy are determined by chemical analysis. The corresponding MnO activity in the slag is obtained from the knowledge of p_{O_2}, K and the previously established manganese activity in the alloy.

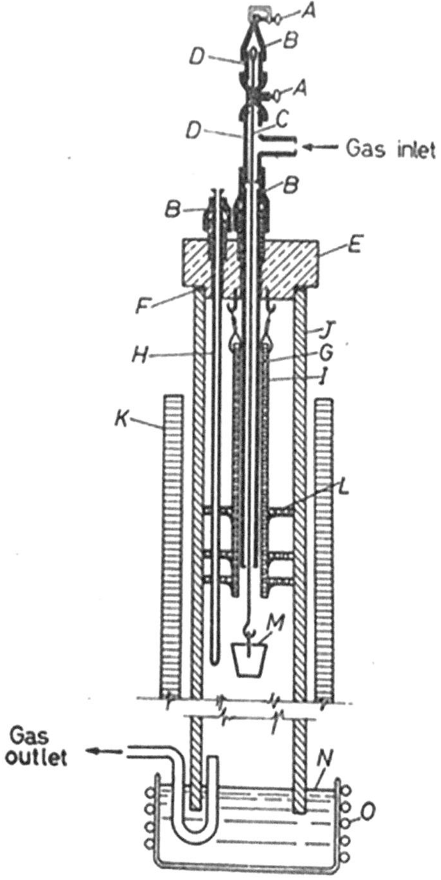

Fig. 5.3 Schematic diagram of experimental set-up for equilibrium studies. (Reproduced from H. Larson and J. Chipman [32], Trans. Amer. Institute of Mining and Metallurgical Engineers, **197** (1953) 1089 with permission of The Minerals, Metals & Materials Society) A – clamp, B – rubber tubing, C – platinum wire, D – glass tubing, E – brass water cooled top, F – silicone rubber gasket, G – alundum tube, H – protection tube for thermocouple, I – alundum tube, J – zircotube, K – cylindrical globar element, L – alundum radiation shields, M – crucible, N – mercury, O – water cooling coils

Larson and Chipman [32] have studied the effect of oxygen potential on the ratio of Fe^{2+}/ Fe^{3+} in CaO-FeO-Fe_2O_3 melts by considering the following equation:

$$2\left(Fe^{2+}\right) + CO_2\,(g) = 2\left(Fe^{3+}\right) + CO\,(g) + \left(O^{2-}\right) \tag{5.20}$$

Figure 5.3 depicts the schematic diagram of the experimental assembly used by them. A refractory zircotube, J is placed inside a cylindrical globar heating element, K. The tube is sealed at the top by a water-cooled brass head, E with a silicon rubber gasket, F. A crucible, M containing the slag sample is suspended by means of a platinum wire, C in the constant temperature zone of the furnace. Both CO_2 and CO driven through flow meters are mixed and then passed on to the furnace containing slag sample.

After attaining the equilibrium the slag sample is quenched by lowering the crucible rapidly to allow it to float on the water cooled pool of mercury, N contained in a trough surrounded with cooling coils, O. The time required for equilibrium varies from 1 to 2 h. The resulting slag is analyzed for metallic iron content and oxides. These investigations based on equilibrium between slags containing iron oxide and gases consisting of air and CO_2-CO gas mixtures at 1550 °C establish the value of the ratio: $(Fe^{3+})/(Fe^{3+} + Fe^{2+})$ over a wide range of oxygen pressures varying from 10^{-1} to 10^{-9} atm. The value of the ratio increases by addition of basic oxides: Bao, CaO, MgO, and MnO and decreases by acid oxides such as SiO_2 and TiO_2.

5.9.5 *Solid-Liquid and Liquid-Liquid Equilibrium*

The experimental techniques employed for investigations of solid-liquid and liquid-liquid equilibria [16] are quite similar to those already described in the preceding paragraphs. In all high temperature studies due consideration has to be given to the choice of the container materials. The container should be inert or react to a minimum extent with the system under investigation. Graphite, magnesia, alumina and beryllia are used as container materials for studies at 1300 °C and above. Slag-metal equilibria happen to be of particular importance in steelmaking. These can be studied by sampling slag and metal at certain time interval in a particular process. Though many studies on slag-metal equilibria have been conducted by equilibrium partitioning study of sulfur between iron and slag, Hatch and Chipman [33] is worth of special attention. They stirred the melt mechanically as well as by bubbling CO gas in order to attain rapid equilibrium. The experiments were carried out on 200 g of metal and 400 g of slag contained in a graphite crucible, heated by induction. A graphite rod rotating at 500 rpm was used as stirrer and CO was introduced through a graphite tube at one atmospheric pressure into the bottom of the melt.

In order to study equilibria involving highly corrosive slags Taylor and Chipman [23] developed a device for rotating the crucible. They used a magnesia crucible to hold about 40 kg of metal under rotation at 200 rpm in an inert atmosphere inside an induction coil. In general, the equilibria between liquid phases in contact are rapidly established. The attainment of equilibrium can be retarded by a gaseous phase which may be required either to remove products or supply reactants. Richardson and Billington [34] while measuring the activity of Cu_2O in molten Cu_2O-Al_2O_3-SiO_2 slags observed that formation of Cu_2O was very slow when oxygen potential is controlled in the gaseous mixture of CO-CO_2 flowing over the crucible containing molten copper covered with slag.

Abraham et al. [35] have used a novel technique of equilibrating a solid with a liquid for activity measurement of a component of the liquid phase. Strips of platinum foil (0.05 mm thick) were wound into spiral and placed in a platinum cup together with a slag containing manganese oxide. The entire mass of the slag and strip was heated in a gaseous mixture (CO_2-H_2-N_2) of known oxygen potential. Equilibrium was achieved in about 6 h. From the knowledge of the thermodynamics of the binary Mn-Pt system and analyses of the slag, activity of MnO was computed.

5.10 Accuracy

In all thermodynamic measurements, it is necessary to estimate the error on the thermodynamic data evaluated. The main sources of error in the measurement of equilibria between gases and condensed phases are listed below. These errors depend on the experimental methods and techniques used.

1. Errors in the preparation of gas mixtures of known chemical potential.
2. Error in temperature measurement.
3. Accuracy of temperature control
4. Error in the chemical analysis of the equilibrated sample.

The error in the temperature measurement may vary from ±0.2 to ±1 °C, depending on the temperature measured, the accuracy being less at higher temperatures (1500 °C) and also on the thermocouples used for measurement. The variation of temperature within the equilibrium zone depends on the temperature control and on the furnace design and could be between ±0.2 and ±3 °C or more depending on the experimental temperature and set up. Reactions which occur on cooling and sampling can cause compositional errors. Interaction of refractories with gases and condensed phases also can give rise to errors in the evaluated data. An estimate will have to be made of the effect of all these errors on the evaluated data.

References

1. F.D. Richardson, C.B. Alcock, in *Physicochemical Measurements at High Temperatures*, ed. by J.O'M. Bockris, J.L. White, J.D. Mackenzie (Butterworth, London, 1959), p. 135
2. R.P. Smith, Trans. Met. Soc. AIME **233**, 397 (1965)
3. R.P. Smith, J. Am. Chem. Soc. **68**, 1163 (1946)
4. R.P. Smith, J. Am. Chem. Soc. **70**, 2724 (1948)
5. S. Wada, J.F. Elliott, J. Chipman, Met. Trans. **2B**, 2199 (1971)
6. M. Busey, E. Giauque, J. Am. Chem. Soc. **75**, 1791 (1953)
7. T. Schafer, N. Kreil, Z. Anorg, Chem **268**, 35 (1952)
8. S. Corney, E.T. Turkdogan, J. Iron Steel Inst. **180**, 344 (1955)
9. P.H. Emmett, E. Brunauer, J. Am. Chem. Soc. **52**, 1456 (1930)
10. M.N. Dastur, J. Chipman, Disc Faraday Soc. **4**, 100 (1946)
11. P.H. Emmett, J.F. Schultz, J. Am. Chem. Soc. **55**, 1376 (1933)
12. L.S. Darken, R.W. Gurry, J. Am. Chem. Soc. **68**, 798 (1946)
13. L.S. Darken, R.W. Gurry, J. Am. Chem. Soc. **67**, 1398 (1945)
14. K.T. Jacob, C.B. Alcock, Acta. Met. **21**, 1011 (1973)
15. P.J. Bowles, H.F. Ramstad, F.D. Richardson, J. Iron Steel Inst. **202**, 113 (1964)
16. K. Schewerdtfeger, E.T. Turkdogan, in *Physiochemical Measurements in Metals Research*, Vol. IV, Part I, ed. by R.A. Rapp (Interscience, New York, 1970) p. 321
17. T. Rosenqvist, B.L. Dunicz, J. Metals Trans., AIME **194**, 604 (1952)
18. A. Fincham, F.D. Richardson, Proc. Roy. Soc. **A213**, 40 (1954)
19. G.R. St Pierre, J. Chipman, J. Metals **8**, 1474 (1956)
20. K.P. Abraham, F.D. Richardson, J. Iron Steel Inst. **196**, 313 (1960)
21. J.B. Bookey, J. Iron Steel Inst. **172**, 61 (1952)

22. E.T. Turkdogan, S. Ignatowiez, J. Pearson, J. Iron Steel Inst. **180**, 349 (1955)
23. C.R. Taylor, J. Chipman, Trans. Met. Soc. AIME **154**, 228 (1943)
24. J.S.L. Leach, in *Physicochemical Measurements in Metals Research*, ed. by R.A. Rapp, vol. Vol. IV, Part I, (Interscience, New York, 1970), p. 197
25. J.L. Curnutt, H. Prophet, R.A. McDonald, A.N. Syverud, *JANAF Thermochemical Tables* (The Dow Chemical Company, Midland, 1975)
26. I. Barin, O. Knacke, O. Kubaschewski, *Thermochemical Properties of Inorganic Substances (Supplement)* (Springer, New York, 1977)
27. E.M. Levin, C.R. Robbins, H.F. McMurdie, *Phase Diagrams for Ceramists* (The American Ceramic Society, Columbus, 1969)
28. F.D. Richardson, T.C.M. Pillay, Trans. Inst. Min. Met., (London) Section C **66**, 309 (1956–57)
29. A. Zupp, D.A. Stevenson, Trans. Met. Soc. AIME **236**, 1316 (1966)
30. J.P. Morris, A.J. Williams, Trans. Am. Soc. Metals **41**, 1425 (1949)
31. P. Diaz, F.D. Richardson, Trans. Inst. Min. Met., (London) Section C **76**, 196 (1966)
32. H. Larson, J. Chipman, J. Metals **5**, 1089 (1953)
33. G.G. Hatch, J. Chipman, J. Metals **6**, 1136 (1954)
34. F.D. Richardson, J.C. Billington, Bull. Inst. Min. Met. **65**, 273 (1955–56)
35. K.P. Abraham, M.W. Davies, F.D. Richardson, J. Iron Steel Inst. **196**, 82 (1960)

Chapter 6
Vapor Pressure Methods

Vapor pressure measurement provides a simple method for determination of activity of any component that is directly related with its partial pressure. If p_i is the partial pressure of the component, i in the solution and p_i^o is the partial pressure of i in pure state at the same temperature, the ratio p_i/p_i^o will give activity, a_i of the component in the solution. The activities are related to the equilibrium constant (Eq. 3.58) that can be used to calculate free energy of a reaction. Heat of reaction can be derived from the variation of the equilibrium constant with temperature (Eqs. 2.94 and 2.95). The experimental methods available for vapor pressure measurement can be broadly classified into three main groups: static, dynamic and effusion.

6.1 Static Methods

In static methods, pressures greater than 1 mm Hg in equilibrium with the sample in an enclosed vessel at a particular temperature and pressure, are measured using appropriate pressure sensing devices. These methods can be further divided into direct and indirect methods. The measurement of the total pressure over the sample at a particular temperature in an enclosed vessel falls in the first group. In such cases pressure is measured by various pressure sensing devices, viz. manometer, Mc Leod gage and isoteniscope. Indirect methods are based on the measurement of the density of vapours in equilibrium with the sample, or on the determination of temperature of phase change for example, as in the dew point and boiling point methods. Measurement of vapor density is again achieved by a variety of instrumental techniques. The isopiestic method using the gas equilibration technique for pressure measurement falls in a separate class.

M. Shamsuddin, *Thermodynamic Measurement Techniques*, The Minerals, Metals & Materials Series, https://doi.org/10.1007/978-3-031-47118-6_6

6.1.1 Direct Methods

Direct static methods [1–9] have been used extensively in determination of vapor pressure of elements, compounds and metallic systems. The construction and operation of different types of manometers and pressure gages will not be discussed here. Only special devices developed under this category will be discussed.

6.1.1.1 Isoteniscope

Among the direct static methods the isoteniscope is most convenient for vapor pressure measurements at elevated temperatures. The method is particularly useful when sample has to be kept anhydrous and when corrosive nature of vapor prevents using simple manometric techniques. The isoteniscope, developed by Smith and Menzies [10] is shown in Fig. 6.1. The apparatus consisting of two parts is joined by a U-tube that contains a fusible eutectic mixture. One part terminates in a small bulb K containing the substance under investigation. On the other hand the second part is connected at O to a pump and an air reservoir fitted with suitable stopcocks and a pressure-sensing device. During operation the entire assembly is evacuated, and the eutectic mixture is first melted by raising a heating bath around the apparatus. After complete melting of the eutectic, the substance is heated. The pressure is measured on the air reservoir side when it is equal on either side of the tube. The isoteniscope, developed by Smith and Menzies [10] for measurement of organic compounds was adapted by Biltz and Meyer [12] for amalgams. This technique was employed for

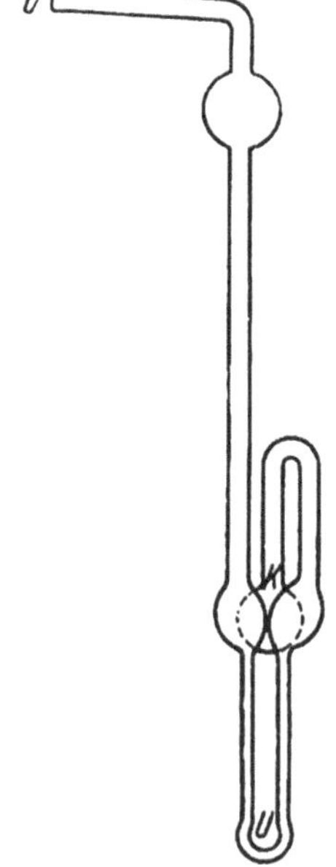

Fig. 6.1 Isoteniscope. (Reproduced from O. Kubaschewski, and C.B. Alcock, Metallurgical Thermochemistry [11] 1967, p 159) [Originally from A. Smith and A. W. D. Menzies [10] J. Amer. Chem. Soc., **32** (1910) 1412]

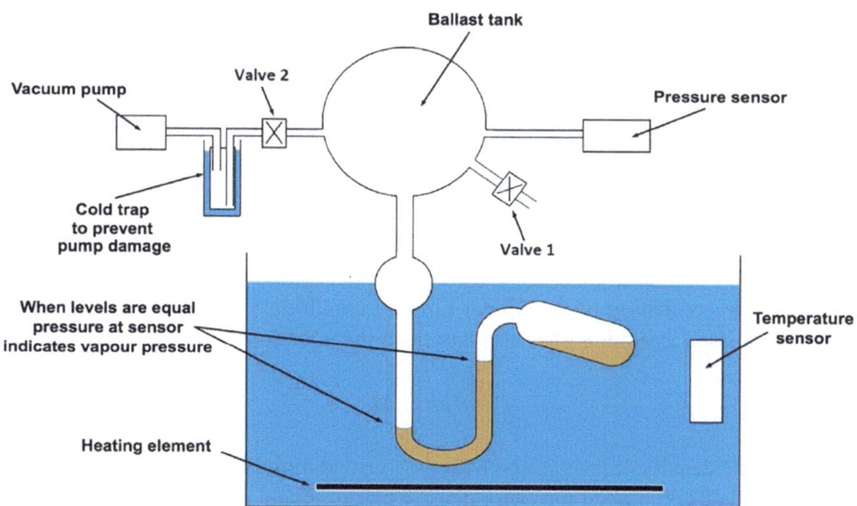

Fig. 6.2 Schematic diagram of isoteniscoppe. (Reproduced with permission from W. Chen et al. [17], J. Chem Education, ACS Publication, DOI: 10.1021/acs.jchemed.5b00990, Copyright 2016 The American Chemical Society and Division of Chemical Education, Inc.)

measuring the vapor pressure of cerium amalgams [11] using the sodium-potassium nitrate eutectic as the monometric liquid. Datz et al. [13] have measured the vapor pressure of alkali halides. Dutrizac and Flengas [14] further modified the design of the isoteniscope for measurement of vapor pressure of alkali and alkaline earth complex chlorides of zirconium and hafnium. Smoronava et al. [15, 16] have measured the vapor of mercury over In-Tl-Hg and In-Bi-Hg alloys. For precise measurements the liquid must be thoroughly degassed by evacuation and a constant temperature is attained before adjusting the liquid levels in the two arms.

In recent years, the isoteniscope technique has been modified to reduce complexity and to improve accuracy. The isoteniscope developed by Chen et al. [17] at the Imperial College, London is depicted in Fig. 6.2. The updated design of the isoteniscope assembly employs a piezo transducer in combination with a microelectromechanical-based high precision piezo sensor and integrated electronics, and a digital pressure display. As shown in figure the major components of the experimental setup include an isoteniscope submerged in a water bath that can be heated to a controlled temperature, a ballast tank, a diaphragm vacuum pump, a piezo pressure transducer with digital display, two valves, and a glass cold trap.

The isoteniscope is connected with the ballast tank, which opens to the atmosphere through valve 1 and is connected to the vacuum pump through valve 2. The pressure inside the ballast tank is measured by the pressure sensor. The pressure read by the pressure sensor is the same as in the bulb of the isoteniscope when the liquid levels are equal on both sides of the U-tube. In order to prevent any entry of vapors of the test liquid, a cold trap (water-ice mixture in a Dewar flask) is placed between valve 2 ad the vacuum pump.

Roche et al. [18] have further modified the technique to measure the vapor pressure of a concentrated solution and to analyse the impact of dissolved solutions in the system.

6.1.2 Indirect Methods

Indirect methods are based on measurement of the density of the vapor in equilibrium over the condensed phase. Under limiting conditions, the total vapor pressure can be calculated assuming ideal gas law from the equation:

$$p = \rho RT/M \tag{6.1}$$

where ρ is the density of the gas and M is the average molecular weight of the vapor. Various methods employed for vapor density measurement include determination of vapor weight, thermal conductivity and ionization of the vapors, optical absorption and tracer techniques. Optical absorption and tracer techniques have been adopted more widely as compared to the other three methods and have been used for vapor pressure measurement over metallic systems at higher temperature. Mass spectrometry may also be considered as a density measurement method though it is always coupled with some other dynamic method.

6.1.2.1 Vapor Weight

The method is restricted to systems that have relatively high vapor pressure and where the molecular weight of the vapor species is well known. Nisel'son [19] measured the vapor density of gaseous $ZrCl_4$ and $HfCl_4$ by the visual–polythermal method which consists of introducing some amount of the tetrachloride into a glass ampoule with a constriction in the middle. The ampoule was evacuated and sealed off, and the chloride was sublimed into the empty end, to separate it from traces of non-volatile product formed by hydrolysis. The ampoule was sealed off at the constriction and placed in a nitrite-nitrate mixture bath, heated by a tubular transparent heater. On gradual heating, the temperature at which the liquid or crystalline phase disappeared was noted. The ampoule was weighed, opened and washed free from chloride. All the pieces of glass were collected and weighed to get the weight of the chloride. The empty ampoule was then filled with water to find its volume. In this way the density of vapor was measured at different temperatures and the vapor pressures were calculated. In some cases where the exact molecular weight of the vapor species is not known, this method can be utilized for the determination of the average molecular weight of the vapor species.

6.1.2.2 Thermal Conductivity

At normal pressure, the thermal conductivity of any gas is independent of pressure. At low pressures where the mean free path of the gaseous molecule is of the order of the transfer distance or larger, the energy loss by conduction of heat increases linearly with the gas pressure. The pressure of perfect gases can thus be determined by measuring thermal conductivity, provided heat losses from radiation can be minimized.

A thermal conductivity cell consists of a glass vessel containing a heated filament similar to that of a Pirani gauge. The power supplied to the filament is maintained constant, and the cooling of the filament by gaseous heat transfer is monitored by a resistance thermometer or a thermocouple. In some cases the filament has been replaced by a resistance thermometer. Though theoretically one can measure the absolute pressure by this method, the instrument is normally calibrated using a Mc Leod gage.

6.1.2.3 Vapor Ionization

The ionization method is again a low pressure technique applied for vapor pressure measurement of easily ionized metals and salts and in particular for the residual gases. A portion of the gaseous phase is ionized and ions are accelerated towards a collector by application of a direct current potential between the collector and the ionization region. Ionization may be introduced either by collision of the molecules with a heated filament or bombardment of the gaseous phase with electrons or bombardment with radioactive decay products. The ion-current established is proportional to the number of molecules per unit volume and therefore a measure of gas pressure. The vapor pressure of liquid and solid mercury has been determined by this method between −79.6 and +19.7 °C assuming mono-atomic vapors. However, this method is rarely used for vapor pressure measurement on account of the following reasons:

(i) Recalibration for each species is necessary.
(ii) Pressures greater than a few micron lead to inefficiency of the ion collector.
(iii) A complex electrical set-up is needed.
(iv) The sensitivity is dependent on the nature of vapor species investigated.
(v) The filament is subjected to burn out, poisoning etc.

The method is however widely used for measurement of residual gas pressure in vacuum system. The gage requires the calibration and is normally carried out with the help of Mc Leod gauge. Vapor ionization technique has been used by Poindexter [20] to measure the vapor pressure of mercury over sodium and potassium amalgams.

6.1.2.4 Tracer Technique

In this method, radioactivity is introduced in the condensed phase either by doping the matrix with radioactive nuclides or by neutron bombardment of the stable nuclides. The gaseous phase in equilibrium with the sample at any temperature is then monitored for radioactivity which is proportional to the partial pressure of the nuclide being studied. Though with proper calibration the absolute pressures can be derived from the vapor phase radioactivity, the method can be used in a better way for determining activity of any component in a multicomponent system. The method is extremely sensitive and very low vapor pressure can be measured. However, in this method the chemical form of the vapor species should be well established. Vapor pressure over solid phosphorus [21, 22] and liquid zinc [23] have been measured with the aid of P_{15}^{31} and Zn_{30}^{65} radioactive tracers.

6.1.2.5 Optical Absorption Technique

Atomic absorption spectroscopy is one of the best methods for determination of vapor density for vapor pressure measurement at high temperature. When vapors contain more than one species, this method can be used to determine concentration of each vapor species independently. Briefly, the method consists of measuring concentration of vapor species by quantity of light absorbed at a frequency characteristic of the species.

This technique is particularly suitable for thermodynamic studies of the multicomponent systems for reasons mentioned below:

(i) It is extremely sensitive, and can be used for vapor pressure measurement in a wide range of pressures: $0.1\text{–}10^{-9}$ torr.
(ii) It is highly selective as interference from the other vapor species is negligible and thus the simultaneous measurements of partial pressures may be made for several vapor species present in a given system.

The following two procedures may be adopted for conducting the experiments:

(i) Light from a continuous spectrum source is passed through the absorption cell (containing the specimen at any temperature) and the ratio of the emergent intensity to the incident intensity is measured with a spectrophotometer.
(ii) The element of interest is incorporated into a hollow cathode lamp to provide a characteristic spectra source of high intensity. These characteristic radiations are then attenuated by atoms of the same element (in the absorption cell) placed in the radiation path.

The latter technique has been preferred over the former by many workers as it gives more accurate results. The relationship between attenuation of resonant wave lengths and vapor pressure has been derived as:

$$I = I_o e^{-Kn} \tag{6.2}$$

where I_o is the incident intensity, I is the emergent intensity, K is the absorption coefficient, and n is the number of absorbing atoms. n is related to the partial pressure by the relationship:

$$n = \frac{pV}{RT} N \tag{6.3}$$

where V is the cell volume and N is the Avagadro number. Thus by introducing the value of n in Eq. (6.3) partial pressure can be obtained as:

$$p = \left[\frac{R}{VNK}\right] Tln\left(\frac{I_o}{I}\right) \tag{6.4}$$

$$p = K' Tln\left(\frac{I_o}{I}\right) \tag{6.5}$$

The constant K' is defined by the cell geometry and the absorption coefficient. In order to determine absolute value of vapor pressures it is necessary to evaluate K' independently. In many cases, however, it may not be necessary to evaluate the value of K' as in the determination of the enthalpy of vaporization or the activity of any component in a multicomponent system. The enthalpy of evaporization, L^e can be determined. by plotting $ln\left[Tln\left(\frac{I_o}{I}\right)\right] vs. \frac{1}{T}$ (instead of $\log p vs. \frac{1}{T}$).

Since the activity of any component, i in the multicomponent system is given as: $a_i = p_i/p_i^o$.

$$\therefore\ lna_i = \ln p_i - lnp_i^o \tag{6.6}$$

Introducing the values of p_i and p_i^o from Eq. (6.5) we get:

$$lna_i = ln\left[\frac{I_o}{I}\right]_{alloy} - ln\left[\frac{I_o}{I}\right]_{pure} \tag{6.7}$$

Thus the a_i and L^e valuess can be determined without determining the value of K'. By measuring the activity at different temperatures and compositions various other thermodynamic properties can be derived.

This technique has been used for investigating various alloys comprising one or more type of vapor species. The optical absorption technique being simple and reliable has been used in study of a number of metallic systems, namely, Pb-Te [24], Sn-Te [25], Ge-Te [26], Cd-Te [27], Hg-Te [28], Hg-Se [29], Ge-Cd [30], Ag-In [31], Ag-Zn-Cd [32], CdTe [33], ZnTe [34], HgTe [35, 36], PbTe-SnTe [37], CdTe-HgTe [38, 39].

6.1.2.6 Phase Change Method

Methods based on phase change are the oldest technique for vapor pressure measurements. Two common phase change methods are based on the determination of (i) the dew point of the vapors in equilibrium with the condensed phase or (ii) the boiling point of a liquid. The boiling point method is used commonly for the measurement of absolute vapor pressure over pure liquid elements, whereas the dew point method is used for the measurement of comparative vapor pressure of the more volatile component over a multicomponent system.

Dew Point Method Dew point is defined as the temperature at which the condensation of vapours takes place in an isothermal reversible way. At any particular pressure, the dew point of the vapor corresponds to the temperature at which the vapor will be in equilibrium over the pure condensed phase. In other words if the vapor of any component is in equilibrium with the pure condensed phase at temperature T and with the multicomponent phase at T, then the temperature T is called the dew point. If the temperature pressure relationship is known for the pure component, the vapor pressure of that component over the multicomponent system at various temperatures can be determined by measuring the dew point at each temperature. The method has been used extensively for variety of materials, viz. intermetallics [40–43], inorganic salts [44] and alloys [45–49] at moderate as well as high temperatures.

Boiling Point Method The boiling point method is applicable mainly for vapor pressure measurement of pure liquid elements and is limited to relatively high pressures. The method is based on the principle that a liquid readily boils at the temperature where the vapor pressure of the liquid just exceeds the external pressure on the liquid. The boiling point may be observed either visually or by some indirect methods, for instance the sudden increase in the rate of weight change of a crucible containing the condensed liquid. Experiments can be carried out either by varying the external pressure at constant temperature of the liquid or by varying the temperature at constant pressure. The vapor pressure of the liquid is then determined by noting the temperature of the condensed phase at the boiling point and the external pressure of the inert gas by some pressure sensing device.

The boiling point method has been used for vapor pressure measurement of inorganic salts [50–53] and liquid metals [54, 55]. While small amount of gaseous impurities do not affect the reliability, as in other static methods, dissolved impurities which cause significant lowering of the boiling point, can be a source of major error. Super heating of the liquid, particularly in case of the metals, can also happen.

6.1.2.7 Isopiestic Method

The isopiestic method may be considered as a special case of static methods. It is one of the most accurate methods for determining comparative vapor pressure at high temperature. The method is based on the principle that if two or more condensed phases are in equilibrium through the vapor species of a common volatile component then the thermodynamic activity of that component will be the same in all the condensed phases. In other words if the vapor pressure of the volatile component is known over one system, the pressures over the other systems can be determined by this technique. The equilibrium is normally achieved under isothermal conditions. However, studies made under non-isothermal conditions have also been found to be very useful and are known as pseudo-isopiestic studies.

The isopiestic method consists of holding two or more samples at the same temperature and equilibriating one volatile component with these samples. Since diffusion of vapour is the limiting factor in determining the duration of the experiment, equilibrium is usually accomplished by first evacuating the system and then heating the samples at the experimental temperature. In some cases radioisotopes have been used for carrying out in-situ analyses as the equilibrium is attained. In general after attaining equilibrium, the results are determined by analyzing the quenched samples.

The isopiestic method has been used by many workers [56–59] for thermodynamic studies of aqueous solutions. For high temperature metallurgical studies the pseudo-isopiestic method is, however, more commonly employed. Komarek and workers have investigated thermodynamic properties of Al-Fe [60], Mg-Sn [61], Mg-Si [62], Al-Cr [63], Al-V [63], Al-Co [64] and many other alloys employing this technique. Thermodynamic properties of Cd-Sn system has been studied by Predel [65].

6.2 Dynamic Methods

Principally, all techniques in which the vapor pressure of a substance is deduced from some time dependent effect may be considered as dynamic methods. In this respect, the effusion methods of Knudsen and Langmuir are also dynamic methods. However, the effusion methods are usually considered as a group of techniques of their own because they are associated with special problems and are applicable in extremely low pressure ranges and at extremely high temperatures. On the other hand boiling point and dew point methods may be considered as static or dynamic, depending upon the method applied for the detection of boiling points and dew points. Generally, dynamic methods are divided into two groups, namely, boiling point and transpiration methods. Dew point methods are not considered as a group of their own because in these methods merely condensation is observed instead of boiling.

6.2.1 Boiling Point Methods

These methods are preferably applied when the boiling point of the substance under investigation is above room temperature at the external pressure of one atmosphere. The measured vapor pressure should be in the high pressure region; greater than 10^{-1} mm Hg, so that the applied external pressure may be controlled and measured by classical manometers. In the static methods employing direct pressure measurements, the considered vaporized substance is in touch with the manometer or at least with some liquid, separating the manometer cell from the thermally equilibrated gas phase, whereas in the boiling point method the vapor pressure is directly exerted to some indifferent gas which acts at some lower temperature upon the manometer. Hence it is possible to employ the boiling point methods at higher temperature as compared to the classical static methods.

The boiling point is the temperature at which vapor pressure of the liquid substance equals the pressure of the surrounding atmosphere. Hence it is necessary to (i) measure the pressure of the atmosphere over the boiling substance or (ii) the temperature of the liquid under investigation or (iii) to detect the onset of boiling. The boiling point may be directed according to the following three effects which characterize the onset of boiling:

1. The continuous enhancement of temperature during heat supply of the liquid ceases at the boiling point; i.e. during boiling the temperature of the liquid remains constant.
2. During boiling the liquid vaporizes not only at the surface but also in the interior; i.e. bubbles may directly be observed.
3. Since the liquid vaporizes during boiling not only at the surface, the onset of boiling is characterized by a discontinuous enhancement of the loss of material due to evaporation; i.e. the boiling point may be detected by continuous weight loss measurement.

At onset of boiling the weight loss vs time plot is characterized by a break. Ruff's method [66] is based on this principle. The substance is enclosed is a crucible suspended from a spring balance, inside a furnace. The boiling point is in effect determined from a discontinuity in the weight–temperature curve. Since the weight change is recorded at changing temperature, the actual temperature of the substance is subject to some error. Additional disadvantage of the method is that the change in slope of the curve at boiling point is sometimes not sharply defined. This method was modified by Fischer and Rahfs [67] who worked at constant temperature and varying pressure to get sharp change in the slope. The application of boiling methods may be hampered by occurrence of delayed boiling. In some cases determination of condensation temperature seems to be more reasonable than boiling temperature measurement. Design of experimental set up varies depending on the above three effects to be adopted in determination of the boiling point.

6.2.2 Transpiration Method

The more general dynamic method for determination of vapor pressure is known as the transpiration method. Since the flow or transpiration method was first introduced by Regnault [68] numerous modifications have been proposed in application of this technique in measurement of vapor pressures exceeding 10^{-6} atm. In this method a steady and measured stream of inert gas is passed over the substance or bubbled through the vaporizing substance under investigation. The gas removes the vapor or volatile component of the substance at a rate which is dependent upon their relative pressures and upon the rate of gas flow. The vapor is condensed or collected by absorption or chemical combination at a cooler portion of the apparatus. The rate of removal of vapor is measured at different rates of gas flow. Theoretically saturation is reached at low yet finite streaming rates and extrapolation may therefore result in an error which can only be avoided by working at very low rates. This may introduce new errors, such as thermal diffusion, due to the long duration of each determination, and care must be taken to reach a satisfactory compromise.

The partial pressure is calculated from the volumes of transported substance, V_s and the transporting gas V_g at STP assuming validity of the gas law according to the expression:

$$p_s = \frac{V_s.760}{V_s + V_g} \text{ mm Hg} \tag{6.8}$$

Design of transpiration apparatus varies drastically from a set of massive bottles filled with solutions for measuring solution vapor pressure to a small ceramic cell used in case of high temperature studies. These apparently different apparatuses, however have some common components, such as a source of carrier gas, a conduction system for this gas into a saturation chamber, a saturation chamber in which the equilibrium partial pressure is maintained over a condensed system, a conduction system to carry the saturated carrier gas away from the saturation chamber and a unit to measure the transpiration rate versus the carrier gas flow rate. The details about various designs of transpiration apparatus may be obtained from Norman and Winchell [69], Merten and Bell [70], Adams and Quan [71], Alcock and Hooper [72] and Schafer [73]. This technique has been used to determine thermodynamic properties of CdS [74], CdSe [75], HgSe [76] and SnS [77].

The shortcomings of the transpiration methods include knowledge about accurate molecular weight of the vaporizing species, uncertainty of the saturation of a carrier gas with the considered vapor, precise measurement of gaseous flow rates, thermal diffusion due to the temperature gradient necessary for condensation of vapor and chemical reaction of the sample with the container and the carrier gas. Gardener and Pang [78] have modified the method to study zinc blende ↔ wurtzite transformation in ZnS. Nakarnura and Fuwa [79] have attached a mass spectrometer to the transpiration apparatus to distinguish the vaporizing species.

6.3 Effusion Methods

The effusion methods include two principal techniques, namely, those owing to Knudsen and Langmuir to measure pressures less than 10^{-3} mm Hg of a substance by its rate of evaporation in vacuum. Such low pressure measurements by these techniques are based on the kinetic theory of gases. The rate of effusion of gaseous molecules per unit area per unit time through a small orifice of a closed vessel containing a gas under thermal equilibrium follows the cosine law [80]. The importance of kinetic theory to pressure measurement was realized by Hertz-Knudsen [81, 82] who established the following equation that relates partial pressure with the mass of vapour m (g), molecular weight M evaporating from an orifice of area A (cm^2) in time t (second) at temperature, T (K):

$$p = \frac{m}{tA}\left[\frac{2\pi RT}{M}\right]^{1/2} = 0.02256\frac{m}{tA}\left[\frac{T}{M}\right]^{1/2} \tag{6.9}$$

However, for the above equation to be applicable following conditions must be fulfilled:

(i) the orifice should small enough to attain proper equilibrium,
(ii) wall should be thin enough to have negligible change of a molecular collision with the edge of the orifice, and
(iii) the pressure should be sufficiently low to manage intermolecular collisions within the orifice.

6.3.1 *Knudsen Effusion Technique*

In Knudsen effusion technique the substance under investigation is enclosed in a gas-tight vessel with a lid drilled with a small orifice. For validity of Eq. 6.9 there are two important conditions as regards to evaporation out of the vessel. Firstly the orifice must be of knife-edged and secondly its diameter must be less than one tenth of the mean free path of the vapor inside the container. If a channel is employed in place of a knife-edged orifice, the length to diameter ratio of the orifice must be known. A continuous weight measuring device is used for estimating the mass of vapor. Sigai and Wiedemeir [83] have investigated CdSe by recording the weight loss measurement with a Cahn microbalance. Bardi and co-workers employed Knudsen effusion apparatus coupled with a vacuum thermobalance for studies of Al_2Se_3 [84], CdTe [85], In_2Se [86], and ZnTe [87]. This technique has also been employed for determination of thermodynamic properties of metallic systems, viz. Al-Mg [88], Cu-Pb [89], Sn-Sb [90], Al-Sb [91], Ga-Sb [92], Fe-Si [93], Au-Te [94], Pd-Cd [95], Hg-Ga [96], Cu-Mn [97], Cd-Bi-Mg [98], Hg-Ga-Al [99], Cu-Pb-Fe [100]. A Knudsen effusion cell [11, 101] with furnace is shown in Fig. 6.3. In order to get best possible sensitivity, use of radioactive isotopes and condensation of vapor

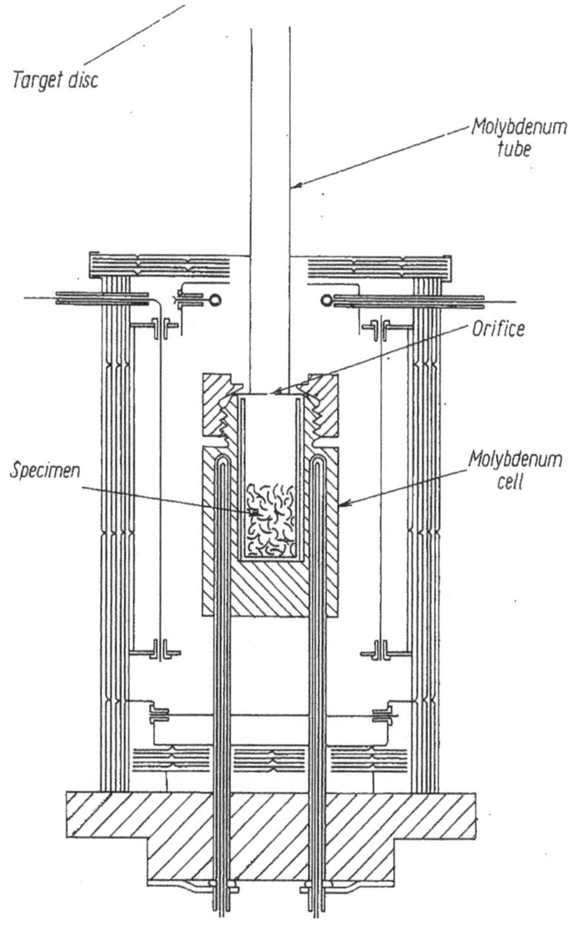

Fig. 6.3 Effusion cell and furnace for the determination of the vapor pressures of metals between 1200 and 1400 °C. (Reproduced from O. Kubaschewski, and C.B. Alcock, Metallurgical Thermochemistry [11] 1967, p 175). (Originally from O. Kubaschewski, and G. Heimer, [101] Acta Metallurgica, **8** (1960) 416)

on a target are advised. The weight loss can then be estimated by radiochemical assay after making sure that a significant fraction of the condensate has not been collected after scattering from the plate defining the solid angle of collection.

The theoretical background and experimental details of Knudsen cell assemblies are available in literature [102–104]. The Knudsen and Langmuir methods for vapor pressure estimation based on the application of Eq. 6.9 can only be employed when the vapor species is simple, chemically well-defined and the molecular weight of the species is known. The interpretation of data is difficult when complex and more species are produced simultaneously in a given vaporization process. Thus, the use of mass spectrometer becomes essential for identifying and studying the equilibria amongst several vapor species. In mass spectrometer the molecules or atoms are ionized with a well-defined electron beam and the resulting, usually positive ions are analysed for charge to mass ratio in the conventional manner by electric and magnetic field deflection or by electric field acceleration and time of flight measurement over a fixed trajectory.

The mass spectrometric method is capable of conducting measurement on several vapour species simultaneously in a single experiment. Absolute vapor pressures can be obtained from the measured ion currents if ionization cross section and detection efficiency are known, alternatively the instrument can determine only relative vapor pressures. Thus by comparing the ion currents obtained for an alloy with that of the pure phase the activity may be estimated accurately. Heat of vaporization can also be determined with high precision. Kubarchewski et al. [105] have measured the vapor pressure of chromium alloys using the combination of effusion and tracer methods. In the experimental set-up, shown in Fig. 6.3, high temperature (about 1400 °C) was generated by resistance heating enclosed in radiation shields. Heating elements at the top and bottom were operated separately to obtain uniform temperature within the molybdenum cell. Mass spectrometer coupled with the Knudsen cell have been used extensively for investigating number of intermetallics, viz. InAs [106], InTe [107], Sb_2Se_3 [108], SnS_2 [109], TlSb [110], PbS-Ga_2S_3 [111] and metallic solutions, viz. La-Ga [112], Fe-Cr [113], Fe-Al [114], Ag-Al [115], Ag-Cu [116], Co-Au [117], Co-Ge [118], Fe-Co [119], Fe-Cr-Co [120], Fe-Cr-Ni [121], Pt-Pd-Ge [122], Ag-In-Sn [123], and In-Sb-Zn [124].

The Knudsen method can be used successfully to measure pressure to a lower limit of about 10^{-4} mm Hg but Langmuir's technique enables measurement of pressure as low as 10^{-7} mm Hg. The latter directly measures the rate of evaporation from metal wire samples heated electrically. The total area of the heated wire is taken as area in Eq. 6.9 which presumes that all vapor molecules deflected back to the metal striking its surface are retained and not reflected. This assumption holds good in Knudsen method but its validity must be tested in Langmuir's technique. Fajans [125] has pointed out that slightest surface contamination lowers the rate of evaporation considerably. There are more chances of errors at low vapor pressure. Hence Lanmuir's method is more reliable at higher temperature. Johnston and Marshall [126] have reported better results by modifying the technique employing metal samples in the form of an annular ring of 22 mm outer and 11 mm inner diameters and 2.5–6 mm thickness. The samples were heated by induction in an evacuated silica tube (Fig. 6.4) and thoroughly degassed. The temperature was measured by an optical pyrometer. While measuring the vapor pressure of liquid copper the metal sample was enclosed in a small trough of molybdenum. In this way errors inherent in Langmuir's method are minimized though not necessarily eliminated as happens in Knudsen technique. Speiser and Johnson [127] and Margrave [128] have critically assessed the effusion techniques for measurements of vapor pressures.

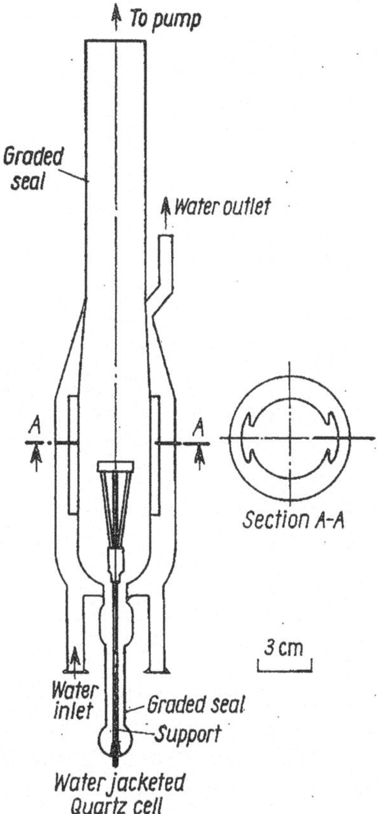

Fig. 6.4 Langmuir type apparatus for determining reactive vapors. (Reproduced with permission from H. L. Johnston and A. L. Marshall [126], J. Amer. Chem. Soc., 62 (1940) 1382. Copyright 1940 American Chemical Society)

6.3.2 Torsion Effusion Technique

Volmer [129] developed torque effusion technique by introducing a very significant modification in Knudsen method. He suggested that pressure can be determined by measuring momentum of the effusing molecules rather their mass in Eq. 6.9:

$$p = \frac{m}{tA}\left[\frac{2\pi RT}{M}\right]^{1/2} = 0.02256\frac{m}{tA}\left[\frac{T}{M}\right]^{1/2} \tag{6.9}$$

Figure 6.5 shows an effusion cell having two orifices located off-centre on opposite sides of the cell suspended from a long suspension fibre of known torsion constant. Suspension fibres made of quartz, tungsten and phosphor bronze may be used, the last being preferred.

The pressure exerted on the wall by a gas is equal to the rate of change of momentum per unit area normal to the wall. The momentum carried away by the effusing molecules through the orifices of different areas of cross section on opposite sides of the walls does not balance the momentum put on the wall. Hence, the cell

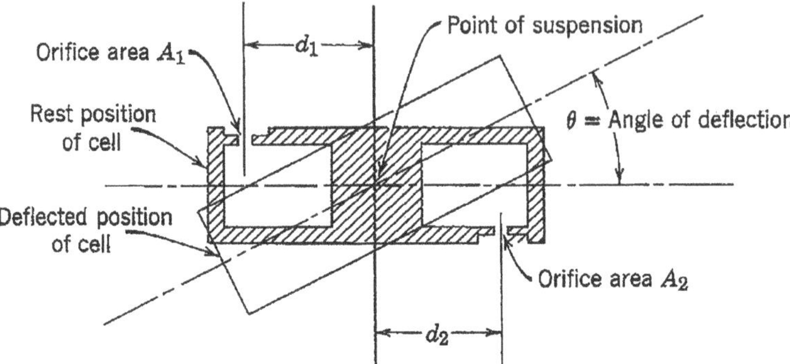

Fig. 6.5 Torsion effusion cell. (Reproduced from E.D. Cater [80] in Physicochemical Measurements in Metals Research, Vol. IV, Part I, R. A. Rapp, ed., Interscience, A division of John Wiley & Sons, 1970, Chapter 2A, p 37)

will rotate against the restraining torque of the suspension wire, which is directly proportional to the pressure in the cell. The pressure is calculated by the expression:

$$p = \frac{2\tau\theta}{(A_1 d_1 + A_2 d_2)} \tag{6.10}$$

where τ and θ are respectively, torsion constant and rotational or angular deflection, d_1 and d_2 are the distances of the two opposed orifices of areas A_1 and A_2 from the axis of suspension. From Eq. 6.10 it is evident that pressure is independent of the molecular weight of the vaporizing species. Thus use of sophisticated and costly mass spectrometer is eliminated because pressure obtained does not indicate the complexity of the mode of vaporization. Further the method is very speedy because vapor pressure can be measured at the instant when the system has reached thermal equilibrium.

The technique can be employed for measurement of pressure between 10^{-4} and 10^{-7} atm in the temperature of 600–2300 °C. However, the following conditions [11] have to be satisfied while adopting this technique:

(i) The mean free path of the vapor inside the cell must be several times larger than the orifice diameter.
(ii) The effective sample area must be very large compared to the orifice.
(iii) The orifice temperature must not be less than that of other portion of the cell.

Various binary and ternary metallic systems, viz. In-Pb [130], In-Mg [131], Tl-Zn [132], Ni-Zn [133], Ag-Tl [134], Ag-Zn-Sn [135] and Ag-Bi-Sn [136] have been investigated by torsion effusion method.

References

1. L.S. Brooks, J. Am. Chem. Soc. **77**, 3211 (1955)
2. F.H. Spedding, J.H. Oye, J. Phys. Chem. **59**, 581 (1955)
3. K. Funagi, K. Uchimura, J. Jpn. Electrochem. Soc. **33**, 167 (1965)
4. R.O. Mac Laren, N.W. Gregory, J. Phys. Chem. **59**, 184 (1955)
5. C.M. Cook, J. Phys. Chem. **66**, 219 (1962)
6. R.J. Sime, J. Am. Chem. Soc. **85**, 501 (1963)
7. R.F. Brebrick, F.T.J. Smith, J. Electrochem. Soc. **118**, 991 (1971)
8. J.H. Greenberg, V.N. Guskov, V.B. Lazrev, J. Chem. Thermody. **17**, 739 (1985)
9. Y.A. Ugai, A.M. Somilov, G.V. Seminova, E.G. Goncharov, Zh. Fiz. Khim. **60**, 25 (1986)
10. A. Smith, A.W.C. Menzies, J. Am. Chem. Soc. **32**, 1412 (1910)
11. O. Kubaschewski, C.B. Alcock, *Metallurgical Thermochemistry*, 4th edn. (Pergamon, Oxford, 1967)
12. W. Biltz, F. Meyer, Z. Anorg. Chem. **176**, 23 (1928)
13. S. Datz, W.T. Smith, E.H. Taylor, J. Chem. Phys. **34**, 558 (1961)
14. F.E. Durizac and S.N. Flengas, in *Advances in Extractive Metallurgy*, Proc. Symp. Inst. Min. Met. (Elsevier, London, 1967), p. 572
15. O.Y. Smoronova, R.S. Nigmetova, L.F. Kozin, Izvestiya Akademii Nauk Kazakhskoi **24**(7), 73 (1974)
16. O.Y. Smoronova, R.S. Nigmetova, Vestnik Akademii Nauk Kazakhskoi **24**(7), 47 (1974)
17. W. Chen, A.J. Haslam, A. Macey, U.V. Shah, C. Brechtelsbauer, J. Chemical Education, Am. Chem. Soc. (2016). https://doi.org/10.1021/acs.jchemed.5b00990
18. P. Roche, R.C. Jones, B. Glennon, P. Donnellan, *AIChE J. Open Access* (2021). https://doi.org/10.1002/aic.17377
19. L.A. Nisel'son, Russian J. Inorg. Chem. **6**, 635 (1961)
20. F.E. Poindexter, Phys. Rev. **28**(2), 208 (1926)
21. F.S. Dainton, H.H. Kimberley, Trans. Faraday Soc. **46**, 912 (1950)
22. F.D. Rosen, W. Davies Jr., Rev. Sci. Instr. **24**, 349 (1953)
23. U. Croatto, L. Riccobani, Chim. Ind. (Milan) **37**, 1955, 538
24. R.F. Brebrick, A.J. Srauss, J. Chem. Phys. **40**, 3230 (1964)
25. R.F. Brebrick, A.J. Srauss, J. Chem. Phys. **41**, 197 (1964)
26. R.F. Brebrick, J. Chem. Phys. **41**, 1140 (1964)
27. R.F. Brebrick, A.J. Srauss, J. Phys. Chem. Solids **25**, 1441 (1964)
28. R.F. Brebrick, A.J. Srauss, J. Phys. Chem. Solids **26**, 989 (1965)
29. R.F. Brebrick, J. Chem. Phys. **43**, 3846 (1965)
30. C.H. Su, T. Tung, A. Mubarak, R.F. Brebrick, High Temp. Sci. **18**, 197 (1984)
31. O.B. Masson, S.S. Pradhan, Met. Trans. **4B**, 591 (1973)
32. G. Scatchard, T. Lin, J. Am. Chem. Soc. **84**, 28 (1962)
33. K. Igaki, M. Nakano, Trans. Jpn. Inst. Met. **20**, 597 (1979)
34. R.F. Brebrick, J. Electrochem. Soc. **116**, 1274 (1969)
35. C.H. Su, P.K. Liao, T. Tung, R.F. Brebrick, High Temp. Sci. **14**, 181 (1981)
36. Y.G. Sha, K.T. Chen, R. Fang, R.F. Brebrick, J. Electrochem. Soc. **136**, 3837 (1989)
37. Y. Huang, R.F. Brebrick, J. Electrochem. Soc. **135**, 486 (1988)., 1547
38. J.P. Schwartz, T. Tung, R.F. Brebrick, J. Electrochem. Soc. **128**, 438 (1981)
39. T. Tung, L. Golonka, R.F. Brebrick, J. Electrochem. Soc. **128**, 451 (1981)
40. V.J. Lyons, J. Phys. Chem. **63**, 1142 (1959)
41. V.J. Lyons, V.J. Silvestri, J. Phys. Chem. **64**, 256 (1960)
42. V.J. Silvestri, J. Phys. Chem. **64**, 826 (1960)
43. L.I. Marina, A.Y. Nashellski, S.V. Yakobson, Russ. J. Phys. Chem. **36**, 575 (1962)
44. L.J. Hawell, R.C. Sommer, H.H. Kellogg, Trans. Met. Soc. AIME **209**, 193 (1957)
45. P. Chiotti, J.T. Mason, K.J. Gill, Trans. Met. Soc. AIME **227**, 910 (1963)
46. A. Yazawa, L.Y. Keun, Trans. Jpn. Inst. Met. **11**, 411 (1970)

47. J.D. Filby, J.N. Pratt, Acta Met. **11**, 427 (1963)
48. G.A. Somorjai, J. Phys. Chem. **65**, 1059 (1961)
49. P. Chiotti, K.J. Gill, Trans. Met. Soc. AIME **221**, 573 (1961)
50. J.W. Johnson, W.J. Silva, O. Cubicciotti, J. Phys. Chem. **69**, 3916 (1965)
51. D.N. Klushin, V.Y. Chernykh, Russ. J. Inorg. Chem. **5**(7), 685 (1960)
52. H. Bloom, J.O.M. Bockris, N.E. Richards, R.G. Taylor, J. Am. Chem. Soc. **80**, 2044 (1958)
53. E.E. Schier, H.H. Clark, J. Phys. Chem. **67**, 1259 (1963)
54. J.F. Walling, J. Phys. Chem. **67**, 1380 (1963)
55. T. Sato, H. Keneko, Technol. Rept. Tohuku Univ. **16**, 18 (1958)
56. W.R. Bousfield, Trans. Faraday Soc. **13**, 401 (1917)
57. D.A. Sinclair, J. Phys. Chem. **37**, 493 (1933)
58. R.A. Robinson, D.A. Sinclair, J. Am. Chem. Soc. **56**, 1830 (1934)
59. O. Scatchard, W.J. Hamer, S.E. Wood, J. Am. Chem. Soc. **60**, 3061 (1938)
60. J. Eldrige, K.L. Komarek, Trans. Met. Soc. AIME **230**, 226 (1964)
61. J. Eldrige, E. Miller, K.L. Komarek, Trans. Met. Soc. AIME **236**, 114 (1966)
62. J. Eldrige, E. Miller, K.L. Komarek, Trans. Met. Soc. AIME **239**, 775 (1967)
63. J. William, K.L. Komarek, E. Miller, US. Govt. Res. Dev. Rept **68**(2), 37 (1968)
64. M. Ettenberg, K.L. Komarek, E. Miller, Trans. Met. Soc. AIME **242**(1968) (1801)
65. B. Predel, Z. Metallk. **49**, 226 (1958)
66. O. Ruff, B. Bergdahl, Z. Anorg. Chem. **106**, 76 (1919)
67. W. Fischer, O. Rahfs, Z. Anorg. Chem. **205**, 1 (1932)
68. H.V. Regnault, Ann. Chim. (Paris) **15**, 129 (1845)
69. J.H. Norman, P. Winchell, in *Physicochemical Measurements in Metals Research*, Vol. IV, Part I, ed. by R.A. Rapp (Interscience, New York, 1970), p. 131
70. U. Merten, W.E. Bell, in *The Characterization of High Temperature Vapours*, ed. by J.L. Margrave (John Wiley, New York, 1967), p. 106
71. C.E. Adams, J.T. Quan, J. Phys. Chem. **70**, 331 (1966)
72. C.B. Alcock, G.W. Hooper, in *Physical Chemistry of Process Metallurgy*, Pt-1, ed. by G.R. St. Pierre (Interscience, New York, 1959), p. 75
73. H. Schafer, *Chemical Transport Reactions* (Academic, New York, 1964)
74. E.I. Boev, L.A. Benderski, N.V. Minaeva, A.M. Bunin, Zh. Fiz. Khim. **43**, 2234 (1969)
75. W.J. Wosten, J. Phys. Chem. **65**, 1949 (1961)
76. A.S. Pashinkin, B. Salamatin, Zhur. Fiz. Khim. **41**, 2395 (1967)
77. H. Rau, Ber. Bunsenges, Physik. Chem. **71**, 716 (1971)
78. P.J. Gardener, P. Pang, J. Chem. Soc. Faraday Trans. I **84**, 1879 (1988)
79. S. Nakamura, A. Fuwa, Nippon Kinzoku Gokkaishi **51**, 124 (1987)
80. E.D. Cater, in *Physicochemical Measurements in Metals Research*, Vol. IV, Part I, ed. by R.A. Rapp (Interscience, A division of John Wiley & Sons, New York, Chapter 2A, 1970), p. 37
81. H. Hertz, Ann. Physik. **17**, 177 (1882)
82. M. Knudsen, Ann. Physik. **29**, 179 (1909)
83. A.G. Sigai, H. Wiedemeir, J. Electrochem. Soc. **119**, 910 (1972)
84. G. Bardi, G. Trionfetti, Thermochim. Acta **136**, 313 (1988)
85. G. Bardi, K. Ieronimakis, G. Trionfetti, Thermochim. Acta **129**, 341 (1988)
86. G. Bardi, V. Piacente, P. Scardala, High Temp. Sci. **25**, 175 (1988)
87. G. Bardi, G. Trionfetti, Thermochim. Acta **157**, 287 (1990)
88. Z. Moser, W. Zakulski, K. Rayman, W. Gasior, Z. Panek, I. Katayama, T. Matsuda, Y. Fukuda, T. Lida, J. Botor, J. Phase Equilibria **19**, 38 (1998)
89. A. Zajaczkowski, Z. Czernecki, J. Botor, Archives Met. **42**(1), 11 (1997)
90. W. Zeng, N. Chen, D. Ye, Jinsu-Xuebo **32**(12), 1233 (1996)
91. A. Zajaczkowski, J. Botor, Z. Metallk. **86**, 590 (1995)
92. A. Zajaczkowski, J. Botor, Z. Metallk. **85**, 472 (1994)
93. J. Botor, A. Zajaczkowski, L. Dziewidek, G. Siderov, Z. Metallk. **82**, 304 (1991)

94. B. Predel, J. Piehl, Z. Metallk **66**, 33 (1975)
95. J.B. Darby, K.M. Myles, J.N. Pratt, Acta Met. **19**, 7 (1971)
96. B. Predel, D. Rothacker, Acta Met. **17**, 783 (1969)
97. R.W. Krenzer, M.J. Pool, I. Kaufman, Trans. Met. Soc. AIME **245**, 91 (1961)
98. L.F. Kozin, R.S. Nigmetova, A.M. Deirova, Zh. Fiz. Khim. **45**(8), 1906 (1971)
99. R.S. Nigmetova, N.V. Boltrukevich, Zh. Fiz. Khim **60**(10), 2613 (1986)
100. A. Zajaczkowski, J. Czernneckj, J. Botor, Z. Metallk **91**, 143 (2000)
101. O. Kubaschewski, G. Heimer, Acta Metallurgica **8**, 416 (1960)
102. W.L. Winterbottom, in *Physicochemical Measurements in Metals Research*, Vol. IV, Part 1, ed. by R.A. Rapp (Interscience, New York, 1970), p. 95
103. G. Bardi, R. Gigli, L. Malapina, V. Piacente, J. Chem. Engg. Data **18**, 126 (1973)
104. V. Piacente, P. Scardala, D. Ferro, R. Gigli, J. Chem. Engg. Data **30**, 372 (1985)
105. O. Kubaschewski, G. Heymer, W.A. Dench, Z. Electrochem. **64**, 801 (1960)
106. M. Tmar, C. Chatillon, J. Chem. Thermodyn. **19**, 1053 (1987)
107. R.S. Srivastava, J.G. Edwards, J. Electrochem. Soc. **131**, 2954 (1984)
108. C.L. Sullivan, J.E. Prusaczyk, R.A. Miller, K.D. Carlson, High Temp. Sci. **11**, 95 (1979)
109. H. Weidemeier, F.J. Csillag, High Temp. Sci. **12**, 277 (1980)
110. G. Balducci, D. Ferro, V. Piacente, High Temp. Sci. **14**, 207 (1981)
111. M.A. Williamson, J.G. Edwards, Thermochim. Acta **107**, 83 (1986)
112. V. Shaji, T. Matsui, K. Nakamura, T. Inoue, J. Nuclear Mater. **247**, 87 (1997)
113. B.R. Reese, R.A. Rapp, G.E.S. Pierre, Trans., Met. Soc., AIME **242**(1968) (1719)
114. N.S. Jacobson, G.M. Mehrotra, Met. Trans. **24B**, 481 (1993)
115. G.R. Belton, R.J. Fruehan, Trans. Met. Soc. AIME **245**, 113 (1969)
116. S.M. Howard, Met. Trans. **20B**, 845 (1989)
117. E.J. Grimsey, J.M. Taguri, Can. Met. Quart. **23**(3), 303 (1984)
118. J. Tomiska, G. Erdelyi, H. Nowtwenty, A. Nickel, High Temp. Sci. **15**, 41 (1982)
119. G.R. Belton, R. Fruehan, J. Richard, J. Phys. Chem. **71**, 1403 (1967)
120. B.B. Argent, M. Ellis, G. Effenberg, High Temp. High Pressure **14**(4), 409 (1982)
121. J. Vrestal, J. Theiner, P. Broz, J. Tomoska, Thermochim. Acta **319**, 193 (1998)
122. R. Chartel, R. Libb, R. Castanet, J. Alloys Compd. **283**, 208 (1999)
123. T. Miki, N. Osawa, T. Nagasaka, M. Hino, Mater. Trans. **42**(5), 732 (2001)
124. J.H. Norman, P. Winchell, H.G. Stalvy, J. Chem. Phys. **36**, 448 (1966)
125. K. Fajans, Z. Electrochem. **69**, 69 (1925)
126. H.L. Johnston, A.L. Marshall, J. Am. Chem. Soc. **62**, 1382 (1940)
127. R. Speiser, H.L. Johnston, Trans. Am. Soc. Metals **42**, 283 (1950)
128. J.L. Margrave, in *Physicochemical Measurements at High Temperatures*, eds. by J.O'M. Bockris, J.L. White, J.D. Mackenzie (Butterworth, London, 1959), p. 225
129. M. Volmer, Z. Physik. Chem., 1931, Bodenstein Festb., 863 (as quoted in Metallurgical Thermochemistry by O. Kubaschewski, and C. B. Alcock, [11])
130. P. Scardal, V. Piacente, D. Ferro, J. Mater. Sci. Lett. **9**, 1197 (1990)
131. G. Chirulli, D. Ferro, V. Piacente, Thermochim. Acta **62**(2–3), 171 (1983)
132. D. Ferro, V. Piacente, B. Nappi, J. Chem. Engg. Data **33**, 3 (1988)
133. J.B. Darby, K.M. Myles, Met. Trans. **3B**, 653 (1972)
134. P.J. Desre, D.T. Hawkins, R. Hultgren, Trans. Met. Soc. AIME **242**, 1231 (1968)
135. P.J. Desre, E. Bonier, J. de Chimie Physique **64**(9), 1243 (1967)
136. B. Brunetti, D. Gozzi, M. Lervolino, A. Latini, V. Piacente, J. Chem. Engg. Data **52**, 1394 (2007)

Chapter 7
Electrochemical Technique

Measurement of emf of a suitable electrochemical cell is one of the most useful methods of obtaining thermodynamic data. In fact this technique is one of the oldest techniques used for this purpose. Based on the combined statement of the first and second laws of thermodynamics the change in free energy accompanying a reversible reaction at constant temperature and pressure, is related to the maximum useful work of the system and is expressed as follows:

$$\Delta G = -\partial w \tag{7.1}$$

where w is the maximum useful work, other than mechanical work which the system is capable of performing. For a finite reversible process, integration of Eq. 7.1, at constant temperature and pressure yields,

$$\Delta G = -w \tag{7.2}$$

The work may be available in electrical form if the system comprises a suitable galvanic cell. In this case the maximum useful available work (w), following a particular chemical reaction is equivalent to the product of potential drop across the cell and the quantity of electricity passing through the cell. Thus for a reaction involving one mole of a substance, $w = nFE$, where E (volt) is the potential drop across the cell, n the number of electrons transferred per mole of the substance, F is the charge on 1 g equivalent (96,494 coulombs per mole).

It is now well established that if the electrochemical cell satisfies the condition of reversibility, reactions at the electrodes are well defined and the electrolyte conducts exclusively through ions, the change in free energy of a reaction on passing one Faraday of electricity is related to the emf of the cell by the expression:

$$\Delta G = -nFE \tag{7.3}$$

M. Shamsuddin, *Thermodynamic Measurement Techniques*, The Minerals, Metals & Materials Series, https://doi.org/10.1007/978-3-031-47118-6_7

Since the developments in electronic circuitry has enabled us to measure the emf very accurately without disturbing the equilibrium, the electrochemical measurements give accurate data about the free energy change associated with the chemical reactions and partial molar free energy of a particular component in solutions. By measuring the temperature coefficient of emf of a cell the information regarding entropy and enthalpy changes can be obtained.

7.1 Classification of Electrochemical Cells

Every galvanic cell consists of two electrodes (anode and cathode) and an electrolyte for conduction of ions. Depending on the overall chemical reaction the cells are categorized into displacement, formation and concentration cells.

7.1.1 *Displacement Cells*

Daniel cell made up of zinc electrode in zinc sulfate solution and a copper electrode in copper sulfate solution is demonstrated in the following manner:

$$Zn/ZnSO_4 \text{ solution} : CuSO_4 \text{ solution}/Cu \tag{7.I}$$

The two solutions are usually separated by means of a porous partition. Neither metal is attacked until the electrodes are connected and a current is allowed to flow. The extent of chemical reaction occurring is proportional to the quantity of electricity passing. Cell reaction is written as:

The electrode on the left hand side is a source of electrons (anodic reaction):

$$Zn = Zn^{2+} + 2e^- \tag{7.4}$$

While these electrons are consumed on the right hand electrode (cathodic reaction):

$$Cu^{2+} + 2e^- = Cu \tag{7.5}$$

The overall cell reaction may be written as:

$$Zn + Cu^{2+} = Zn^{2+} + Cu$$

or

$$Zn + CuSO_4 = ZnSO_4 + Cu \tag{7.6}$$

The cell emf depends on the concentration of zinc and copper ions in the solutions.

Similarly reactions for the cell:

$$Zn/ZnCl_2(aq) : KCl\ (aq)AgCl\ (s)/Ag \tag{7.II}$$

may be written as:

On the left hand electrode:

$$Zn = Zn^{2+} + 2e^- \tag{7.7}$$

On the right hand electrode:

$$AgCl = Ag^+ + Cl^- \quad (Ag^+ + e^- = Ag) \tag{7.8}$$

The net reaction is

$$AgCl + e^- = Ag + Cl^- \tag{7.9}$$

Hence, the overall cell reaction on passage of two Faradays is then

$$Zn + 2\ AgCl\ \ (s) = Zn^{2+} + 2Cl^- + 2\ Ag$$

or

$$Zn + 2\ AgCl\ \ (s) = ZnCl_2 + 2\ Ag \tag{7.10}$$

A special case of this type arises when both electrodes are of the same metal:

$$Ag/AgCl(s)KCl(aq) : AgNO_3(aq)/Ag \tag{7.III}$$

By convention the reaction at the left hand electrode is the reverse of that at the right hand electrode of the previous cell: $Ag + Cl^- = AgCl + e^-$.

And on the right hand electrode the reaction is $Ag^+ + e^- = Ag$

Hence, the net reaction in the cell on the passage of one Faraday is represented as:

$$Ag^+ + Cl^- = AgCl \tag{7.11}$$

7.1.2 Formation Cell

This type of cell can be used to determine the free energy of formation of a compound from its constituent elements. The free energy of formation of the compound, AB from its constituents, A and B can be obtained by measuring the open circuit emf (E) of the cell:

$$-\,\mathrm{A}\ (s,l)/\mathrm{AX}/\mathrm{AB}\ (s)+\mathrm{B}\ (s,l) \tag{7.IV}$$

Where A and (AB + B) are the two electrodes and AX is the electrolyte. The two half-cell reactions can be written as:

(i) anodic reaction at the left hand electrode:

$$\mathrm{A}=\mathrm{A}^{+}+\mathrm{e}^{-} \tag{7.12}$$

(ii) cathodic reaction at the right hand electrode:

$$\mathrm{B}+\mathrm{e}^{-}=\mathrm{B}^{-} \tag{7.13}$$

The overall cell reaction is

$$\mathrm{A}+\mathrm{B}=\mathrm{AB} \tag{7.14}$$

The free energy of formation of AB is given by Eq. 7.3: $\Delta G = -nFE$

By convention cell reaction involves passage of positive ions through the cell from left to right and negative ions from right to the left.

Since the electrons are generated at the electrode located on the left hand side, it will be negatively charged whereas the electrons being consumed on the right hand electrode will be positively charged. With these conventions the open circuit emf (*E*) generated across the cell will be positive.

7.1.3 Concentration Cell

This type of cell is used to determine the partial molar free energy of mixing of the component *i*, $\overline{G}_i$ which is related to activity by the expression:

$$\overline{G}_i = -nFE = RT\ln a_i \tag{7.15}$$

For example, cell of the type:

$$-\,A(l)/AX/A - B - C\ (l)+ \tag{7.V}$$

where pure A and alloy A–B–C are the two electrodes and AX is an electrolyte. The two half cell reactions may be written as

Anodic reaction on the left hand electrode:

$$A = A^{+} + e^{-} \tag{7.16}$$

Cathodic reaction on the right hand electrode:

$$(A\text{–}B\text{–}C)^{+} + e^{-} = A\text{–}B\text{–}C \tag{7.17}$$

The overall cell reaction for dissolution or mixing of pure A in the alloy, A–B–C is given as:

$$A = (A\text{–}B\text{–}C)\ \ (l) \tag{7.18}$$

7.2 Theoretical Considerations and Criteria of Reversibility

Although the above cells appear to be simple and straight forward there are several experimental difficulties which have to be overcome before one can rely on the correct emf to get accurate thermodynamic data. This needs very careful planning keeping in mind a number of theoretical considerations [1–8]. Even if the electrodes are in equilibrium with the electrolyte, one has to make sure that the activity measured is not only the activity at the surface of the multicomponent alloy. In order to avoid this problem one must choose systems in which the diffusion rate of the metal in the alloy is very high. The activity gradient can be minimized by operating the cell at temperatures 0.3–0.4 times the melting point of the metals and alloys forming the electrodes [1].

The emf of the above cells must be reversible and true representative of the correct cell reaction. To ascertain the correct and accurate value of the emf the following criteria of reversibility must be applied on the cell:

(i) Time independent emf at a constant temperature i.e. the cell emf must be stable for fairly long time at a particular temperature.
(ii) Reproducibility of the emf, irrespective of whether it is approached from the higher or lower temperature side.
(iii) Recoverability of the same emf after passage of a small amount of impressed current through the cell in either direction.

The testing of the above criteria of reversibility is usually carried out by disturbing the equilibrium by passing Faradic current in either direction or changing the temperature to a higher or to a lower value, and observing that the original emf value is obtained soon after removing the disturbance. While measuring the equilibrium emf values one must use electrometer with input impedance many orders of magnitude higher than that of the galvanic cell. Although the random variation of emf is tolerable because it would only affect the mean deviation from the equilibrium value, the variation in emf continuously in one direction is a sure sign of ideal behavior of a cell even if the variation is only couple of millivolts in one day.

Another important requirement for the satisfactory performance of a cell is that the value of n in Eq. 7.3 must be unequivocally defined. The participating ion should be present in one well-defined valence state. Presence of oxidizing and / or reducing agent in the cell must be avoided. If necessary, one should perform separate experiment to confirm the value of n. It is important to note that during the slag-metal reaction involving silicon transfer, Toporischev et al. [9] reported that the reaction: $Si = Si^{2+} + 2e^-$, was more probable at the slag-metal interface instead of the general reaction: $Si = Si^{4+} + 4e^-$. Occurrence of the former reaction has been confirmed by actually measuring the silicon transferred in a suitable electrochemical cell [10] by passing known current.

Similarly one must have complete knowledge regarding the conductivity of the electrolyte. The electrolyte must show exclusively ionic conduction. Significant electronic conductivity of the electrolyte would give lower values of emf because of the short circuiting effect. In addition to this, electrodes may become polarized. Herzog and Klemm [11] reported that the transport number of electrons in the cell: W/Cd/ $CdCl_2$/Cd-Bi/W having molten $CdCl_2$ in contact with Cd, was found to be nearly 0.12. This transport number gave inaccurate value of the emf. Hence, the activity data in Cd-Bi system based on this cell emf was wrong. This difficulty was overcome by dissolving 3–5% $CdCl_2$ in the solvent melt of LiCl-KCl. Fused salts which dissolve appreciable quantities of metal, may be expected to depart widely from purely ionic behavior.

Before planning to construct a galvanic cell it is advisable to make sure from the available thermodynamic data that no reaction other than the electrochemical reaction of interest is feasible. Particularly the following exchange reaction replacing A^+ ions from the electrolyte AX by B or C must be avoided:

$$\text{AX (electrolyte)} + \text{B or C (in alloy)} = \text{BX (or CX)} + \text{A (alloy)} \tag{7.19}$$

This would limit the choice of electrolyte which should be far more stable than compound of the element(s) in contact with it. In order to avoid the exchange reaction and for precise measurement of emf, metal A, should be less noble than alloy components B and C. A qualitative measure of the relative nobility of various metals is given by their position in the electromotive series. Further it can be estimated by the difference of the heat of formation and free energy of formation of the corresponding salts. Wagner [2] has discussed such exchange reaction between alloys and molten oxides, sulfides and chlorides. Due to the displacement reaction in case of single phase alloy the concentration of A in the alloy is increased and the concentration of A at the right hand electrode is decreased. Both changes tend to decrease the emf of the cells (7.IV) and (7.V) below a normal value. Wagner and Werner [3] have estimated the magnitude of error due to the exchange reaction in galvanic cells and tested the basic assumption for the estimate by measuring the emf with certain alloy systems.

No thermal gradient should exist in the cell. If the dimensions of the electrochemical cell are large or if the furnace is not properly constructed to give uniform heating zone it is possible to have appreciable temperature in-homogeneities throughout the cell. This would give rise to a thermoelectric emf for which the correction becomes necessary. These errors are more pronounced if there are several junctions of dissimilar metals. Precise temperature control as well as ensuring the uniformity of temperature at least in the region occupied by the cell becomes essential for accurate measurements.

Transfer of material from one part of the cell to another by a flow of current, by diffusion or by physical mixing can cause the cell to have "mixed potential" due to local shifts in electrolyte composition. This transfer can also change conditions in the cell by altering the total or surface composition of electrodes. In the cell: Ag/AgCl (l) /Pt, $Cl_2(g)$, Grube and Rau [4] used glass as a solid intermediate electrolyte to minimize the transfer of chlorine to the left hand electrode:

$$Ag(s)/AgCl(l)/glass/AgCl(l)/graphite, Cl_2(g) \qquad (7.VI)$$

Although use of various types of diaphragms [12] have been suggested, the expression for the diffusion potential becomes complicated. The method of calculation for such potentials has been illustrated elsewhere [5–8].

Theoretically the lead wires must be identical. In case of dissimilar metals/ materials, a correction for compensating the extra emf between them must be applied. The correction factor must be determined in a separate experiment without electrolyte in the temperature range of operation. The leads should be insoluble in the electrode materials to avoid altering the composition. A liquid metal with high surface tension may not have good contact with the lead and hence an error of several millivolts in the cell emf may result.

The presence of concentration gradient or liquid junction potential in the electrolyte can alter the cell emf because of the irreversible nature of diffusion occurring in the electrolyte. Hence the Daniel cell of the type Pb/$PbCl_2$: AgCl /Ag, cannot be used to obtain accurate thermodynamic data because of the uncertain magnitude of the liquid junction potentials. This may vary from a few millivolts to a considerable value depending on the salts of similar and dissimilar metals. For example, values up to 200 mV have been reported [13] for the junction between KCl and $CdCl_2$.

7.3 Eelectrolytes

Choice of an electrolyte for determining thermodynamic properties of systems is a very tricky problem and need careful considerations and planning. In fact one can say that one of the major limiting factors for the electrochemical technique is the selection of an electrolyte. The available electrolytes can be classified into three categories: aqueous, fused or molten salt and solid electrolytes.

7.3.1 Aqueous Electrolyte

Use of aqueous electrolytes based on water as a solvent is limited to low temperature applications only. Goates et al. [14] have measured the thermodynamic properties of HgS at 298 K making use of the following cells:

$$\mathrm{Hg + HgS/H_2S\ (atm)/HCl\ (0.1\ molal)\ H_2\ (atm)\ Pt} \tag{7.VII}$$

and

$$\mathrm{Hg + HgS/S^{-2}/normal\ calomel} \tag{7.VIII}$$

Kutsenok and Geiderikh [15] have studied In-Sb system at 298 K using aqueous solution of $In(NO_3)_3$. But complications due to liquid junction potential and slow attainment of equilibrium make it sometimes difficult to obtain reliable results. Thus aqueous electrolyte is not suitable for measurements at higher temperature (e.g. metallurgical systems) because of evaporation of water. The method has been extended to the use of organic solvents to increase the operating temperature to a limited extent. Terpilowski et al. [16] have employed the following cell to determine the thermodynamic properties of Tl-Te system in the temperature 398–468 K:

$$\mathrm{Tl/glycerol + NaCl + TlCl/Tl\text{-}Te} \tag{7.IX}$$

In a similar way organic electrolytes with suitable conducting cations have been used to investigate the thermodynamic properties of HgSe [17], HgTe [18], TlSe [19] and ZnSe [20] by Terpilowski and coworkers. Vasil'ev and coworkers have studied Bi-Se [21], Tl-S [22], Tl-Se [23] and Tl-Te [24] systems. Babanly and coworkers have extended the application of glycerol with KCl + TlCl in the study of ternary systems, Tl-Ge-Se [25] and Tl-Bi-Se [26]· Although the results based on these electrolytes seem to be consistent, the mode of conduction in the electrolyte is not very clear.

7.3.2 Fused or Molten Salt Electrolytes

Prior to 1960 most of the electrochemical measurements were based on fused salts electrolytes. The choice of the electrolyte depends entirely on the system to be investigated. However some general remarks are made here. The electrolytes for formation and concentration cells normally contain only a few percent of the compound of the least noble metal in the system to be investigated. The solvent compounds in the electrolyte have a common anion and they must be appreciably more stable than the compound of the less noble metals of the cell to avoid exchange reactions. The auxiliary electrolytes help in reducing the tendency of side reactions and also minimize the liquid junction potentials.

These auxiliary electrolytes (solvents) can be divided into two groups: halides and oxides. Following points should be considered while selecting a solvent:

(i) There should not be any polymeric anion formation between salts in the melt.
(ii) Extreme care should be taken to avoid the presence of moisture and oxygen in the electrolyte. The melt should be dehydrated and prepared under vacuum or dry argon, free from oxygen.
(iii) The electrolyte should have low melting and high boiling points and low vapor pressure and must be chemically stable and neutral towards the system to be investigated.

7.3.2.1 Fused Halides as Solvent

From literature it is clear that the mixtures of fused LiCl-KCl, LiCl-KCl-NaCl, KCl-NaCl and KCl-NaCl-$ZnCl_2$ have been most extensively used for dissolving other chlorides without exhibiting liquid junction potential and electronic conduction. Chlorides of many metals especially the transition metals exhibit appreciable electronic conductivity, especially in contact with an element of its cation. When this is dissolved in the above mixture of monovalent chlorides the electronic conductivity is appreciably suppressed. Ordinarily, $AlCl_3$ in contact with aluminum exhibits electronic conductivity due to the dissolution of aluminum but when it is dissolved in KCl-NaCl mixture solubility of the electrode material becomes negligibly small. Wilder and Elliott [27, 28] have used this electrolyte to determine the activity of aluminum in Al-Ag and Al-Bi-Pb systems in the temperature range: 700–980 °C. For large solubility of electrodes in the electrolyte and for the exchange reaction between metals of the alloy, whose difference in electrode potential is not very large Wagner and Warner [3] have estimated the errors in the measurement of activity.

Major difficulty in the use of chloride electrolytes at higher temperatures is that due to slightest contamination with water vapor they form volatile complexes in addition to reacting with the container. Silica and glass have been used up to 1100 °C and 600 °C, respectively by completely removing water vapor from the electrolyte. Table 7.1 lists the electrolyte mixture used so far for determination of thermodynamic properties of various systems [27–108]. This table only shows the type of

Table 7.1 Fused salt electrolytes and the related systems investigated therewith

Auxiliary electrolyte	Systems studied	Temperature range, K	Reference
LiCl-KCl	Cd-Sb	663–713	[29]
	CdTe, PbTe, ZnTe		[30]
	Ga-Sb		[31]
	PbS, PbSe, PbTe		[32]
	Ag_2Se		[33]
	Bi-Te		[34]
	Sn-Te		[35]
	Tl-Bi		[36]
	Ni-Te		[37]
	PbTe-PbSe		[38–40]
	ZnTe-CdTe		[41, 42]
	CdTe-HgTe		[43]
	CdTe-CdSe		[44, 45]
	CdTe		[46]
	CdSe		[47]
	ZnTe		[48]
	ZnSe		[49]
	Bi-Sn		[50]
	Ag-In		[51]
	Fe-Sb		[52]
	Ga-Cd		[53]
	In-Tl		[54]
	Ag-Tl		[55]
	Pb-Tl		[56]
	Sb-Te		[57]
	As-Cd-Zn		[58]
	Bi-In-Sb		[59]
	Cd-In-Sb		[60]
	Ge-Sb-Zn		[61]
	In-Pb-Sb		[62]
	Cd-Tl-Te		[63]
	Pb-Au-Bi		[64]
	La-U-Ga-X		[65]
	Gd-Al		[66]
	Ag-Cd-Bi		[67]
	Cu-Dy intermetallics		[68]
	Tl-In-Te		[69]
	Sc-Al intermetallic		[70]
	Sn-Ag-Bi		[71]
	Gd-Bi		[72]
	Zn-In-Cd		[73]
	Ag-Bi-In		[74]
	Al-Cu-Zn		[75]

(continued)

Table 7.1 (continued)

Auxiliary electrolyte	Systems studied	Temperature range, K	Reference
	Ag-Sb-Zn		[76]
	Ag-Sb-Sn		[77]
	Al-Sn-Zn		[78]
	La-Ga-Al/U-Ga-Al		[79]
	Cd-Pr		[80]
	Cu-Ga/Ge/Sn-Zn	1170–1270	[81]
NaCl-KCl	Al-Ag	970–1250	[27]
	Zn-In		[82]
	Zn-Sn, Zn-Cd		[83]
	Zn-Bi-Pb		[84]
	Zn-Cd-Sn		[85]
	Zn-Pb-Cd		[86]
	Zn-Pb-In		[87]
	Zn-Bi-Sn-Cd		[88]
	Bi-In	950–1200	[89]
	Bi-Pb		[90]
	Sb-Zn-Cd	823–1023	[91]
	Al-Au	973–1153	[92]
	Ag_2S, PbS	770–1070	[93]
	Sn-Zn-Cd		[94]
KCl-NaCl-$ZnCl_2$	PbSe	700–850	[95]
	ZnTe		[96]
	Se-Te		[97]
	Pb-As		[98]
NaCl-$AlCl_3$	Ag-Al, Al-Bi-Pb	870–1170	[28]
	Al-Zn, Al-Ga	870–1170	[99]
	Al-Sb, Al-Au	930–1420	[100]
NaCl-CuCl	Cu-Pb	1370–1470	[101]
$CaCl_2$-$MgCl_2$	Mg-Al	970–1070	[102]
$CaCl_2$-CaC_2	Graphite in austenite	1070–1270	[103]
$SnCl_2$ (l)	Sn-Cu, Sn-Ag,	520–570	[104]
	Sn-Au, Sn-Hg		
$SbCl_3$-KCl	HgTe	340–410	[105]
LiBr-KBr	Bi-In	671–773	[106]
	In-Sb	637–873	[107]
	In-Pb	773–873	[108]
CaO-SiO_2	Fe-Si, Co-Si	1720–1885	[109]
CaO-P_2O_5	Phosphides	–	[110]
CaO-SiO_2-B_2O_3	Ag-Si	1670	[111]
Na_2O-B_2O_3	H-O		[112]
PbO-SiO_2	Pb-O		[113]
PbO-B_2O_3	Pb-O		[114]
PbO-Na_2O-P_2O_5	Pb-O		[115]
NiO-Na_2O-K_2O-SiO_2	Ni-O		[116]

electrolytes which can be used for various systems. Advantageous characteristics of LiCl-KCl and LiCl-KCl-NaCl solvents are good thermal conductivity, wide span of electrolytic decomposition, high conductivity and fluidity, commercial availability and absence of strongly acidic or basic properties. Further, the melting point is low and its chemical reactivity is such that pyrex/corning glass can be used as material of construction.

7.3.2.2 Oxide Melts as Solvents

Another group of molten electrolytes which are commonly used are mixture of oxide melt. Schwerdtfeger and Engell [109] determined the free energy of formation of SiO_2 using the following cell:

$$\mathrm{Pt},\ \ \mathrm{Si}(l)/\mathrm{CaO.SiO_2}\ \ (l).\mathrm{SiO_2}(s)/\mathrm{O_2}(g),\mathrm{Pt} \tag{7.X}$$

in the temperature range 1450–1615 °C. This is a formation type cell where electrolyte is saturated with SiO_2. When four Faradays are transported through the cell the overall cell reaction is represented as:

$$\mathrm{Si}\ \ (l) + \mathrm{O_2}\ \ (g) = \mathrm{SiO_2}\,(s) \tag{7.20}$$

If Si in the electrode of the above cell is substituted by silicon alloys like Fe-Si or Co-Si then the activity of silicon can be obtained in terms of emf of the cell by the expression:

$$lna_{Si} = \frac{4EF}{RT} + \Delta G^{o}_{SiO_2}/RT \tag{7.21}$$

They have also used $CaO.P_2O_5$ as an electrolyte for studies of phosphides [110]. Tupkary [111] determined the activity of silicon in Ag-Si alloys at 1150 °C using $CaO\text{-}B_2O_3\text{-}SiO_2$ melt in a cell of the type:

$$\mathrm{Pt/Ag\text{-}Si/CaO\text{-}B_2O_3\text{-}SiO_2/Pt},\ \ \mathrm{O_2}\ \ (g) \tag{7.XI}$$

Similarly application of a number of oxide melts is available in literature, for example, $Na_2O\text{-}B_2O_3$ [112], $PbO\text{-}SiO_2$ [113], $PbO\text{-}B_2O_3$ [114], $PbO\text{-}Na_2O\text{-}P_2O_5$ [115] and $NiO\text{-}Na_2O\text{-}K_2O\text{-}SiO_2$ [116]. However, the result in all the cases is not very promising because of the corrosive nature of the melt and reaction with the container material.

Use of molten oxide electrolyte reduced extensively since Kiukkola and Wagner [117] introduced the utility of oxide solid electrolytes for thermodynamic measurements in 1957.

7.3.3 Solid Electrolytes

Solid electrolytes have some distinct advantages over fused salts. Due to the refractory nature of many solid electrolytes these can be used at much higher temperatures. The construction and operation of a galvanic cell with solid electrolyte is relatively less complicated. Often electrolyte itself acts as a container especially if the metals and alloys are in molten state. There are, however serious limitations restricting the use of solid electrolyte. The main one is the occurrence of electronic conductivity and the subsequent polarization due to the slow diffusion. Only solids exhibiting predominantly ionic conductivity ($t_{ion} \geq 0.99$) over a useful range of temperature and chemical potential are considered to be a good electrolyte. Thus it becomes essential to know the extent of contribution of ionic and electronic conductivity in a solid under the changing experimental conditions to consider their use in the formation of a galvanic cell.

A predominantly ionic conduction in a solid can be obtained if the concentration of ionic defects is at least 1000 times larger than the electronic defects [118]. Thus a typical solid electrolyte is a pure solid having a band gap energy of larger than 3 eV, or a solid solution in which the number of ionic defects has been greatly increased by doping. Solid halides such as NaCl, KCl and AgCl are examples of pure solid electrolytes. Calcia stabilized zirconia and yttria doped thoria are examples of solid electrolytes in which the number of ionic defects has been greatly increased by doping. A polycrystalline solid should have a density of ≥95% of the theoretical value to be used as a solid electrolyte.

In order to avoid complications in the measurement of emf of a galvanic cell based on solid electrolytes, it is not only essential that the solid exhibits predominantly ionic conductivity but also it should have high total conductivity. The ionic conductivity is dependent on temperature along with other factors. These solid electrolytes may be classified in three different categories on the basis of their conductivity. There are excellent reviews [119–124] on solid electrolytes and their applications. Table 7.2 lists known solid electrolytes [119, 125]. A brief account on these three groups of solid electrolytes is given below:

7.3.3.1 Halide Solid Electrolytes

The first group includes electrolytes based on silver halides and their modifications, such as AgI, $RbAg_4I_5$, and $Ag_6I_4WO_4$. These compounds exhibit remarkably high conductivity at low temperatures. α- AgI has a very high conductivity in the temperature range of 146–556 °C. $RbAg_4I_5$ has very high conductivity but its thermodynamic stability is poor. $Ag_6I_4WO_4$ is very useful below 293 °C.

Lead chloride and fluorides of barium, stroncium, calcium, and magnesium exhibit ionic conduction and thus serve as solid electrolytes. These compounds conduct both anions as well as cations and are good ionic conductors above 500 °C. However, in fluorides the conduction is mainly due to F^- ions [126].

Table 7.2 List of known solid electrolytes [119, 125]

Conducting ion	Compounds
Br^-	$BaBr_2$, KBr, NaBr and $PbBr_2$
Cl^-	$BaCl_2$, KCl, NaCl, $PbCl_2$ and $SrCl_2$
F^-	BaF_2, CaF_2, MgF_2, NaF, Na_3AlF_6, PbF_2 and SrF_2
I^-	KI and PbI_2
O^{2-}	$Zr_{1-x}M_x^{2+}O_{2-x}, Zr_{1-x}M_x^{3+}O_{2-(x/2)}, Th_{1-x}M_x^{2+}O_{2-x}, Th_{1-x}M_x^{3+}O_{2-(x/2)};$ $Hf_{1-x}M_x^{2+}O_{2-x}, Hf_{1-x}M_x^{3+}O_{2-(x/2)}, La_2O_3(CaO); B_2O_3, 3B_2O_3.WO_3,$ $LaAlO_3(CaO \text{ or } BaO), CaTiO_3(Al_2O_3), SrTiO_3(Al_2O_3), PbO,$ $Bi_2O_3 - SrO, Ce_{1-x}M_x^{2+}O_{2-x}, Ce_{1-x}M_x^{2+}O_{2-(x/2)}; SrZrO_3(Sc_2O_3);$ $CaZrO_3, Sm_2O_3, Gd_2O_3, Dy_2O_3, \text{and } Y_2O_3.$
S^{2-}	$CaS - Y_2S_3$
Al^{3+}	Al_2O_3 (purox)
NH_4^+	$(NH_4)_2O.11Al_2O_3$
Be^{2+}	BeO
Cu^+	β and γ CuBr, CuCl, β CuI; $3[C_6H_{12}N_2.2(H \text{ or } C_2H_5)$ Br or Cl or I] 17Cu, $HgCu_2I_4$
Li^+	LiH;Li_2SO_4, Li_2WO_4, $LiAgSO_4$, $LiNaSO_4$, $LiNaSO_4$, $Li_{1.72}$ $Mg_{0.14}SO_4$; Li_4SiO_4-Li_3PO_4, $LiAlCl_4$, Li_5AlO_4, Li_5CaO_4, Li_6ZnO_4 LiI, LiI.NH_4I and β $Li_2O.11Al_2O_3$.
Mg^{2+}	MgO
K^+	KBr, KCl, KI and β $1.3K_2O.0.2Li_2O.10Al_2O_3$
Rb^+	β $Rb_2O.11Al_2O_3$
Ag^+	α and β AgI, AgCl, AgBr, Ag_3SBr, Ag_3SI, Ag_2HgI_4, KAg_4I_5, $RbAg_4I_5$, $NH_4Ag_4I_5$; $Ag_7I_4PO_4$, $Ag_{19}I_{15}P_2O_7$, $Ag_6I_4CrO_4$, $Ag_6I_4MoO_4$, $Ag_6I_4WO_4$,$Ag_7I_4AsO_4$,$Ag_7I_4VO_4$,$(C_5H_5NH)Ag_5I_6$; $(C_5H_5NH)_5Ag_{18}I_{23}$; $Ag_5I_3SO_4$, β$Ag_2O.11Al_2O_3$;QAg_6I_7 and $Q_2Ag_{13}I_{15}$ (where Q is a tetramethyl or tetraethyl radical)
Na^+	NaF, NaCl, NaBr, β $Na_2O.11Al_2O_3$; β" $Na_2O.5Al_2O_3$ and β" $Na_2O.MgO.5Al_2O_3$
Tl^+	β $Tl_2O.11Al_2O_3$

Electronic conduction is almost negligible even up to the decomposition pressure [125]. CaF_2 functions satisfactorily in the temperature range: 400–1000 °C. Wagner [127] has estimated the limit of calcium activity in CaF_2 above which $t_{ion} \geq 0.99$ to be $a_{Ca} \leq 10^{-5}$ at 600 °C and $\leq 6 \times 10^{-6}$ at 840 °C.

7.3.3.2 β-Alumina

The second group of solid electrolytes are based on β alumina type structure with a general formula M_2. n Al_2O_3, where M is either Li, Na, K, Rb, Ag or Tl and n varies from 5 to 11. These compounds have hexagonal unit cell in which M ions are situated exclusively in planes which contain in loose packing an equal number of M and oxygen ions. These planes are perpendicular to c-axis with aluminum ions at both octahedral and tetrahedral positions. This particular structure imparts high M ion conduction in the direction perpendicular to c-axis [118]. The specific conductivity of sodium and silver β aluminas vary between 10^{-1} and 10^{-2} $ohm^{-1}.cm^{-1}$ at 25 °C.

7.3.3.3 Oxide Solid Electrolyte

Zirconia and thoria based electrolytes are most often used for high temperature studies. Zirconia or thoria doped with 15 mole % dopant shows maximum ionic conductivity [125]. These solids are useful electrolytes because the electronic contribution to the total conductivity is <1% over a wide range of temperature and pressure.

Conduction Mechanisms in the Functioning of Oxide Solid Electrolytes In the late nineteen fifties, Wagner [117] showed the utility of solid electrolytes for thermodynamic measurements at high temperature. Initially solid glass and halide electrolytes were used, but it was established that oxide electrolytes were more promising. The most widely used solid electrolyte is calcia doped with zirconia. A large number of metallic and oxide systems have been studied since Wagner demonstrated the role of oxide electrolytes for thermodynamic measurements.

These solid oxide electrolytes are mainly ionic conductors. With solid oxides three types of conduction, namely, ionic or electrolytic conduction, electronic conduction (n-type) and electron hole conduction (p-type) occur. In order to visualize the concept of ionic conductivity consider an oxide crystal MO where M is divalent. In this oxide some R^+, a monovalent ion is dissolved. Thus for every R^+ replacing two M^{2+} ions there are two extra negative charges in the lattice. One way of accommodating this solute ion in the lattice of MO is to balance these two extra charges by creating oxygen ion vacancy. This phenomenon is represented in Model – A. At high temperature these oxygen vacancies can migrate under the influence of potential gradient. This results in a conduction which is predominantly ionic.

M^{2+}	O^{2-}	M^{2+}	O^{2-}	M^{2+}
O^{2-}	R^+	O^{2-}	M^{2+}	O^{2-}
M^{2+}	O^{2-}	M^{2+}	O^{2-}	M^{2+}
O^{2-}	M^{2+}		M^{2+}	O^{2-}
M^{2+}	O^{2-}	M^{2+}	O^{2-}	M^{2+}

Model – A

Ni^{3+}	O^{2-}	Ni^{2+}	O^{2-}	Ni^{2+}
O^{2-}	Li^+	O^{2-}	Ni^{2} + −	O^{2-}

Model – B

P-type or electron hole conduction is found with non-stoichiometric oxides e.g., FeO_{1-x}, NiO etc. Here charges can be balanced over the lattice by changing one valency of cations. The extra charge on cations will be moving in the lattice causing the flow of current. When manovalent Li^+ is added to NiO to compensate the charge more Ni^{3+} are produced resulting in the hole conduction as represented in Model –B.

The predominant ionic conduction is found with zirconia and thoria [120, 121]. When ZrO_2 is doped with divalent Ca^{2+}, to maintain electroneutrality in the lattice, oxygen ion vacancies are created. The same phenomenon occurs when thoria is doped with yttria (Y_2O_3). The electronic hole conduction is also function of oxygen pressure. At relatively low oxygen partial pressures, *n*-type or conduction due to electron results while at high pressures, *p*-type conduction results, which varies slightly with temperature. Ionic conductivity is observed with oxides like ZrO_2, ThO_2 etc.

Figure 7.1 shows the typical behavior of conductivity with varying partial pressure of oxygen. A-type behavior is shown by zirconia doped with CaO or Y_2O_3 (Fig. 7.1a). In this case electronic conductivity predominates at lower partial pressures of oxygen. Thoria doped with CaO or Y_2O_3 exhibits predominantly B-type behavior (Fig. 7.1b) where ionic conduction is found at lower oxygen partial pressures and at higher pressure range it is mainly electron hole conduction. The general shape of the conductivity curve will be according to Fig. 7.1c. The only

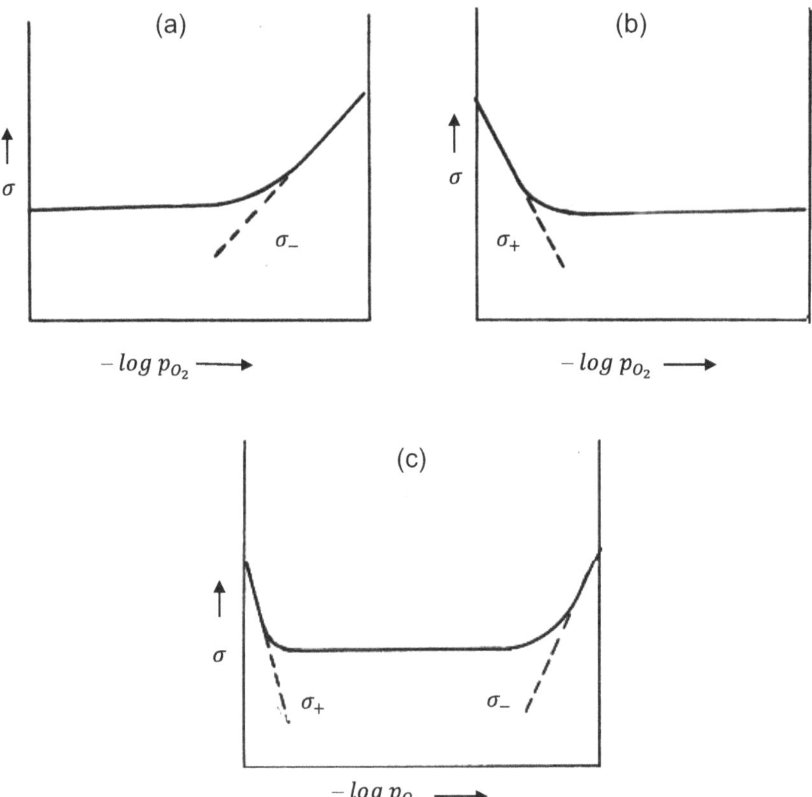

Fig. 7.1 Conductivity of different solid electrolytes as a function of partial pressure of oxygen

range in which conductivity is independent of oxygen partial pressure is the ionic conductivity range. In this range only the material can be used as a solid electrolyte. The commonly known oxide electrolytes are calcia-stabilized-zirconia (CSZ) and yttria-doped-thoria. The former exhibits ionic conduction between 1 and 10^{23} atmosphere at 1600 °C [128, 129] while the latter at the same temperature between 10^{-6} and 10^{-27} atmosphere [130]. At higher pressure thoria based electrolytes exhibit hole conduction.

Construction of a Galvanic Cell with Oxide Solid Electrolytes The schematic cell diagram is shown in Fig. 7.2. Generally, commercially available solid electrolytes, namely, zirconia doped with 15 mole% calcia or thoria doped with 7–8 mole % yttria is used. The ionic mobility changes considerably by solute addition. Figure 7.3 shows the range of oxygen pressures in which various oxide solid solutions can be used as solid electrolytes. In these pressure ranges the transference number 't_{ion}' is >0.99 which can be assumed to be unity for practical purposes. The other requirements of the cell are: (i) auxiliary contact electrode and (ii) standard electrode. Being inert to the cell atmosphere generally platinum is used as auxiliary electrode. Moreover it does not react with the electrolyte.

Metal and metal oxide mixture of a known oxygen partial pressure is employed to prepare the standard electrode. A few such mixtures with the equilibrium oxygen pressure at 1000 °C are listed in Table 7.3:

All the three types of cells, viz. formation, displacement and concentration can be constructed using solid electrolytes for determination of free energy of formation or reaction and activity. For example, free energy of formation of calcium chromate, $CaCrO_4$ has been obtained by measuring the open circuit emf of the following cell [131]:

$$\mathrm{Pt, O_2/CaO/CaF_2/Cr_2O_3 + CaCrO_4/Pt, O_2} \qquad (7.XII)$$

The cell reaction:

$$\mathrm{CaO + ½\, Cr_2O_3 + 3/4O_2 = CaCrO_4} \qquad (7.22)$$

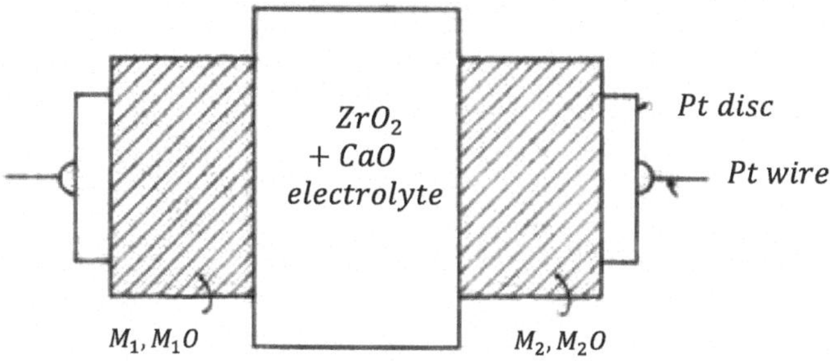

Fig. 7.2 Schematic diagram of the solid electrolyte cell

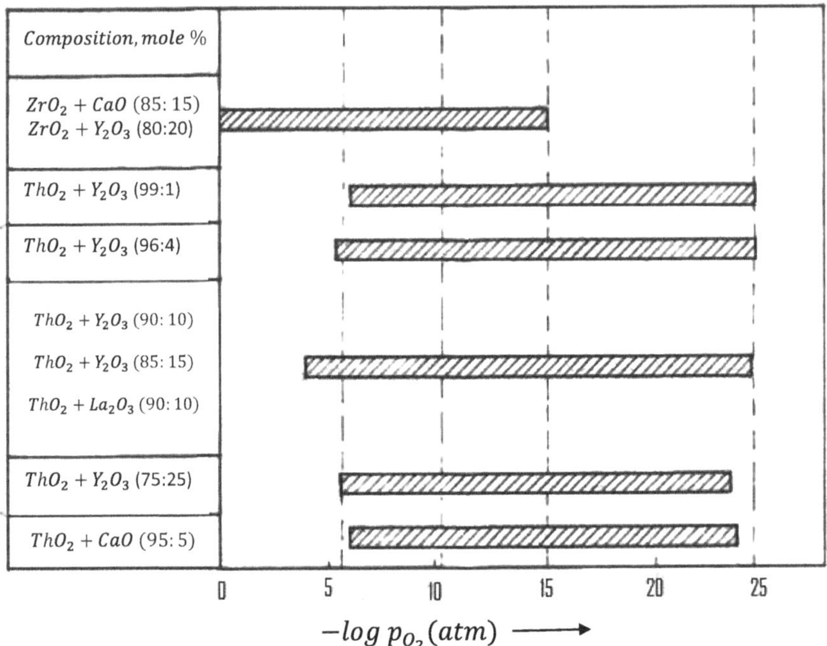

Fig. 7.3 Utility ranges of various solid electrolytes at 1273 K

Table 7.3 Metal and metal oxide mixtures of known oxygen partial pressures

System	Equilibrium p_{O_2} (atm)
Cu-Cu_2O	$10^{-6.3}$
Ni-NiO	$10^{-10.4}$
Fe-FeO	$10^{-14.9}$
Nb-NbO	$10^{-24.8}$

From Eq. 7.22 it is clear that the experiment requires maintenance of an atmosphere of pure and dry oxygen for successful establishment of the equilibrium.

Activity of silver in solid silver-tellurium alloys has been determined by Kiukkola and Wagner [117] using the cell:

$$\text{Pt/Ag/AgI/Ag-Te} \qquad (7.\text{XIII})$$

In the following cell based on calcia-stabilized-zirconia (CSZ):

$$\text{Pt, Fe/FeO/ZrO}_2\text{-CaO/NiO, Pt} \qquad (7.\text{XIV})$$

the virtual cell reaction involves displacement of oxygen ions from the right hand side to the left hand side, expressed as:

$$Fe + NiO = FeO + Ni \tag{7.23}$$

The cell (XIV) is thus a displacement type cell with co-existing electrodes [119, 122]. The emf for the above cell is given by the expression:

$$E = -(\Delta G^o{}_{FeO} - \Delta G^o{}_{NiO})/2\,F \tag{7.24}$$

Hence by measuring the open circuit E and from the value of the free energy of formation of FeO (or NiO), the free energy of formation of NiO (or FeO) can be obtained.

Activity of a component in alloys and oxide solid solutions can also be determined. For example, Rapp and Maak [132] determined the activity of copper in solid copper-nickel alloys by constructing the following cell:

$$Pt/Ni\text{-}NiO/ZrO_2\text{-}CaO/(Cu\text{-}Ni), NiO/Pt \tag{7.XV}$$

The open circuit emf of this cell is given by

$$E = -RT/2F \ \ln a_{Ni} \tag{7.25}$$

Kachhawaha et al. [133] have used the cell:

$$Pt/Ni\text{-}NiO/ZrO_2\text{-}CaO/(NiO\text{-}ZnO), Ni/Pt \tag{7.XVI}$$

to determine activity of nickel oxide in NiO-ZnO solid solutions.

Seetharaman and Abraham [134] have measured the activity of FeO in FeO-MnO solid solutions by setting up the cell based on thoria doped with yttria as the solid electrolyte:

$$Pt/Fe + (Fe, Mn)O/ThO_2\text{-}Y_2O_3/(Ni + NiO), Ni/Pt \tag{7.XVII}$$

Principle of Activity Measurement with Oxide Solid Electrolyte As oxide solid electrolyte conducts through the migration of O^{2-} ions principle of measurement of activity in metallic systems needs a brief discussion:

If we have an alloys, A–B, where oxide of A, AO is more stable than BO, i.e., ΔG^o (AO) < ΔG^o (BO) the activity of oxygen in A–B and AO would be different and will depend on the concentration of A in the alloy A–B. Thus following concentration type galvanic cell based on calcia-stabilized-zirconia may be constructed to determine the activity of A in the alloy, A–B:

$$Pt/AO, A/ZrO_2.CaO/(A\text{–}B), AO/Pt \tag{7.XVIII}$$

To formulate virtual cell reaction in the above cell one has to consider the migration of oxygen ions in zirconia and migration of electrons in the leads. Since oxide electrolytes work due to the difference in p_{O_2} on two sides, the half-cell reactions can be written as follows:

On the left hand side electrode with low p_{O_2} electrons are generated as per the reaction:

$$O^{2-} = \frac{1}{2}O_2(g) + 2e^- \quad \text{(hence a negative electrode)} \tag{7.26}$$

On the right hand electrode with high p_{O_2}these electrons are consumed according to the reaction:

$$\frac{1}{2}O_2(g) + 2e^- = O^{2-} \quad \text{(hence a positive electrode)} \tag{7.27}$$

By convention left hand side has low oxygen partial pressure.

Thus the overall cell reaction is:

$$O_2\left(p'_{O_2}\right) = O_2\left(p''_{O_2}\right) \tag{7.28}$$

Accordingly, the cell reaction may be formulated as the transfer of one g atom of oxygen from the right hand to the left hand side of the cell on passing two Faradays. Hence,

$$\mu'_O - \mu''_O = 2FE \tag{7.29}$$

Where μ'_O *and* μ''_O are the respective chemical potentials of oxygen on the right hand and left hand sides. Since AO of the same chemical potential is present on both the sides of the cell, the following reactions **apply for the chemical potentials of A, and O:**

On the right hand electrode:

$$\mu'_O + \mu''_A = \mu^o_{AO} \tag{7.30}$$

On the left hand electrode:

$$\mu'_A + \mu''_O = \mu^o_{AO} \tag{7.31}$$

From Eqs. 7.29, 7.30, and 7.31, we get:

$$\mu'_A - \mu''_A = -2FE \tag{7.32}$$

As pure A is present on the left hand side of the cell:

$$\mu''_A = \mu^o_A \text{ (standard state, i.e. unit activity)}$$

For the alloy electrode on the right hand side we have:

$$\mu'_A = \mu^o_A + RT\ ln a_A \tag{7.33}$$

Hence, from Eqs. 7.32 and 7.33, we can write:

$$RT\ \ ln a_{A\ (alloy)} = -2\ FE \tag{7.34}$$

From the variation of the activity coefficient of A with atom fraction, the activity coefficient of B can be calculated by Gibbs–Duhem integration.

7.4 Galvanic Cell Assembly

In this Section construction of a few selected galvanic cell assemblies will be discussed.

7.4.1 Galvanic Cell Assembly for Fused Salt Electrolytes

The cell assembly usually made of glass consisting of a simple H-shaped [135] tube or two beakers [136] connected by a salt bridge containing the electrolyte solution was employed while using aqueous or organic liquid electrolytes. Richards and Forbes [137] constructed a cell with several compartments to load a number of samples simultaneously. There was provision to introduce amalgams with special pipette.

A number of cell designs have been developed for molten or fused electrolytes. Reactivity of the fused salt causes problem in taking out lead wires for emf measurement. The leads may be sealed at the bottom of the electrode compartments made of glass. More often leads are taken out from top of the assembly through protection tubes of ceramic or glass. Graphite rod, tungsten, tantalum, iridium, platinum, molybdenum, iron and nichrome wires have been successfully used. Generally the cells are operated under the cover of an inert gas.

McAteer and Seltz [30], Weibks and Quadt [138] and Elliott and Chipman [139] have employed glass assembly of different designs for thermodynamic investigations of tellurides, alloys and binary/ternary cadmium solutions, respectively. Seigle, Cohen and Averbach [140] constructed a quartz assembly for studying solid alloys. The electrode was suspended by a gold lead wire in the liquid electrolyte contained in a nickel cup. At top of the assembly there were arrangements for evacuation and insertion of inert gas. Egan and Wiswall [141] used molybdenum and tantalum crucibles to hold the electrolyte and electrodes.

In order to protect the molten electrolyte (LiCl-KCl) from atmosphere, Oriani [1] fabricated the cell and the filling vessel into different compartments. During operation the vessel was filled with the salt mixture and interconnected with the cell. The entire assembly was connected to a vacuum system. The filling vessel and the cell were heated under vacuum in separately controlled furnaces. The salt mixture was maintained liquid at 500 °C for a long time. The electrode was transferred to the cell by first sealing off the vertically connected tube between the cell and the vacuum system. Pure helium was then admitted over the liquid in the filling vessel to force the liquid up and over into the cell. After transferring the electrolyte, more helium was admitted and the cell was sealed off. Following this method of preparation of electrolyte and electrode Oriani [1] operated a galvanic cell for nine months without any sign of attack on silica.

In the cell assembly used by Wilder and Elliott [27] the metal electrodes were contained in doubly recrystallized pure alumina crucibles. The large cup containing crucibles and sheathing for tantalum lead wires were of 97% vitreous alumina. Elliot and Lemons [142] and Belton and Rao [102] modified their cell assemblies according to the need of investigations. Thompson and Flengas [93] fabricated their apparatus in two compartments to determine the free energy of formation of PbS and Ag_2S using sulfur vapor electrode. The upper portion of the assembly was made of pyrex glass to contain sulfur whereas the lower portion containing fused salt was of quartz. In order to control the temperature of sulfur and the electrolyte independently the entire cell was located in a furnace divided into two compartments.

For determination of activity of silicon in molten Ag-Si alloys using moten oxide electrolyte Tupkary [111] employed a cell assembly consisting of a silica tube narrowed down at one end which was connected to a fine bore capillary to form a crucible. The Ag-Si alloy electrode under the cover of the slag was contained in the crucible. A tungsten lead wire in a silica sheath was inserted through the capillary for electrical connection to the molten metal electrode. Another electrode of oxygen was prepared by passing oxygen gas at one atmosphere pressure over a platinum wire sheathed in a silica tube.

Most of the cell designs discussed so far are complicated and needs excellent glass blowing facilities in the laboratory for smooth conduction of experiments. However with careful planning a simpler apparatus can be fabricated at places with limited facilities. One such cell assembly made of borosil glass, designed in our laboratory for determination of thermodynamic properties of binary and ternary alloys and chalcogenides, is shown in Fig. 7.4.

The assembly used for lead and zinc chalcogenides, CdTe, CdSe and HgTe had two lower limbs (6 mm diameter) below a tubular electrolyte compartment of 35 mm diameter. At the bottom of each limb a tungsten lead wire of 0.4 mm diameter and 20 cm length was sealed. One of the two limbs contained lead, zinc, cadmium or mercury as one electrode and another was meant for the electrode of the intermetallic compound under study. The top of the cell assembly was covered with a rubber cork

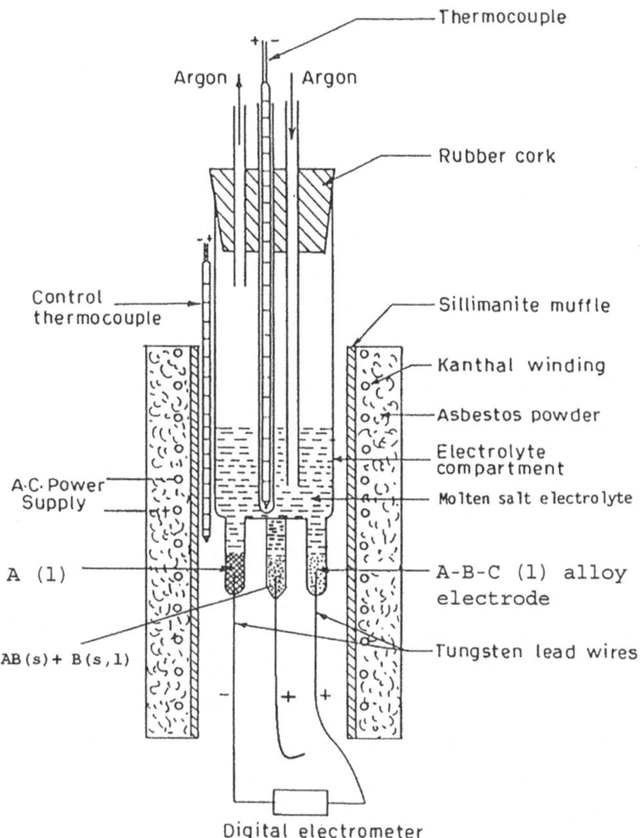

Fig. 7.4 Fused salt galvanic cell assembly

having three holes, meant for the passage of argon inlet tube, a thermocouple well (one end-closed borosil glass tube of 5 mm diameter) and an argon outlet tube. The thermocouple as well as the inlet tube reached up to the bottom of the cell assembly.

The cell used for activity measurement was made of 45 mm diameter borosil glass tube containing four or five lower limbs of 6 mm diameter each. In this case one limb contained pure metal A, and others had different compositions of solid solutions or alloy electrodes. Plan of the cell assembly is show in Fig. 7.5.

Shamsuddin and coworkers [32, 37–45] have used a eutectic melt of LiCl and KCl (43 wt% LiCl, eutectic point: 360 °C) with 5 wt% $PbCl_2$, $CdCl_2$ or $ZnCl_2$ as the fused salt electrolyte to determine free energy of formation of PbS, PbSe, PbTe, CdSe, CdTe, ZnSe, or ZnTe and activity of PbTe in PbTe-PbSe, CdTe in CdSe-CdTe and ZnTe in CdTe-ZnTe systems. For the determination of free energy of formation of HgTe [105] in the low temperature range the eutectic mixture of $SbCl_3$ and KCl (89.7 wt% $SbCl_3$, eutectic point: 55 °C) was used. Free energy of formation of PbSe [32] has also been determined by using eutectic melt of $ZnCl_2$-KCl-NaCl

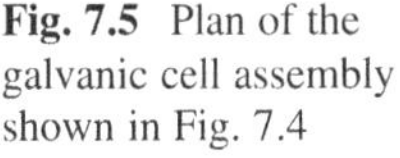

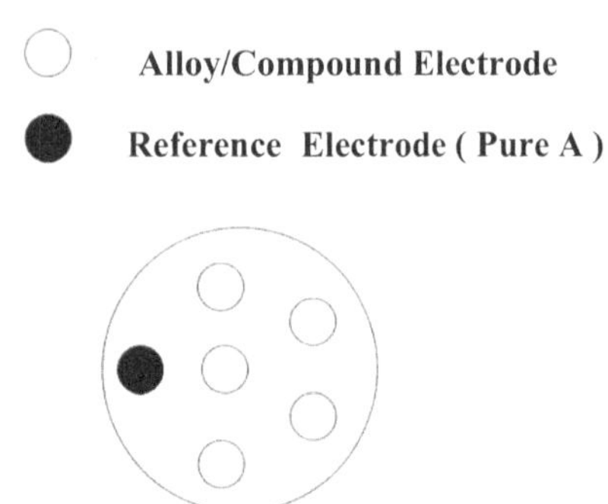

Fig. 7.5 Plan of the galvanic cell assembly shown in Fig. 7.4

with 5 wt% $PbCl_2$. LiCl-KCl fused melt containing small quantity of $GdCl_3$ or $LaCl_3$ or UCl_3 has been used for thermodynamic studies of Gd-Bi [72], La-Ga-Al [79] and U-Ga-Al [79] alloys, respectively.

7.4.2 Galvanic Cell Assembly for Solid Electrolytes

The schematic diagram of a cell assembly [125] is shown in Fig. 7.6. It consists of a one-end closed re-crystallized alumina tube A (450 mm long and 25 mm diameter). The open ends of this tube is closed tightly with two brass collars (C-1 and C-2) having neoprene ‘O’ ring in between them. The ‘O’ ring is close-fitted in a groove, made in the lower brass collar. The lower brass collar has two holes- one at the center and the other slightly away from it. A single-bore sheath (S-1) and a two-bore sheath (S-2) pass respectively, through the side hole and the central hole. On top of the central hole an alumina tube (A-1, 350 mm long and 8 mm diameter) is fixed with an adhesive. Two copper tubes (CT-1 and CT-2) are soldered respectively on the other side of the hole and over the hole. A platinum disc (P-1), the reference electrode (R-1), the solid electrolyte pellet (SE), the electrode pellet under investigation (R-2) and another platinum disc (P-2) are stacked one above the other and rested on the top end of the alumina tube (A-1). A long platinum wire passing through the sheath (S-1) is welded to the platinum disc (P-2) and the tip of a Pt-Pt-10% Rh thermocouple which passes through the sheath (S-2) is welded to the platinum disc (P-1). The cell assembly is held tightly together with the help of an alumina disc (D) with two slots in it, two alumina rods (A-2 and A-3) with grooves round them, matching with the slots in the alumina disc and two springs (S-3 and S-4). One end of the springs are attached to hooks, brazed to the lower brass collar and the other end is attached to the alumina rods. The exit ends of the wires are sealed with an adhesive. Copper tubes (CT-1 and CT-2) act as gas inlet and outlet, respectively. An earthed Kanthal wire is wound over the alumina tube (A) to prevent any inductive electrical pick-up. The ‘O’ ring is protected by cooling the brass heads by flow of water. The entire assembly is introduced in a muffle furnace in such a way that the stacked pellets lie in the constant temperature zone of the furnace.

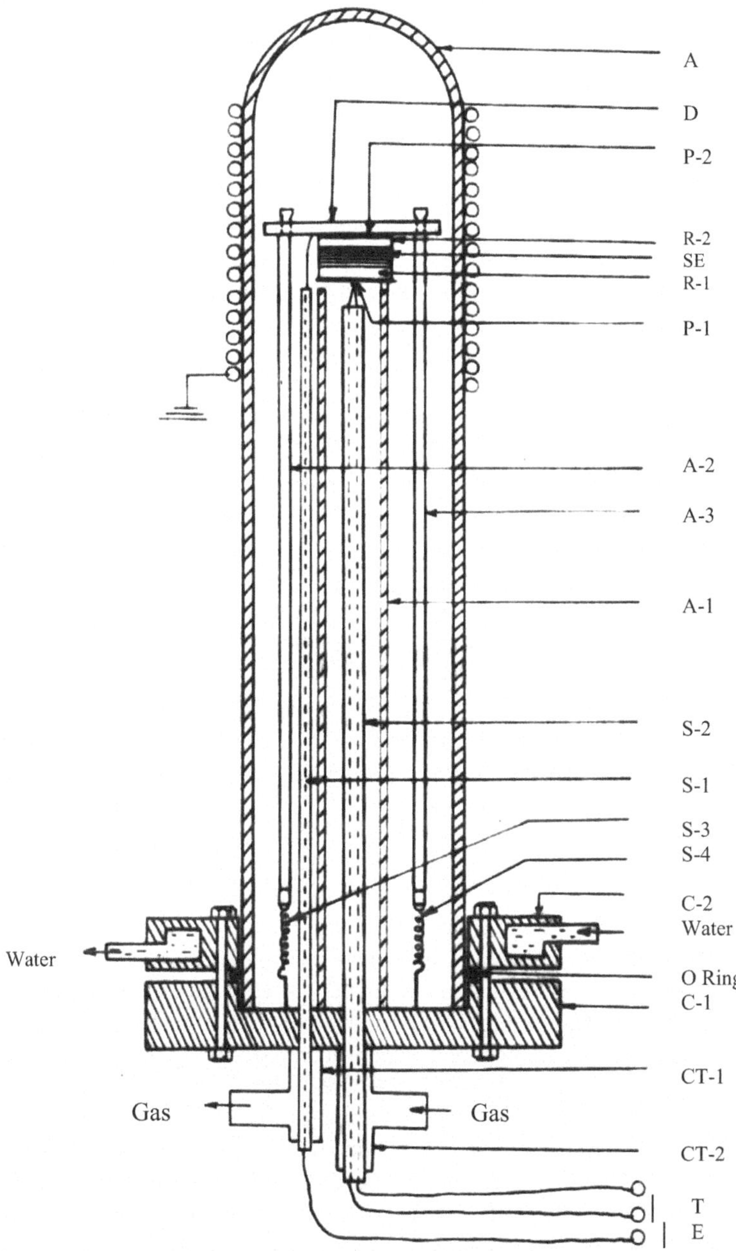

Fig. 7.6 Schematic cell diagram using a solid electrolyte. (Reproduced from J. S. Kachhawaha [125] Ph D Thesis, 1979, Banaras Hindu University). A and A-1 – alumina tubes, A-2 and A-3 – alumina rods, D – alumina disc, P-1 and P-2 – platinum discs, R-1 and R-2 – electrodes, SE – solid electrolyte pellet, S-1 – two-bore sheath, S-2 – one-bore sheath, S-3 and S-4 – springs, C-1 and C-2 – brass collars, CT-1 and CT-2 – copper tubes, T – cell temperature, E – cell emf

The cell assembly is first evacuated to 0.01 mm Hg and filled with purified argon. It is evacuated again and then a constant stream of purified argon is flown throughout the experiment. The furnace is then heated to the desired temperature and the cell is allowed to attain equilibrium. The emf of the cell is measured by connecting the two platinum leads (E in Fig. 7.6) whereas the cell temperature by connecting the thermocouple leads (T in Fig. 7.6). Before noting the emf one must ensure that it is not influenced by the rate of flow of the gas. The reversibility of the cell should be checked by disturbing the equilibrium by impressing an external emf or by passing small current (1–5 μ A) through the cell. The emf was considered to be reliable only if it returned to the initial value after removing the disturbance. The cell is held at a particular temperature for a long period. The experiment has to be discontinued in case the cell emf shows a continuous drift in one direction. The measurements have to be made at randomly varying temperatures.

A large number of metallic and oxide systems have been studied using solid electrolytes. Schmalzried and coworkers [143, 144] have investigated the thermodynamic properties of a number of ternary oxides with the aid of both halide and oxide electrolytes. No attempt has been made to list all the system investigated. However, in order to emphasize the importance of the technique some important investigations are mentioned here.

Nigl et al. [145] have studied the thermodynamic properties of alloys of alkaline earth metal (Ba, Sr, Ca) with Bi and Sb using the electrochemical cell based on a CaF_2 solid-state electrolyte in the temperature range: 723–1123 K. The thermodynamic properties of Sr-Pb alloys at 773–1073 K have also been investigated by Nigl and coworkers [146]. Reddy et al. [147] have investigated the thermodynamic properties of Ti-Al intermetallics using CaF_2 solid electrolyte and Al-$CaAlF_7$ reference electrode in the temperature range: 820–900 K. Nanko et al. [148] have measured the activity of aluminum in Pt-Al solid solutions in the temperature range: 820–900 K using CaF_2 as solid electrolyte. A galvanic cell: $AlF_3/CaF_2/Co, CoF_2$ was employed to determine activity of aluminum in solid Co-Al alloys (0–48.7 at % Al) [149] in the temperature range 900–1050 K. Paasch and Schaller [150] have determined activity of Cd, Y and Ce in Pd-rich alloys in the temperature range 973–1073 K using CaF_2. Alqasmi and Egan [151] have measured free energy of mixing of liquid Na-Sn alloys whereas Katayama et al. [152, 153] have determined the thermodynamic properties of Mn-In and Mn-Bi alloys in the temperature range: 800–1100 K using CaF_2 as the solid electrolyte.

Solid elelctrolytes based on calcia-stabilized-zirconia have been extensively used for activity measurement in binary metallic solutions: Ga-Te [154], Se-Te [155], Ni-In [156], Ga-Te [157], Cu-In [158], Cu-Te [159], Al-Mg [160], Zn-Cd [161], Ga-Sn [162], Pt-Ga [163], Ga-Bi [164], Te-Bi [165], Tl-Bi [166], Pb-Tl [167], Ag-Bi [168], Te-In [169], Sb-In [170], Bi-Sb [171], Pb-Sn [172] and Pb-Ag [173] as well as ternary metallic systems: In-Bi-Pb [174], In-Pb-Sn [175], Ga-Sb-Tl [176], Ga-In-Te [177], Ga-In-Tl [178], In-Sb-Ag [179], In-Sb-Cu [180], Ga-Sb-Ge [181], Ga-Sb-Sn [182], In-Pb-Ag [183], In-Bi-Ag [184], Ga-Sb-Bi [185] and Ga-Sb-In [186].

The yttria doped thoria solid electrolyte has been used for the activity measurement in many binary and ternary systems, viz. Ga-Te [187], Ni-Mn [188] Co-Mo [189] and Ag-Cu-Ti [190].

References

1. R.A. Oriani, J. Electrochem. Soc. **103**, 194 (1956)
2. C. Wagner, *Thermodynamics of Alloys* (Addition-Wesley, Massachusetts, 1962)
3. C. Wagner, A. Warner, J. Electrochem. Soc. **110**, 32 (1963)
4. G. Grube, E.A. Rau, Z. Elektrochem. **40**, 252 (1934)
5. P.B. Taylor, J. Phys. Chem. **31**, 1478 (1927)
6. L. Bass, Trans. Faraday Soc. **60**, 19 (1964)
7. A. Guggenheim, Proc. Phys. Soc. **85**, 393 (1965)
8. A. Guggenheim, J. Phys. Chem. **33**, 844 (1929)
9. G.A. Toporischev, O.A. Esin, V.N. Kalugin, Izv. Vyssh. Ucheb. Zaved. Chern Met. **7**(2), 19 (1964)
10. R.A. Tupkary, Indian J. Tech. **5**, 14 (1967)
11. W. Herzog, A. Klemm, Z. Naturforsch. **15**, 1100 (1960)
12. C. Solomons, in *Physicochemical Measurements in Metals Research*, Vol. IV, Part 2, ed. by R.A. Rapp (Interscience, New York, 1970), p. 35
13. J.W. Tomlinsen, in *Physicochemical Measurements at High Temperature*, ed. by J.O'M Bockris, J.L White, J.D. Mackenizie (Butterworth, London, 1959), p. 247
14. J.B. Goates, A.G. Cole, E.I. Gray, J. Amer, Chem. Soc. **73**, 3596 (1951)
15. I.B. Kutsenok, V.A. Geiderikh, Fiz. Khim. Electrokhim. Redk. Met. Solevykh Rasplavakh., P.T. Stangrit, ed., Akad. Nauk- SSSR, Kol'sk. Fil. Aptity, USSR, 1984, Chem. Abs.**101** (1984) 200098 d
16. J. Terpilowski, E. Zaleska, W. Gawel, Rocz Chem. **39**, 1337 (1965)
17. E. Ratajczak, J. Terpilowski, Rocz Chem. **42**, 433 (1968)
18. E. Ratajczak, J. Terpilowski, Rocz Chem. **43**, 1609 (1969)
19. J. Terpilowski, E. Zaleska, W. Gawel, Rocz Chem. **42**, 845 (1968)
20. J. Terpilowski, E. Ratajczak, Rocz Chem. **41**, 429 (1967)
21. V.P. Vasil'ev, A.P. Somov, A.V. Nikol'skaya, Y.I. Gerasimov, Zh. Fiz. Khim **42**, 675 (1968)
22. V.P. Vasil'ev, A.V. Nikol'skaya, Y.I. Gerasimov, Dokl. Akad. Nauk, SSSR **199**, 1094 (1971)
23. V.P. Vasil'ev, A.V. Nikol'skaya, Y.I. Gerasimov, Zh, Fiz. Khim. **45**, 2061 (1971)
24. V.P. Vasil'ev, A.V. Nikol'skaya, Y.I. Gerasimov, A.F. Kuzmetsov, Izv. Akad. Nauk SSSR, Neorg. Mater. **4**, 1040 (1968)
25. M.B. Babanly, N.A. Kulieva, Zh. Neorg. Khim **31**, 2365 (1986)
26. M.B. Babanly, I.S. Zamani, A. Azizuland, A.A. Kulieva, Zh. Neorg. Khim. **35**, 1285 (1990)
27. T.C. Wilder, J.F. Elliott, J. Electrochem. Soc. **107**, 628 (1960)
28. T.C. Wilder, J.F. Elliott, J. Electrochem. Soc. **111**, 352 (1964)
29. H. Seltz, J.C. DeHaven, Amer. Inst. Min. Met. Engg. Tech. Publ., 622 (1935)
30. J.H. McAteer, H. Selts, J. Am. Chem. Soc. **58**, 2081 (1936)
31. A.S. Abbasov, A.V. Nikol'skaya, Y.I. Gerasimov, V.P. Vasil'ev, Dokl. Akad. Nauk SSSR **156**, 1399 (1964)
32. M. Shamsuddin, S. Misra, in *Chemical Metallurgy – A Tribute to Carl Wagner*, ed. by N.A. Gockcen, (TMS – AIME, Warrendale, 1981), p. 241
33. A. Nasar, M. Shamsuddin, Metall. Mater. Trans. **28 B**, 519 (1997)
34. C.H. Liu, J.C. Angus, J. Electrochem. Soc. **116**, 1054 (1969)
35. J. Rakotomavo, M.C. Baron, C. Petot, Met. Trans. **12B**, 461 (1981)
36. K. Kameda, Met. Trans. Jpn. Inst. Met. **31**, 129 (1990)

37. R.K. Singh, M. Tech. Disserttion, Banaras Hindu University, 2005
38. M. Shamsuddin, S. Misra, Z. Metallk. **70**, 541 (1979)
39. M. Shamuddin, A. Nasar, T.K.S.P. Gupta, J. Alloys Compd. **199**, 189 (1993)
40. M. Shamuddin, A. Nasar, T.K.S.P. Gupta, Thermochim. Acta **232**, 303 (1994)
41. M. Shamsuddin, T.K.S.P. Gupta, A. Nasar, P. Ramachandrarao, Thermochim. Acta **246**, 213 (1994)
42. T.K.S.P. Gupta, A. Nasar, M. Shamsuddin, P. Ramachandrarao, Thermochim. Acta **247**, 415 (1994)
43. A. Nasar, M. Shamsuddin, J. Less Common Met. **171**, 83 (1991)
44. M. Shamsuddin, A. Nasar, Met. Trans **23 B**, 467 (1992)
45. M. Shamsuddin, A. Nasar, Metall. Mater. Trans. **26 B**, 569 (1995)
46. M. Shamsuddin, A. Nasar, High Temp. Sci. **28**, 245 (1990)
47. A. Nasar, M. Shamsuddin, J. Less Common. Metals **158**, 131 (1990)
48. A. Nasar, M. Shamsuddin, J. Less Common. Metals **161**, 193 (1990)
49. A. Nasar, M. Shamsuddin, Z. Metallk. **81**, 244 (1990)
50. N.A. Asryan, A. Mikula, Inorg. Mater. **40**, 366 (2004)
51. D. Jendrzejczyk, K. Fitzner, Thermochim. Acta **433**, 66 (2005)
52. A.A. Vechev, V.A. Geidrikh, Y.I. Gerasimova, Zh. Fiz. Khim. **35**, 1578 (1961)
53. M. Shamsuddin, S.B. Singh, A. Nasar, Thermochim. Acta **316**, 11 (1998)
54. E. Kundys, J. Terpilowski, Arch. Hutnic **9**, 333 (1964)
55. J. Terpilowski, E. Kundys, Arch. Hutnic **6**, 137 (1961)
56. E. Kundys, J. Terpilowski, Arch. Hutnic **7**, 238 (1962)
57. E. Kundys, J. Terpilowski, Arch. Hutnic. **7**, 39 (1962)
58. A. Mikula, K.L. Komarek, Z. Metallk. **81**, 209 (1990)
59. V.V. Egorkin, A.V. Ermakov, Zavod. Lab. **48**, 48 (1982)
60. A.N. Momedov, Z.B. Bagiorov, R.A. Karimova, Zh. Fiz. Khim. **61**, 597 (1987)
61. V. Gerling, R. Luck, B. Predel, Z. Metallk. **78**, 13 (1987)
62. Z.B. Bagirov, A.N. Mamedov, Zh. Fiz. Khim. **59**, 2839 (1985)
63. M.M. Asadov, M.B. Babanly, F.K. Guseinov, A.A. Kuliev, Zh. Fiz. Khim. **57**, 2865 (1983)
64. P. Rebouillen, S. Spas, J.M. Maine, J. Riou, R. Baret, J. Less Common Met. **160**, 61 (1990)
65. E.V. Raguzina, D.S. Maltsev, V.A. Volkovich, Russian Forum of Young Scientists, Volume 2018 in *Russian,* KnE Engineering, p. 72–82. https://doi.org/10.18502/keg.v3i4.2227
66. C. Caravaca, G. De C'ordoba, Z. Naturforsch. **63a**, 98 (2008)
67. R.A. Vecher, V.V. Tsyganok, Vestn. Beloruss University **2**(1), 6 (1971)
68. W. Han, Z. Li, M. Li, Y. Gao, X. Yang, M. Zhangab, Y. Sun, R. Soc. Chem. Adv. **8**, 8118 (2018)
69. M.B. Babanly, Z.A. Guseynov, Z. Metallk **92**, 451 (2001)
70. Y. Castrillejo, A. Vega, M. Vega, P. Hernándezb, J.A. Rodriguezb, E. Barradoa, Electrochim. Acta **118**, 58 (2014)
71. I. Katayama, T. Tanaka, S.I. Aka, K. Yamazaki, T. Lida, Mater. Sci. Forum, 502 (2005)
72. S.C. Baldivieso, N.D. Smith, S. Im, H. Kim, J. Alloys Compd. **886**, 161229 (2021). https://doi.org/10.1016/j.jallcom.2021.161229
73. J. Terpilowski, Arch. Hutnic **11**, 63 (1966)
74. V. Vassilev, M. Alaoui-Elbelghiti, A. Zrinesh, M. Gambino, J.P. Bros, J. Alloys Compd. **265**, 160 (1998)
75. D.V. Tan, L. Segers, R. Winand, J. Electrochem. Soc. **141**, 927 (1994)
76. A. Mikula, Met. Trans. **23B**, 601 (1992)
77. C.S. Oh, J.H. Shim, B.J. Lee, D.N. Lee, J. Alloys Compd. **238**, 155 (1996)
78. S. Knott, A. Mikula, Mater. Trans. Jpn. Inst. Met. **43**, 105 (2002)
79. A. Novoselova, V. Smolenski, V.A. Volkovich, A.B. Ivanov, A. Osipenko, T.R. Griffiths, J. Nucl. Mater. **466**, 373 (2015). https://doi.org/10.1016/j.jnucmat.2015.08.010
80. T. Yin, Y. Liu, L. Wang, Y. Yan, G. Wang, Z. Chai, W. Shi, J. Nucl. Mater. **523**, 16 (2019). https://doi.org/10.1016/j.jnucmat.2019.05.042

81. B. Predel, V. Schallner, Z. Metallk. **63**(119), 341 (1972)
82. Z. Moser, Bull. Acad. Polon. Sci. **18**, 481 (1970)
83. Z. Moser, Bull. Acad. Polon. Sci. **17**, 467 (1969)
84. Z. Moser, Arch. Hutnic **15**, 385 (1970)
85. W. Ptak, Z. Moser, Arch. Hutnic **15**, 375 (1970)
86. Z. Moser, Arch. Hutnic **15**, 125 (1970)
87. Z. Moser, Met. Trans. **2**, 2175 (1971)
88. W. Ptak, Z. Moser, Bull. Acad. Polon. Sci. **19**, 71 (1970)
89. K. Kameda, Mater. Trans. Jpn. Inst. Met. **30**, 523 (1989)
90. W. Jin, J. Weiliang, K. Qian, F. Xiaohong, X. Saijun, Z. Jun, Int. J. Electrochem. Sci. **15**, 3836 (2020). https://doi.org/10.20964/2020.05.12
91. A.L. Alabyshev, M.F. Lantratov, Tr. Leningr. Politekhn. Inst. **223**, 55 (1963)
92. Y.K. Lee, A. Yazawa, Tohuku Daigaku Senko Seiren Kenkyusho Iho **25**(1), 1 (1969)
93. W.T. Thompson, S.N. Flengas, J. Electrochem. Soc. **118**, 419 (1971)
94. W. Ptak, Z. Moser, Bull. Acad. Polon. Sci **18**, 383 (1970)
95. M. Shamsuddin, S. Misra, Scripta Met. **7**, 547 (1973)
96. L.A. Zabdyr, J. Electrochem. Soc. **131**, 2157 (1984)
97. N. Moulondj, G. Petot- Ervas, C. Petot, Thermochim. Acta **136**, 87 (1988)
98. M.M. Asadov, M.B. Babanly, F.K. Guseinov, A.A. Kuliev, Zh. Fiz Khim **57**, 2865 (1983)
99. B. Predel, V. Schallner, Z. Metallkd. **60**, 869 (1969)
100. B. Predel, V. Schallner, Mater. Sci. Engg. **5**, 210 (1969/70)
101. V.I. Deev, V.I. Pybnikov, V.P. Goldobin, V.I. Smirnov, Russ. J. Phys. Chem. **45**, 1730 (1971)
102. G.R. Belton, Y.K. Rao, Trans. Met. Soc. AIME **245**, 2189 (1969)
103. G.R. Belton, D.R. Morris, Trans. Met. Soc. AIME **242**, 108 (1968)
104. G.R.B. Elliott, J. Electrochem. Soc. **115**, 1143 (1968)
105. A. Nasar, M. Shamsuddin, J. Less Common Met. **161**, 87 (1990)
106. Z. Terpilowski, Arch. Hutnic **3**, 226 (1958)
107. Z. Terpilowski, Arch. Huntic **4**, 355 (1959)
108. Z. Terpilowshi, Z. Gregorczy, Arch. Huntic **6**, 197 (1961)
109. K. Schwerdtfeger, H.J. Engell, Arch. Eisenhutten. **35**, 533 (1964)
110. K. Schwerdtfeger, H.J. Engell, Arch. Eisenhutten. **34**, 64 (1963)
111. R.H. Tupkary, Ind. J. Tech. **6**, 132 (1968)
112. V.P. Csaki, A. Dietzel, Glasstech. Ber. **18**, 33 (1940)
113. H. Ito, T. Yanagase, Trans. Jpn. Inst. Metals **1**, 115 (1960)
114. B.M. Lepinskikh, O.A. Esin, Zh. Neorg. Khim **6**, 1223 (1961)
115. Y.K. Delimarski, V.N. Andreeva, Zh. Neorg. Khim **5**, 1800 (1960)
116. R.L. Ranford, S.N. Flengas, Can. J. Chem. **43**, 2879 (1965)
117. K. Kiukkola, C. Wagner, J. Electrochem. Soc. **104**(308), 379 (1957)
118. J.T. Kummer, in *Progress in Solid State Chemistry*, ed. by H. Reiss, J.O. Mc Caldin, vol. 7, (Pergamon, Oxford, 1972), p. 141
119. R.A. Rapp, D.A. Shores, in *Physicochemical Measurements in Metals Research*, Vol. IV, Pt. 2, ed. by R.A. Rapp (Interscience, New York, 1970), p. 123
120. H. Schmalzried, in '*Metallurgical Chemistry Symposium, 1971*' Proceedings of the Symposium, 1972, ed. by O. Kubaschewski (Brunel University and NPL, London, 1972), p. 39, HMS0
121. C. Wagner, in *Advances in Electrochemistry and Electrochemical Engineering*, ed. by P. Delahay, vol. 3, (Interscience, New York, 1966), p. 1
122. B.C.H. Steel, in *Electrochemical Measurement in High Temperature System*, ed. by C.B. Alcock (Inst. Min. Met., London, 1965), p. 132
123. T.H. Estell, S.N. Flengas, Chem. Rev. **70**, 339 (1970)
124. W.L. Roth, J. Solid State Chem. **4**, 60 (1972)
125. J.S. Kachhawaha, Ph. D. Thesis, *Thermodynamic Investigations on Oxide Systems*, Banaras Hindu University, 1979

126. R.W.J. Ure, J. Chem. Phys. **26**, 1363 (1957)
127. C. Wagner, J. Electrochem. Soc. **115**, 933 (1968)
128. K. Schwerdtfeger, Trans. Met. Soc. AIME **239**, 1276 (1967)
129. R.J. Fruehan, Trans. Met. Soc. AIME **242**(1968) (2007)
130. P.C. Lidster, H.B. Bell, Trans. Met. Soc. AIME **245**, 2273 (1969)
131. K.K. Prasad, H.S. Ray, K.P. Abraham, *Chemical and Metallurgical Thermodynamics* (New Age International Publishers, New Delhi, 2007)
132. R.A. Rapp, F. Maak, Acta. Met. **10**, 63 (1962)
133. J.S. Kachhawaha, M.P. Ganu, V.B. Tare, Scripta Met. **7**, 311 (1973)
134. S. Seetharaman, K.P. Abraham, Ind. J. Technol. **6**, 123 (1968)
135. T.W. Richards, C.P. Smith, J. Am. Chem. Soc. **44**, 524 (1922)
136. O. Kubaschewski, C.B. Alcock, *Metallurgical Thermochemistry*, 4th edn. (Pergamon, Oxford, 1967)
137. T.W. Richards, G.S. Forbes, Z. Phys. Chem. **58**, 683 (1907)
138. F. Weibke, V. Quadt, Z. Elektrochem. **45**, 715 (1939)
139. J.F. Elliott, J. Chipman, J. Am. Chem. Soc. **73**, 2683 (1951)
140. L.L. Seigle, M. Cohen, B.L. Averbach, Trans. Met. Soc. AIME **104**, 308 (1952)
141. J.J. Egan, R.H. Wiswall, Nucleonics **15**, 105 (1957)
142. G.R.B. Elliott, J.F. Lemons, J. Electrochem. Soc. **114**, 935 (1967)
143. R. Taylor, H. Schmalzried, J. Phys. Chem. **68**, 2444 (1964)
144. Y.D. Tretyakov, H. Schmalzried, Ber. Bunsenges, Physik Chem. **73**, 396 (1969) and **70** (1966) 180
145. T.P. Nigl, N.D. Smith, T. Lichtenstein, J. Gesualdi, K. Kumar, H. Kim, J. Vis. Exp. **129**, e56718 (2017). https://doi.org/10.3791/56718
146. T.P. Nigl, T. Lichtenstein, N.D. Smith, J. Gesualdi, Y. Kong, H. Kim, J. Electrochem. Soc. **165**(14), H991 (2018)
147. R.G. Reddy, A.N. Yahya, L. Brewer, J. Alloys Compd. **32**, 223 (2001)
148. M. Nanko, Y. Kishi, T. Maruyama, Mater. Trans. Jpn. Inst. Met. **39**, 1238 (1998)
149. H.J. Schaller, T. Breischneider, Z. Metallk. **76**, 143 (1985)
150. S. Paasch, H.J. Schaller, Ber. Bunsenges. Phys. Chem. **87**, 812 (1983)
151. R. Alqasmi, J.J. Egan, Ber. Bunsenges. Phys. Chem. **87**, 815 (1983)
152. I. Katayama, S. Matsushima, Z. Kozula, Mater. Trans. Jpn. Inst. Met. **32**, 943 (1991)
153. I. Katayama, S. Matsushima, Z. Kozula, Mater. Trans. Jpn. Inst. Met. **31**, 789 (1990)
154. I. Katayama, J. Nakayama, T. Nakai, Z. Kozula, Trans. Jpn. Inst. Met. **28**, 129 (1987)
155. N. Mouloudj, G. Petot- Ervas, C. Petot, Thermochim. Acta **136**, 87 (1988)
156. I. Katayama, Y. Suzuki, Y. Yamamoto, T. Oishi, Monatsh. Chemie **136**(11), 1955 (2005)
157. I. Katayama, K. Shimazawa, D. Zikovic, D. Manasijevic, Z. Zivkovic, T. Lida, Z. Metallk. **94**, 1296 (2003)
158. I. Katayama, K. Miyakusu, T. Lida, J. Min. Met. **38**(1–2), 23 (2002)
159. P. Wai, I. Jie, Mater. Sci. Eng. A **269**(1–2), 104 (1999)
160. G. Lu, Z. Qiu, Trans. Nonferrous Metals Soc. China **8**(1), 109 (1998)
161. I. Katayama, K. Maki, Y. Fukuda, A. Ebara, T. Lida, Mater. Trans. Jpn. Inst. Met. **38**, 119 (1997)
162. I. Katayama, K. Maki, M. Nakano, T. Lida, Mater. Trans. Jpn. Inst. Met. **37**, 988 (1996)
163. I. Katayama, T. Makino, T. Lida, High Temp. Mater. Sci. **34**, 127 (1995)
164. I. Katayama, J. Nakayama, T. Ihura, Z. Kozuka, T. Lida, Mater. Trans. Jpn. Inst. Met. **34**, 79 (1993)
165. K. Kameda, K. Yamaguchi, H. Horie, Nippon Kinzoku Gakkacshi **57**, 158 (1993)
166. K. Kameda, K. Yamaguchi, T. Kon, Nippon Kinzoku Gakkaishi **56**, 900 (1992)
167. K. Kameda, K. Yamaguchi, Nippon Kinzoku Gokkaishi **55**, 951 (1991)
168. K. Kameda, K. Yamaguchi, Nippon Kinzoku Gakkaishi **55**, 536 (1991)
169. K. Kameda, K. Yamaguchi, Nippon Kinzoku Gakkaishi **54**, 1222 (1990)
170. K. Kameda, J. Tanabe, Nippon Kinzoku Gokkaishi **51**, 1174 (1987)

171. S. Itoh. And T. Azakami, Nippon Kinzoku Gakkaishi, 48 (1984) 239
172. E. Sugimoto, S. Kutowa, Z. Kozuka, Nippon Kogyo Kaishi **98**, 429 (1982)
173. E. Sugimoto, S. Kutowa, Z. Kozuka, Nippon Kogyo Kaishi **97**, 1199 (1981)
174. M. Zeng, Z. Kozuka, Nippon Kinzoku Goakkaishi **51**, 44 (1987)
175. M. Zeng, Z. Kozuka, Nippon Kinzoku Goakkaishi **51**, 666 (1987)
176. I. Katayama, Y. Seudai, D. Zivkovic, D. Manasijevic, Z. Zivkovic, H. Yamashita, Diffusion Defect. Data Solid State Pt. B **127**, 71 (2007)
177. I. Katayama, Y. Seudai, D. Zivkovic, D. Manatijevic, Z. Zivkovic, H. Yamashita, Thermochim. Acta **431**, 138 (2005)
178. I. Katayama, K. Yamazaki, M. Nakano, T. Lida, Scand. J. Met. **32**, 1 (2002)
179. S. Itabashi, K. Yamaguchi, K. Kameda, Nippon Kinzoku Gakkaishi **65**, 888 (2001)
180. S. Itabashi, K. Kameda, K. Yamaguchi, T. Kon, Nippon Kinzoku Gakkaishi **63**, 817 (1999)
181. I. Katayama, Y. Fukuda, Y. Hattori, Berichte der Bunsen- Gesellschaft **102**, 1235 (1998)
182. I. Katayama, Y. Fukuda, Y. Hattori, T. Maruyama, Thermechim. Acta **314**, 175 (1998)
183. K. Kameda, K. Yamaguchi, T. Kon, Nippon Kinzoku Gakkaishi **61**, 444 (1997)
184. K. Kameda, K. Yamaguchi, T. Gonii, Nippon Kinzoku Gakkaishi **60**, 452 (1996)
185. I. Katayama, J. Nakoyama, T. Ikura, Z. Zozuka, T. Lida, Maler. Trans. Jpn. Inst. Met. **34**, 792 (1993)
186. I. Katayama, J. Nakayoma, T. Ikura, Z. Kozuka and T. Lida, J. Non Cryst. Solids, 156–158 (1) (1993) 393
187. D.E. Grima, Trans. Met. Soc. AIME **233**, 1442 (1965)
188. K.T. Jacob, Met. Trans. **13B**, 283 (1982)
189. V.N. Drobyshev, T.N. Rozukhina, L.A. Tarasova, Zh. Fiz. Khim. **39**(1), 141 (1965)
190. J.J. Pak, M.L. Santella, R.J. Fruehan, Met. Trans. **21B**, 349 (1990)

Chapter 8
Phase Diagram Analyses

An equilibrium phase diagram represents the geometrical loci [1] of thermodynamic parameters when equilibrium between different phases under a given set of conditions is established. Thus phase diagram and thermodynamic data of a system are interrelated. Heumann and Predel [2] have reviewed many aspects of inter-relationship between phase diagram and thermodynamic properties of phases of a system. Many thermodynamic and physical properties are at least qualitatively inferred from the shapes of the phase diagrams. For example, in isomorphous phase diagrams a depression indicates higher stability of the liquid compared to that of the solid. This tendency in gradually accentuated with liquidus and solidus having a common minimum in the phase diagrams and further on the solid state miscibility gap and the eutectic system, showing tendency of un-mixing of the components. The reverse is true of systems showing elevation of solidus, a common maximum in the solidus and liquidus, and finally compound formation. Miller and Komarek [3] have studied the retrograde solubility in PbS, PbSe and PbTe and correlated the point of retrograde solubility to the inflexion point in the corresponding liquidus.

8.1 Phase Diagram Analyses in Light of Thermodynamic Data

It is well known that whereas the liquidus in many phase diagrams can be determined fairly accurately by cooling curves in thermal analysis, a similar determination of solidus involves significant errors due to segregation in the solid phase. The determination of melting temperatures of homogeneous solid alloys by heating curves in thermal analysis is relatively more reliable but usually requires long annealing times due to slow diffusion rates in the solid state. The ideal liquidus and ideal solidus curves of binary and pseudo-binary alloy systems exhibiting complete miscibility in

M. Shamsuddin, *Thermodynamic Measurement Techniques*, The Minerals, Metals & Materials Series, https://doi.org/10.1007/978-3-031-47118-6_8

the liquid and/or solid phases can be calculated from the heats of fusion (ΔH^f) of the pure components by the following relations [4]:

$$ln\,\frac{x_B^s}{x_B^l} = \frac{\Delta H_B^f}{R}\left(\frac{1}{T} - \frac{1}{T_B^f}\right) \tag{8.1}$$

and

$$ln\,\frac{x_A^s}{x_A^l} = \frac{\Delta H_A^f}{R}\left(\frac{1}{T} - \frac{1}{T_A^f}\right) \tag{8.2}$$

Where, subscripts A and B denote the components of the binary system A–B, T and T^f are any temperature (between T_A^f and T_B^f) and melting point, respectively, and x^s and x^l are mole fractions of solid and liquid, respectively.

Steininger [5] has derived the following general equation for the thermodynamic relationship between solidus and liquidus at equilibrium for systems which are nonideal but homogeneous and monotonic.

$$RT\left[\ln\left(\frac{x_A^s}{x_B^s}\right) - \left(\frac{x_A^l}{x_B^l}\right)\right] = \Delta H_A^f\left[1 - \frac{T}{T_A^f}\right] - \Delta H_B^f\left[1 - \frac{T}{T_B^f}\right] - D \tag{8.3}$$

where, D accounts for nonideality and is defined as;

$$D = \left(\frac{\partial G^{xs\,(l)}}{\partial x_A^s}\right) - \left(\frac{\partial G^{xs\,(l)}}{\partial x_B^l}\right) \tag{8.4}$$

Here, $G^{xs(l)}$ is the excess molar free energy of mixing of pure liquid A and pure (supercooled) liquid B. There are many systems where the liquidus has been determined fairly accurately, but little or no reliable solidus data are available. There are a few systems where the solidus has been determined but not the liquidus because of unattainable temperature or due to the decomposition or sublimation of the constituents at higher temperatures. In such cases, a quick and reliable method of calculating the two phase gap based on Eq. 8.3 could be used as an alternative method or as a guide for experimental investigations.

Steininger [5] has pointed out that for ideal solutions, the nonideal term (D) in Eq. 8.3 is negligible indicating thereby that the following equation of the ideal form may be used to calculate one of the boundaries of the two-phase field when the other one and heats of fusion of the pure components are known.

$$RT\left[\ln\left(\frac{x_A^s}{x_B^s}\right) - \left(\frac{x_A^l}{x_B^l}\right)\right] = \Delta H_A^f\left[1 - \frac{T}{T_A^f}\right] - \Delta H_B^f\left[1 - \frac{T}{T_B^f}\right] \tag{8.5}$$

8.2 Evaluation of Thermodynamic Functions from Phase Diagrams

The thermodynamic functions such as heat of fusion, excess free energy of formation, activity and solubility can be calculated from the phase diagram. Analysis of phase diagram can further yield information regarding factors governing theories of alloying e.g. strain energy, size, valence, and electron concentration effects. However, reliable thermodynamic data are rarely obtained by these analyses due to the approximate and inaccurate nature of phase diagrams which have been determined by measurements of intensive physical properties and thermal analysis. Nevertheless a number of thermodynamic parameters can be obtained by referring to the phase diagrams. A few are mentioned below:

8.2.1 *Heat of Fusion*

When the heat of fusion of a component is not available, it can be estimated from its phase diagram with another solute by knowing the degree of lowering of the melting point of the solvent for a given mole fraction of the solute. The heat of fusion can be calculated by the following relation assuming negligible solid solubility:

$$\Delta H_A^f = x_B\left(RT^2/\Delta T\right) \quad (8.6)$$

Where, ΔH_A^f is the heat of fusion of A, x_B the mole fraction of B, ΔT, the depression in the melting point of pure A and T is the liquidus temperature. The relation is of particular use for calculating the melting point of metals which sublime e.g. arsenic. Also the change in fusion point or liquid or liquid solubility of the alloy can be obtained if heat of fusion of the solvent is known.

8.2.2 *Activity*

The activity of the component i, in a binary liquid solution can be calculated from the location of the liquidus and solidus phase boundaries in the phase diagrams. The principle underlying the method is simple. At equilibrium, the chemical potential of components A and B are the same in both the liquid and solid phases. The activity is given by the expression [6]:

$$\Delta \overline{G}_i^l = RT \ ln\, a_i = -\Delta S_i^f\left(T_i^f - T\right) \quad (8.7)$$

Where, a_i, T_i^f and ΔS_i^f stand for the activity, melting point, and entropy of fusion of the solute i and T is the liquidus temperature.

8.2.3 Free Energy of Formation

The excess free energy of the liquid phase and the free energy of formation of a congruently melting compound can be calculated from the liquidus data and the heats of fusion and formation of the compound. The methods of analyses depend on the shape of the liquidus in the vicinity of the compound. Brief accounts on such analyses suggested by Wagner and Jordan are given below:

8.2.3.1 Wagner's Analysis

For a regular solution of a system A–B having a minimum of the liquidus and solidus, Carl Wagner [7] suggested a method for calculating the difference in the free energies between the liquid and solid phases $\left[G_A^{xs(s)} - G_A^{xs(l)}\right]$ at the minimum composition. At this minimum (common for the liquidus and solidus) the following relations hold:

$$G_A^{xs\ (s)}(T, x_B) - G_A^{xs\ (l)}(T, x_B) = \left(T_A^f - T\right)\Delta S_A^f \tag{8.8}$$

and

$$G_B^{xs\ (s)}(T, x_B) - G_B^{xs\ (l)}(T, x_B) = \left(T_B^f - T\right)\Delta S_B^f \tag{8.9}$$

where x_B is the mole fraction of B, T is the temperature at the minimum of the liquidus and solidus, and T^f and ΔS^f stand for the melting point and entropy of fusion, respectively. On multiplying Eqs. 8.8 and 8.9 by $(1 - x_B)$ and x_B, respectively and adding corresponding sides, the difference of the excess molar free energies in the solid and liquid states at the mole fraction x_B and temperature T, we get the following expression:

$$\begin{aligned} G^{xs\ (s)} - G^{xs\ (l)} &= \left[(1 - x_B)G_A^{xs\ (s)} + x_B G_B^{xs\ (s)}\right] - \left[(1 - x_B)G_A^{xs\ (l)} + x_B G_B^{xs\ (l)}\right] \\ &= (1 - x_B)\left(T_A^f - T\right)\Delta S_A^f + x_B\left(T_B^f - T\right)\Delta S_B^f \end{aligned} \tag{8.10}$$

This difference in the molar free energies of the solid and liquid phases at the minimum has been correlated to the size of disparity of the component elements.

Wagner [7] has also suggested a method for calculating the free energy of formation of a congruently melting line compound at its melting point from the knowledge of its heat of fusion and the liquidus phase boundary. The analysis, which assumes the melt to be regular solution, has been applied to a number of intermetallic compounds [8, 9]. The excess free energy G^{xs} of a liquid phase from the Wagner's Eq. (61) [7] is expressed as:

$$G^{xs}\left(T^f, n_B\right) = -\left(1-n_B\right)\ x_B \int_0^{n_B} I_1\left(n_B\right)\, dn_B - n_B\left(1-x_B\right) \int_{n_B}^{1} I_1\left(n_B\right)\, dn_B \quad (8.11)$$

where, n_B and x_B are the atom fraction of B in the liquid, and in the compound, respectively, and the integrand $I_1(n_B)$ is defined as

$$I_1\left(n_B\right) = \Delta S_c^f\left[T^f - T\left(n_B\right).\text{Ø}\left(n_B\right)\right] / \left(n_B - x_B\right)^2 \quad (8.12)$$

In the above equation, T^f, $T(n_B)$ and ΔS_c^f respectively, stand for the melting point of the compound, the liquidus temperature and the entropy of fusion of the compound. The auxiliary function $\text{Ø}(n_B)$ is defined as;

$$\text{Ø}\left(n_B\right) = 1 + \frac{R}{\Delta S_c^f}\left[\left(1-x_B\right)\ln\left(\frac{1-x_B}{1-n_B}\right) + x_B\, ln\, \frac{x_B}{n_B}\right] \quad (8.13)$$

where, R is the gas constant.

Following Wagner's Eq. (63) [7] the integrand may also be written as:

$$I_1\left(n_B\right) = \Delta S_c^f\left[T_{id}\left(n_B\right) - T\left(n_B\right)\right]\text{Ø}\left(n_B\right) / \left(n_B - x_B\right)^2 \quad (8.14)$$

where, $T_{id}(n_B) = T^f / \text{Ø}\ (n_B)$ is temperature of the liquidus for a hypothetical ideal liquid.

The integrand $I_1(n_B)$ calculated from Eq. 8.14 is plotted as a function of n_B and the integration in Eq. 8.11 is performed graphically. From Wagner's Eq. (35) [7] the standard free energy of formation of the compound AB at its melting point is obtained as:

$$\begin{aligned} \Delta G^o &= \left[\frac{1}{2}\ A\ (l) + \frac{1}{2}\ B\ (s) = \frac{1}{2}\ AB\ (s),\ \ T^f\ K\right] \\ &= -RT^f\ ln\ 2 + G^{xs}\left(n_B = 0.5\right) - \frac{1}{2}\left(T^f - T_B^f\right)\ \Delta S_B^f \text{ cal g atom}^{-1} \end{aligned} \quad (8.15)$$

where, T_B^f and ΔS_B^f are respectively the melting point and entropy of fusion of B. The uncertainty in ΔG^o at T^f is nearly equal to the uncertainty in G^{xs}, which in terns depends on the heat of fusion of *AB* and the values of the liquidus.

The free energy of formation of the compound *AB* from the solid components at $T_o = 298$ K may be calculated by the Gibbs–Helmholtz equation:

$$\frac{d\,(\Delta G^o/T)}{dT} = -\frac{\Delta H^o}{T^2}$$

Integration of the above equation between $T_o = 298\ K$ and the melting point T^f of the compound, yields:

$$\Delta G^o(T_o) = \frac{T_o}{T^f}\,\Delta G^o(T^f) + \left(1 - \frac{T_o}{T^f}\right)\Delta H^o\ (T_o) + T_o \int_{T_o}^{T^f} \frac{\Delta H^o(T) - \Delta H^o\ (T_o)}{T^2}\mathrm{dT} \tag{8.16}$$

where, ΔG^o and ΔH^o denote, respectively the free energy and heat of formation of the compound AB from the components in their most stable state.

8.2.3.2 Jordan's Analysis

In Wagner's analysis the melt is assumed to be regular and applicable to systems showing parabolic and symmetric variation of the liquidus. The regular solution behavior is not obeyed where the liquidus shows a sharp peak near the equiatomic composition. Jordan [10] has derived expressions for such liquidus curves of a binary system, A–B with a congruently melting compound, AB by assuming association in the liquid phase with species, A, B and AB being present. This type of solution has been designated as a 'Regular Associated Solution'. For a regular solution the Guggenheim interchange energy (α) is given as:

$$\alpha = -\frac{1}{2}\,\frac{RT\ln 4x_A x_B - \Delta H^f + T\Delta S^f}{(x_B - 0.5)^2} \tag{8.17}$$

where, x_A *and* x_B are the mole fractions of solvent and solute, respectively and T is the liquidus temperature. Jordan [10] has derived the following equation for the liquidus incorporating the interchange energy ω_{12} for A-AB interaction, and β as the degree of dissociation:

$$\omega_{12} = \frac{\frac{RT}{2}\ln\frac{x_B}{x_A} + \Delta H^f + T\Delta S^f}{(x_B - 0.5)^2} = \alpha + \frac{RT\beta}{(x_B - 0.5)^2} \tag{8.18}$$

The values of α and β are calculated from the intercept and slope, respectively of the plot of ω_{12} vs $(x_B - 0.5)^2$. The free energy of formation of the compound AB at the melting point T^f is given by the relation:

$$\Delta G\left(T^f\right) = \frac{1}{2}RT\ln\left[\frac{\overline{\beta}}{1 + \overline{\beta}}\right] + \frac{1}{4}\overline{\alpha} \tag{8.19}$$

where, $\overline{\alpha}$ and $\overline{\beta}$ are the average values of α and β, respectively for the different composition ranges *A*-*AB* and *AB*-*B*. The theory of regular associated solution was first extended by Jordan to derive thermodynamic *data* for Zn-Te and Cd-Te systems [10].

8.3 Application of Wagner's and Jordan's Approach in Some Intermetallic Compounds

Chou et al. [11] have conducted extensive studies on the retrograde solubility of PbS, PbSe, and PbTe. They have reported maximum deviation from stoichiometry on the Pb-rich and the chalcogen-rich side of the phase fields as: PbS – 2.5×10^{-4} at % excess Pb at 1075 °C and 1×10^{-4} at % excess S at 850 °C; PbSe – 4×10^{-4} at % excess Pb at 925 °C, and 3×10^{-4} at % excess Se at 800 °C, and PbTe – 2.8×10^{-4} at % excess Pb at 780 °C, and 10^{-4} at % excess Te at 720 °C. Wagner and Jordan analyses have been applied to evaluate certain thermodynamic properties of PbS, PbSe, PbTe, and SnTe because they are congruently melting line compounds.

8.3.1 Lead Sulfide

Since PbS melts congruently and has virtually no solid solubility for Pb or S, the free energy of formation of PbS at its melting point has been calculated according to the standard analysis due to Wagner [7]. Calculations by this analysis according to the procedure outlined above yield a value of ΔG at the melting point (1386.3 K) as −4.6 kcal mol^{-1} which is not in agreement with the extrapolated value of −9.72 kcal mol^{-1} from the data of Thomson and Flengas [12]. Hence, it is concluded that Pb-S melts do not follow the regular solution behavior.

The Wagner's integrand calculated from the liquidus [7] according to the Eq. 8.12 shows sharp peaks near the equiatomic composition. Thus Pb-S melt may behave like a regular associated solution as suggested by Jordan [10] for the melts in the system Zn-Te and Cd-Te. The free energy of formation of PbS at the melting point calculated on the basis of Jordan's analysis is −9.3 kcal mol^{-1} which is in reasonable agreement with the extrapolated value of −9.72 kcal mol^{-1} from the data of Thomson and Flengas [12]. Thus the regular associated solution model seems to be valid for Pb-S melts.

Sharma et al. [13] have employed the concept of the regular solution model to describe the thermodynamic properties of the liquid phase in Pb-S system by assuming the existence of PbS species in addition to Pb and S. Treating the system as a pseudoternary solution of Pb, S and PbS, they have discussed the thermodynamic properties of the intermediate phase, PbS, by a point–defect model.

8.3.2 Lead Selenide

The free energy of formation of PbSe at its melting point can be calculated by phase diagram analysis due to Wagner [7], since this compound melts congruently and has a negligible range of solid solubility. Calculations by this analysis yield ΔG [Pb (l) + Se (l) = PbSe (s) at 1353.7 K] as -13.2 kcal mol^{-1} which does not compare with an extrapolated value of -18.4 kcal mol^{-1} based on our experimental data [8]. ΔG at the melting point of PbSe calculated by Jordan's analysis [10] is -11.6 kcal mol^{-1} which is also not in agreement with the above extrapolated value of -18.4 kcal mol^{-1}. The estimated values of ΔG from phase diagram by the analyses due to Wagner and Jordan are only approximate in view of several simplified assumptions involved in these analyses. Part of the discrepancy may be due to the large temperature range of extrapolation applied to our investigation.

8.3.3 Lead Telluride

The free energy of formation of PbTe at its melting point (1197 K) calculated according to Wagner's analysis [7] is -9.8 kcal mol^{-1}. This compares well with the extrapolated value of -10.1 kcal mol^{-1} based on our investigations [8].Thus, in conformity with the assumption in the Wagner's analysis, liquid Pb-Te alloys may behave like a regular solution. The free energy of formation of PbTe at the melting point calculated by Jordan's phase diagram (Fig. 8.1 [14]) analysis [10] is -7.6 kcal mol^{-1} which is 2.5 kcal less exothermic as compared to the value reported in reference [8]. Hence, the regular associated model may not be applicable to the Pb-Te melts.

8.3.4 Tin Telluride

The excess free energy of liquid SnTe calculated from Wagner's analysis by using the value of heat of fusion and the liquidus from the phase diagram (Fig. 8.2 [14]) is highly exothermic (-4.0 kcal mol^{-1}). This indicates that atomic distribution in Sn-Te melt may be non-random. Calculations from Wagner's analysis yield a value for the free energy of formation, [ΔG, Sn (l) + Te (l) = SnTe (s), 1079 K] as -7.0 kcal mol^{-1}. The corresponding value from Mcteer and Seltz [15] is -10.8 kcal mol^{-1}. From the wide discrepancy from the above two values it may be concluded that the assumption of regular behavior for the melts adopted in Wagner's analysis may not be valid for Sn-Te liquid alloys.

In the system Sn-Te, both the liquidus and the Guggenheim interchange energy show sharp peaks near the equiatomic composition. Following the similar procedure as adopted by Jordan [10] for Cd-Te and Zn-Te melts, it may be postulated that

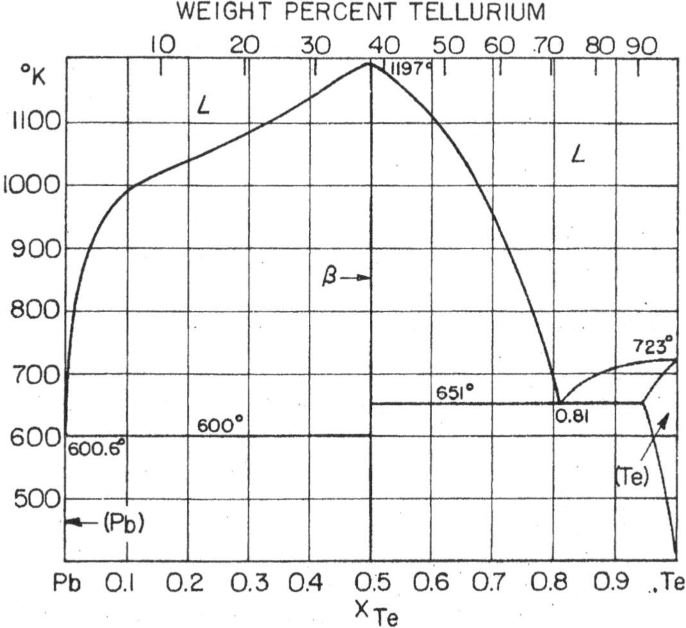

Fig. 8.1 Phase diagram of Pb-Te system. (Reproduced from *Selected Values of Thermodynamic Properties of Metals and Alloys* by R. Hultgren et al. [14], John Wiley, 1963)

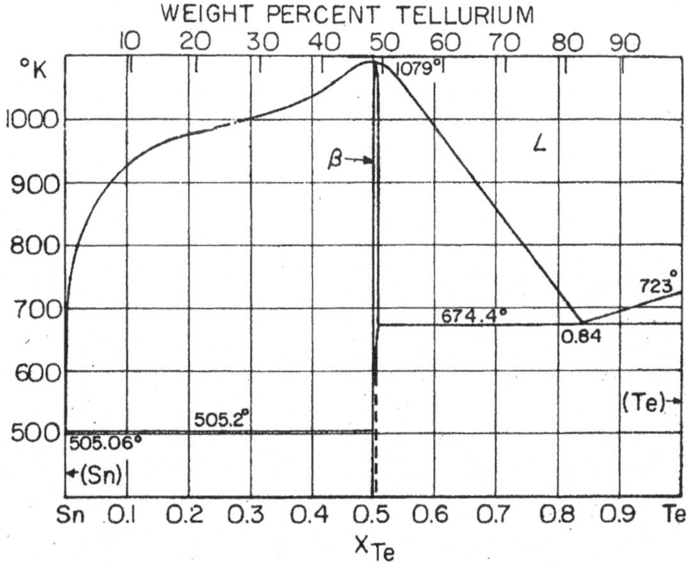

Fig. 8.2 Phase diagram of Sn-Te system. (Reproduced from *Selected Values of Thermodynamic Properties of Metals and Alloys* by R. Hultgren et al. [14], John Wiley, 1963)

Sn-Te melt may behave like a regular associated solution. The free energy of formation of SnTe at the melting point according to this analysis based on the procedure outlined in Sect. 8.2.3.2 is -10.6 kcal mol^{-1} which is in good agreement with the extrapolated value of -10.8 kcal mol^{-1} from the data of Mc Ateer and Seltz [15].

The regular associated behavior of molten Sn-Te melt is also reflected in its activity-composition relationship. The activity of tellurium in tellurium rich melt of Sn-Te may be calculated from the depression in the melting point of tellurium (Fig. 8.2) according to the Eq. 8.7. The calculated activity values exhibit a negative deviation from Raoults law. This is consistent with the postulates of existence of association in the melt as discussed in the preceding paragraph.

8.3.5 Ternary Melts

The theory of regular associated solution has been extended by Tung et al. [16] to the ternary system, Cd-Hg-Te. Brebrick [17] has developed thermodynamic equations governing the equilibrium in ternary solutions of Hg-Cd-Te and Hg-Zn-Te. Together with the associated solution model for the liquid with species Hg, Cd, Te, HgTe and CdTe (and similarly in the other system) the equations have been used to calculate thermodynamic properties of the system.

8.4 Relative Merits and Accuracy of Different Techniques

The error limit in the data obtained by vapor pressure and chemical equilibrium techniques is high and especially the vapor pressure technique is unreliable in the absence of the knowledge of the correct molecular weight of the vaporizing species. Other limitations of the chemical equilibrium technique such as immediate analysis of the quenched samples, container material problem have already been discussed. For the vapor pressure technique one needs mass spectrometer as an expensive accessory. The indirect calculation of heats of formation and fusion from the free energy of formation obtained by mass spectrometric technique involves very high error limit [18], up to $\pm 20\%$. The gas equilibration technique suffers from the lack of thermal equilibrium between the gas and solid or liquid phase due to the slow diffusion of species in the condensed phases. It has also been mentioned that only approximate thermodynamic data and information can be estimated from phase diagrams. The heats of formation, fusion and dissolution can be accurately measured by calorimetric techniques but the value of free energy of formation calculated from calorimetric data is inaccurate mostly due to the non-availability of reliable heat capacity data in the required temperature range.

The electrochemical technique using solid or fused salt electrolytes measures directly the free energy of formation. The accuracy of the free energy of formation of the compound or the partial molar free energy of a component in the solution depends on the accuracy of measurement of the open circuit emf of the reversible galvanic cell. A variation of 1 mV in emf would mean an accuracy of ±96 J in the value of ΔG^o *or* $\overline{G}_i$ for a chemical reaction involving transfer of 1 electron. However, calculation of entropy of formation from the slope of the plot of emf vs. temperature may not be reliable because a slight change in (*dE/dT*) can cause a large error in entropy value. Similarly the heat of formation calculated from the equation:

$$\Delta H = -nF[E - T(dE/dT)] \tag{8.20}$$

will involve higher error limits. Schmalzried [19] has estimated this error to be 2.1 kJ/mole or more in the temperature range of 100–200 K.

In the light of the above discussion it is advisable to determine the enthalpy and the free energy values by calorimetry and electrochemical technique respectively, and combine them, wherever possible, to arrive at the reliable values of entropies.

References

1. A.D. Pelton, H. Schmalzried, Met. Trans. **4**, 1395 (1973)
2. T. Heuman, B. Predel, Arch. Eisenhutten. **39**, 783 (1960)
3. E. Miller, K. Komarek, Trans. Met. Soc. AIME **236**, 832 (1966)
4. R.A. Swalin, *Themodynamics of Solids* (John Wiley, New York, 1970)
5. J. Steininger, J. Appl. Phys. **41**, 2713 (1970)
6. B. Predel, W. Schwermann, Z. Metallk. **61**, 58 (1970)
7. C. Wagner, Acta Met. **6**, 309 (1958)
8. M. Shamsuddin, S. Misra, in *Chemical Metallurgy – A Tribute to Carl Wagner*, ed. by N.A. Gockcen, (TMS – AIME, Warrendale, 1981), p. 241
9. W.F. Schottky, M.B. Bever, Acta Met. **6**, 320 (1958)
10. A.S. Jordan, Met. Trans. **1**, 239 (1970)
11. N. Chou, K.L. Komarek, E. Miller, Trans. Met. Soc. AIME **245**, 1553 (1969)
12. W.T. Thompson, S.N. Flengas, J. Electrochem. Soc. **118**, 419 (1971)
13. R.C. Sharma, J.C. Lin, Y.A. Chang, Met. Trans. **18B**, 237 (1987)
14. R. Hultgren, R.L. Orr, P.D. Anderson, K.K. Kelley, *Selected Values of Thermodynamic Properties of Metals and Alloys* (John Wiley, New York, 1963)
15. J.H. McAteer, H. Selts, J. Am. Chem. Soc. **58**, 2081 (1936)
16. T. Tung, L. Golonka, R.F. Brebrick, J. Electrochem. Soc. **128**, 451 (1981)
17. R.F. Brebrick, J. Crystal Growth **86**, 39 (1988)
18. O. Kubaschewski, C.B. Alcock, *Metallurgical Thermochemistry*, 4th edn. (Pergamon, Oxford, 1967)
19. H. Schmalzried, in *'Metallurgical Chemistry Symposium, 1971'* Proceedings of the Symposium, 1972, ed. by O. Kubaschewski (Brunel University and NPL, London, 1972), p 39, HMS0

Chapter 9
Miscellaneous Category

In this chapter the calorimetric–equilibrium technique based on the combination of the two techniques has been discussed in Sect. 9.1 whereas in Sect. 9.2, another combination of vapor pressure and electrical conductivity has been presented.

9.1 Calorimetric–Equilibrium Technique

Since despite the enormous literature on palladium-hydrogen system, the nature of the hydrogen interaction in palladium and its alloys was not well understood, Kleppa [1, 2] developed the calorimetric–equilibrium technique at the James Franck Institute, The University of Chicago to understand the behavior of hydrogen in palladium alloys. Historically, palladium-hydrogen happens to be the most widely studied system since Graham [3] discovered the absorption of hydrogen in palladium and palladium-silver alloys. In addition, the high solubility and mobility of hydrogen in the f.c.c. palladium lattice [4] as well as the superconducting [5] nature of the hydrogen-rich β-phase attracted researchers for detailed investigations in palladium and its alloys.

In Chaps. 4 and 5, calorimetry and chemical equilibria have been discussed in detail. The salient features of the equilibrium apparatus and the microcalorimeter used in the calorimetric–equilibrium technique are briefly given below.

9.1.1 Metal-Hydrogen Equilibrium Apparatus

The equilibrium apparatus, shown in Fig. 9.1 is based on the design of Veleckis and Edwards [6]. It was operated under vacuum of the order of 10^{-4} mm Hg or better, generated by a combination of mechanical fore-pump and an oil diffusion pump through a liquid nitrogen trap. Purified hydrogen after titanium sponge treatment

M. Shamsuddin, *Thermodynamic Measurement Techniques*, The Minerals, Metals & Materials Series, https://doi.org/10.1007/978-3-031-47118-6_9

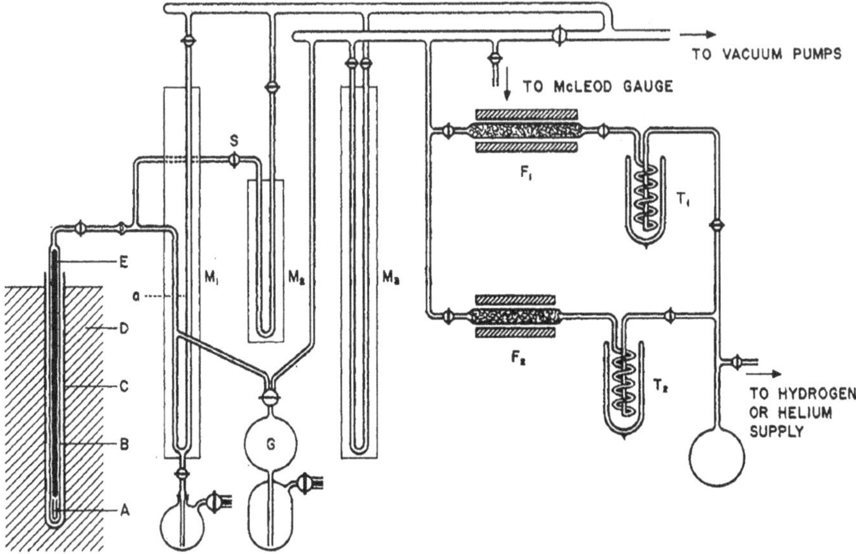

Fig. 9.1 Schematic diagram of hydrogen-metal equilibrium apparatus. (Reproduced from E. Veleckis and R. K. Edwards [6], J. Phys. Chemistry, 73 (1969) 683). A – porcelain specimen container; B – vycor tube; C – protection tube; D – liquid metal bath; E – gas displacement bell; F_1 – purification and storage furnace for hydrogen; F_2 – Ti sponge furnace; G – calibrated bulb for gas measurement; M_1 – mercury monometer; M_2/M_3 – oil monometer; T_1/T_2 – cold trap

collected in a 2 dm^3 bulb at near atmospheric pressure was delivered to the metal-hydrogen apparatus (calorimeter + equilibrium apparatus) in amounts ranging from 0.1 to 2.0 m.mol. The amount of hydrogen added in each experiment is based on the volume of a calibrated glass bulb (522.6 cm^3 at 294.85 K). The temperature of the gas in the bulb is noted by a Hg-in-glass thermometer and the pressure by a large octoil or a regular mercury manometer.

9.1.2 Calorimeter

It is a Calvet type twin micro-calorimeter (Fig. 4.5) suitable for work at temperature below 770 K. The apparatus consists of two nearly identical differential calorimetric units located in two cylindrical wells in a heavy aluminum jacket. In each of these twin units a temperature difference between the calorimeter proper and the surrounding jacket gives rise to an emf in a suitably constructed 96 couple thermopile. The two thermopiles are connected in series, bucked against each other. In this manner it is assured that small drifts in the temperature of the aluminum jacket will affect the ways of the two thermopiles in essentially the same way, but in the opposite sense.

Only one of the two calorimeters is involved in each experiment, the other one serves as a "dummy".

As long as no heat is generated in the calorimeter, a net emf is zero with no drift. If the heat effect is of short duration, the emf will first rise quite fast and will then decay exponentially to a near zero value in a period of the order of one hour or less. For heat effects of larger duration this period will be correspondingly longer.

Since the emf generated in the thermocouple system is directly proportional to the temperature difference between the calorimeter and the surrounding jacket. It is also proportional (according to Newton's Law) to the rule of heat transfer. The total area between the emf vs. time curve and the apparatus zero line accordingly will be proportional to the total heat effect associated with the process in the calorimeter. The calorimeter can be calibrated either electrically or by a 'drop' method i.e. an inert substance of known heat contribution can be dropped into the calorimeter, and the corresponding area determined. The results of these two independent calibration agreed to better than 0.5%. The output of the thermopiles of the twin calorimeter was fed to a Leeds an Northrup 9835 – B DC amplifier and from this in parallel circuits to a Leeds and Northrup Type G Azar recorder and to an electronic integrator. It was connected to a Hewlett–Packard Digital voltmeter, connector and printer.

9.1.3 Thermodynamic Investigations and Interpretation Based on the Above Combination

With the aid of the above calorimetric–equilibrium technique Kleppa and coworkers [7–14] studied the thermodynamics of solubility of hydrogen in alloys of palladium with silver, gold, copper, iron, nickel, cobalt and manganese. From the equilibrium pressure of hydrogen p over an alloy, the partial molar free energy is obtained from the relation:

$$\overline{G}_H = \frac{1}{2} RT \ln p_{H_2} \tag{9.1}$$

The relationship between the solubility of hydrogen in the alloy sample x and the equilibrium pressure p may be represented by Sieverts' law according to the equation:

$$\frac{10^3 x}{p^{1/2}} = a + bp^{1/2} \tag{9.2}$$

where, $x = \frac{n_H}{n_{Pd} + n_M}$ (M = Au, Ag, Cu, Ni, Co, Fe, Mn)

The coefficients a and b in Eq. 9.2, known as Sieverts' constants [15] may be used to calculate the excess partial molar free energy $\overline{G}_H^{xs}$ and its derivative with x:

$$\overline{G}_H^{xs}\,(x \to 0) = -RT \ln a\,(760)^{1/2} \tag{9.3}$$

$$\frac{\partial \overline{G}_H^{xs}}{\partial x} = -RT\left(1 + \frac{b}{a^2}\right) \tag{9.4}$$

The observed partial molar enthalpies are represented by the linear relation:

$$\overline{H}_H = c + dx \tag{9.5}$$

The interaction between two or more dissolved solutes at dilute concentrations in palladium can be examined through the effect of one on the activity coefficient of the other. Wagner [16] first introduced the concept of an interaction parameter, which for the effect of an alloying element M on the activity coefficient of hydrogen is expressed as

$$\varepsilon_H^M = \frac{\partial \ln \gamma_H}{\partial x_M} \tag{9.6}$$

where, γ_H and x_M are respectively, the activity coefficient of hydrogen and the atomic fraction of the alloying element M in the Pd-H-M system. A negative value ε_H^M means attraction between M and H atoms, whereas a positive value is indicative of M-H repulsion. The self-interaction parameter of hydrogen ε_H^H is defined as $\frac{\partial \ln \gamma_H}{\partial x_H}$, in which x_H refers to the atomic fraction of hydrogen in the alloy.

When palladium is alloyed with gold, silver or manganese (upto 33 at % Mn) the lattice parameter is increased, and when it is alloyed with copper, nickel, cobalt or iron, the lattice parameter is reduced. This correlates with the fact that alloying palladium with gold, silver and manganese makes the enthalpy of solution more exothermic, whereas alloying it with copper, nickel, cobalt and iron makes the enthalpy of solution less exothermic or indeed endothermic, depending on the alloy composition.

The solubility of hydrogen in palladium alloys which obeys Sieverts' law, expressed by Eq. 9.2, is shown in Fig. 9.2. In this expression, a and b are constants at constant temperature. The coefficient a is always positive and its numerical value is a direct measure of the strength of the interaction between hydrogen and the metallic solvent at high dilution; a low value shows a weak interaction and hence low solubility. Numerically, this will be reflected both in a correspondingly large positive value of $\overline{G}_H^{xs} = -RT \ln a \sqrt{760}$, (when $x \to 0$) and in a small negative or positive value of $\overline{H}_H$ $(x \to 0)$. The second coefficient b reflects the effective hydrogen-hydrogen interaction in the solution. This coefficient may have a positive, negative or zero value. A positive b indicates hydrogen-hydrogen attraction, a negative value repulsion and a value of zero means that the solution is ideal according to Henry's law. The values of the coefficients a and b for palladium and its alloys at 555 K and 700 K have been listed in Table 9.1 and also plotted in Figs. 9.3a and 9.3b.

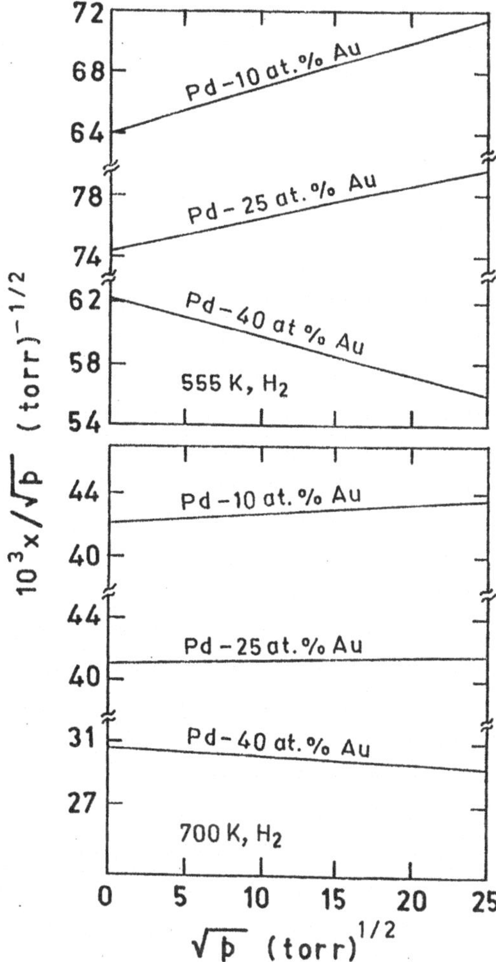

Fig. 9.2 $x/\sqrt{p}$ vs $\sqrt{p}$ for hydrogen in Pd-Au alloys at 555 and 700 K. (Reproduced from M. Shamsuddin and O. J. Kleppa [11] J. Chem. Phys., **71** (1979) 5154)

From Table 9.1 and Fig. 9.3a it is seen that the coefficient a is positive for palladium and all its alloys. The value is larger for Pd-Au, Pd-Ag and Pd-Mn (up to 33 at % Mn) than for pure palladium. This correlates with the fact that when palladium is alloyed with gold, silver and manganese (up to 33 at %), the lattice expands and more hydrogen enters the lattice, exhibiting thereby more exothermicity in the partial molar enthalpy of solution of hydrogen. By contrast a decrease in the value of a for Pd-Cu, Pd-Ni, Pd-Co and Pd-Fe alloys is consistent with the contraction of the palladium lattice and hence low solubility and less exothermicity in the partial molar enthalpy. In all the cases the solubility of hydrogen (i.e. a) decreases with an increase in temperature (Table 9.1).

The values of the interaction parameters listed in Table 9.2 show the effect of the alloying elements, i.e. gold, silver, copper, nickel, cobalt, iron and manganese (dilute concentration) on the activity coefficient of hydrogen in palladium.

Table 9.1 Summary of thermodynamic data[a] for dilute solutions of hydrogen in palladium and its alloys

	555 K.H_2		700 K.H_2		
Solvent	10^3 *a* (Torr $^{-1/2}$)	10^6 *b* (Torr $^{-1}$)	10^3 a (Torr$^{-1/2}$)	10^6 b (Torr $^{-1}$)	555 K, H_2 ε_H^H
Pd	0.500	2.60	0.373	0.90	−8.86
$Pd_{0.90}Au_{0.10}$	0.640	3.00	0.422	0.54	−4.03
$Pd_{0.75}Au_{0.25}$	0.743	2.15	0.410	0.02	−2.48
$Pd_{0.60}Au_{0.40}$	0.624	−2.74	0.306	−0.50	+9.13
$Pd_{0.90}Ag_{0.10}$	0.914	7.8	0.572	2.0	−6.89
$Pd_{0.75}Ag_{0.25}$	2.042	21.9	0.964	3.7	−2.68
$Pd_{0.60}Ag_{0.40}$	3.12	−15.9	1.19	−1.4	+20.49
$Pd_{0.90}Cu_{0.10}$	0.404	0.99	–	–	−1.79
$Pd_{0.75}Cu_{0.25}$	0.269	0.23	–	–	−0.06
$Pd_{0.60}Cu_{0.40}$	0.139	0.05	–	–	(0)
$Pd_{0.90}Ni_{0.10}$	0.229	0.52	0.192	0.08	−1.07
$Pd_{0.75}Ni_{0.25}$	0.068	0.00	0.069	0.00	0.00
$Pd_{0.90}Co_{0.10}$	0.201	0.30	–	–	(−14)
$Pd_{0.90}Fe_{0.10}$	0.177	9.30	–	–	(−8)
$Pd_{0.90}Mn_{0.10}$	0.503	1.716	0.342	0.629	−1.05
$Pd_{0.82}Mn_{0.18}$	0.574	3.34	0.310	0.695	−10.04
$Pd_{0.67}Mn_{0.33}$	0.537	- 8.82	0.154	0.00	+30.15
$Pd_{0.60}Mn_{0.40}$	0.033	0.26	0.021	0.00	(−5)
$Pd_{0.50}Mn_{0.50}$	0.008	0.00	0.004	0.00	0.00

[a]Reproduced from M. Shamsuddin [17] J. Less Common Metals, **154** (1989) 285

In terms of inter atomic attraction the negative values of ε_H^{Ag}, ε_H^{Au} and ε_H^{Mn} may be understood on the basis of a very simple model, illustrated by the palladium-hydrogen-gold or silver or manganese solutions. In Pd-H solutions each hydrogen atom is surrounded by a number of palladium atoms which share the Pd-H bonding energy. The bonding of gold, silver and manganese with hydrogen is stronger than the Pd-H bonding and this leads to two results: (i) the ratio Au: Pd, Ag: Pd or Mn: Pd is greater in the nearest neighbourhood of hydrogen atoms than in the bulk of the solution and (ii) the hydrogen atoms become more firmly bonded as the gold or silver or manganese concentration increases. The values of the interaction parameters (Table 9.2) demonstrate that the effect of silver on hydrogen is greater than that of gold while that of manganese is much lower. The extent of this effect depends on the amount of expansion on alloying (Table 9.2) which is due to the repulsion of palladium and gold or silver or manganese (around 33 at %) atoms in the nearest neighborhood.

Copper, nickel cobalt and iron show a positive effect, i.e. the activity coefficient of hydrogen in palladium increases on alloying with these elements. Each atom of copper, nickel, cobalt or iron is probably bonded, like hydrogen, to a nearest-neighbor shell of palladium atoms. Bonding of palladium atoms to copper, nickel, cobalt or iron leaves less palladium to be bonded with hydrogen and raises the

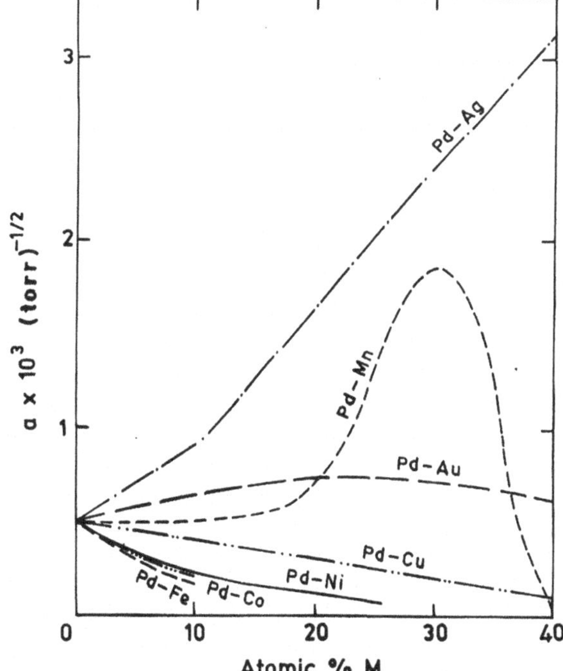

Fig. 9.3a *a* vs. alloy composition for solutions of hydrogen in palladium alloys at 555 K. (Reproduced from M. Shamsuddin [17] J. Less Common Metals, **154** (1989) 285)

activity coefficient of the hydrogen. This competition for palladium atoms increases the activity coefficient of each competing species so long as the effect is not overshadowed by stronger intersolute attractive forces as in Pd-H-Cu, Pd-H-Ni, Pd-H-Co and Pd-H-Fe solutions.

Figure 9.3b very clearly demonstrates that the coefficient b (i.e. slope of plots of $x/\sqrt{p}$ vs $\sqrt{p}$) decreases and then changes its sign from positive to negative with an increase in gold content from 10 to 40 at % at both the temperatures 555 and 700 K. A similar trend has been observed in $\Delta\overline{H}_H$ vs. x plots (Fig. 9.4). From Fig. 9.3b it is evident that b changes its sign from positive to negative at around 30–40% Au, Ag or Mn. This means that the hydrogen-hydrogen interaction changes from attractive to repulsive in this concentration range. The numerical values of b for Pd-Cu, Pd-Ni and Pd-Co at 555 and 700 K are quite low, indicating that these solutions are nearly ideal. In these alloys b approaches zero with an increasing quantity of alloying elements and an increase in temperature (Table 9.1). In the Pd-Mn system the nature of the interaction changes again from repulsive to attractive near 40 at % Mn. The successive changes from attractive to repulsive and then repulsive to attractive in Pd-Mn alloys indicate the complex nature of the system in the composition range 30–40 at % Mn, due to the order-disorder transformations.

The change in the nature of the hydrogen-hydrogen interaction from attractive to repulsive in palladium alloys at around 30–40 at % Au, Ag and Mn is also evident from the change in sign of the self-interaction parameters of hydrogen, ε_H^H. From

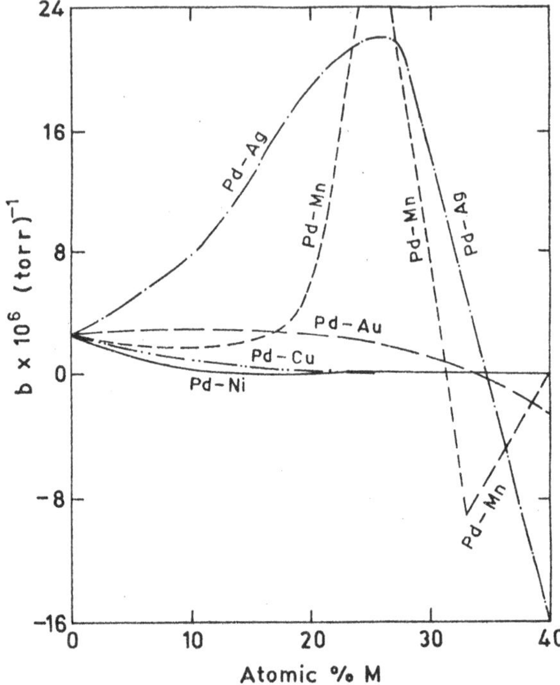

Fig. 9.3b b vs. alloy composition for solutions of hydrogen in palladium alloys at 555 K. (Reproduced from M. Shamsuddin [17] J. Less Common Metals, **154** (1989) 285)

Table 9.2 Atomic interaction in palladium alloys[a]

Interaction Parameter	Alloying additions (M)						
	Au	Ag	Cu	Ni	Co	Fe	Mn
ε_H^M	−2.36	−6.42	2.63	8.58	9.70	10.96	−0.18

[a]Reproduced from M. Shamsuddin [17] J. Less Common Metals, **154** (1989) 285

Table 9.1 it is clear that ε_H^H is negative or positive respectively for the positive or negative values of b. A positive value of b or negative value of ε_H^H indicates hydrogen-hydrogen attraction. A negative b or positive ε_H^H means repulsion. A zero value of b or ε_H^H shows that the solution is ideal in the sense of Henry's law. This ideal behavior has been observed for the solubility of hydrogen in $Pd_{0.75}Ni_{0.25}$ and $Pd_{0.5}Mn_{0.5}$ at 555 K and in $Pd_{0.75}Ni_{0.25}$, $Pd_{0.67}Mn_{0.33}$, $Pd_{0.6}Mn_{0.4}$ and $Pd_{0.5}Mn_{0.5}$ at 700 K (Table 9.1).

Thus, the data generated by the calorimetric–equilibrium technique in Kleppa's laboratory [7–14] at the James Franck Institute, University of Chicago have been quite useful in analyzing the nature of interaction of hydrogen in palladium alloys. A detailed analysis of the significance of the calorimetric measurements and discussions of the special experimental problems associated with calorimetric–equilibrium

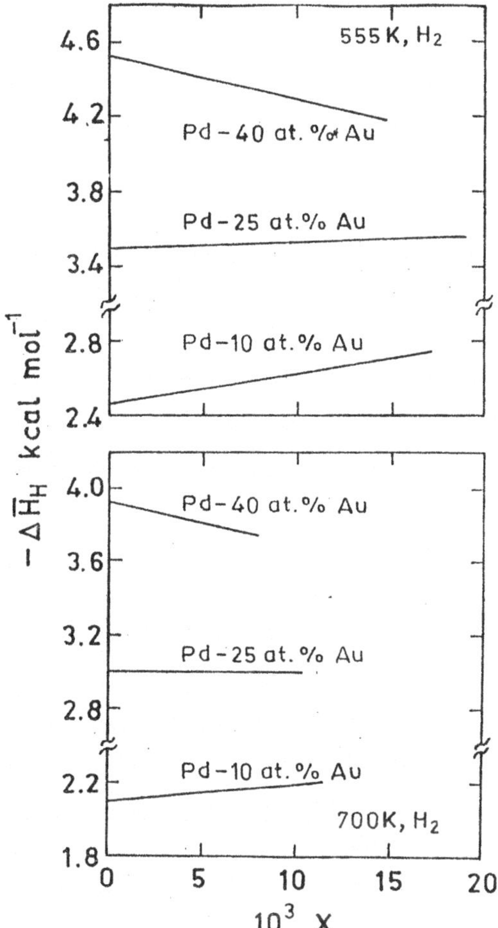

Fig. 9.4 $\Delta \overline{H}_H$ vs. x for hydrogen in Pd-Au alloys at 555 and 700 K. (Reproduced from M. Shamsuddin and O. J. Kleppa [11] J. Chem. Phys., **71** (1979) 5154)

studies of the palladium-hydrogen systems, are given elsewhere [2]. The following important conclusions [17] have been drawn from these studies on palladium alloys:

1. The more exothermic heat of solution of hydrogen in Pd-Au, Pd-Ag and Pd-Mn (upto 33 at % Mn) is in conformity with the expansion of palladium lattice on alloying with gold, silver and manganese and the negative values of the interaction parameters, ε_H^{Au}, ε_H^{Ag} and ε_H^{Mn}.
2. The less exothermic or endothermic values of the heat of solution of hydrogen in Pd-Cu, Pd-Ni, Pd-Co and Pd-Fe alloys are in conformity with the contraction of palladium lattice on alloying with copper, nickel, cobalt or iron and the positive values of the ε_H^{Cu}, ε_H^{Ni}, ε_H^{Co} and ε_H^{Fe} interaction parameters,
3. The hydrogen-hydrogen interaction in Pd-Au, Pd-Ag and Pd-Mn alloys changes from attractive to repulsive at around 30–40 at % Au, Ag and Mn. The nature of interaction in Pd-Mn alloys is more complex due to the order-disorder transformation in the system.

9.2 Vapor Pressure and Electrical Conductivity

Cadmium telluride (CdTe) is the only II-VI compound which can be prepared as n- as well as p-type semiconductor easily by controlling the non-stoichiometry. Since point defects associated with the deviation from stoichiometry strongly affect the electrical and optical characteristics of a compound, it is essential to know the defect structure of both n- and p-type CdTe. There are plenty of literature on electrical properties of this compound but the defect structure has not been systematically studied. The major investigations regarding its defect structure has been restricted to the electrical characterization of n-type CdTe in equilibrium with Cd vapor [18–26]. Smith [18] has measured the electrical conductivity of CdTe by controlling the vapor pressure of either Cd or Te but he was not successful in investigating the effect at lower pressure of tellurium. In order to measure the electrical conductivity of polycrystalline CdTe in the very low ambient tellurium pressure ranges varying from 1.0×10^{-15} to 2.0×10^{-5} atm in the temperature range 675–975 K an especially designed experimental assembly [27] was designed.

9.2.1 Experimental Assembly

A sample holder assembly shown in Fig. 9.5, made of two piston type silica tubes is housed in the especially designed experimental assembly, schematically shown in Fig. 9.6. A CdTe pellet (cylindrical rod), placed between two platinum disks was pressed by two silica tubes open at both ends, with the aid of nichrome springs (Fig. 9.5). To each of the disks one Pt/Pt-13% Rh thermocouple was welded. The

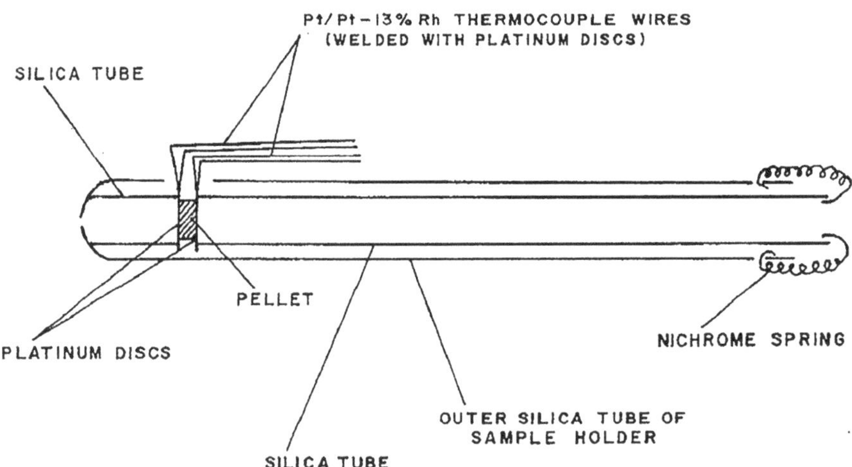

Fig. 9.5 Schematic illustration of the experimental sample holder assembly. (Reproduced from M. Shamsuddin et al. [27] J. Appl. Phys. **74** (1993) 6208)

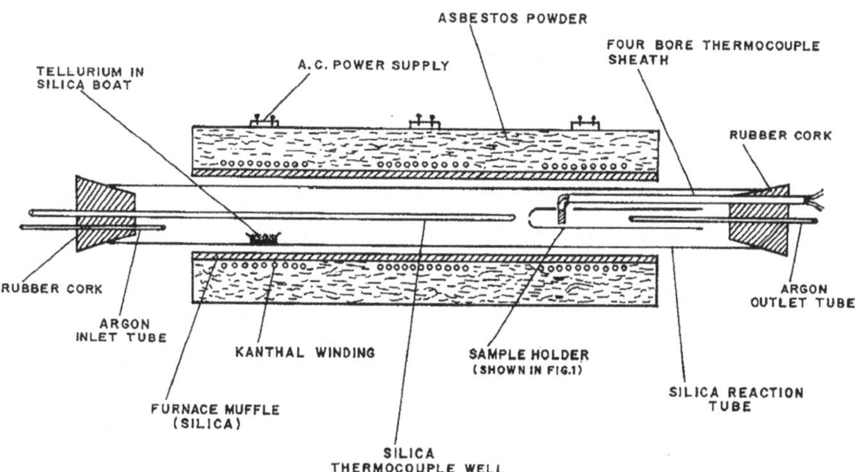

Fig. 9.6 A three zone furnace housing the sample holder assembly. (Reproduced from M. Shamsuddin et al. [27] J. Appl. Phys. **74** (1993) 6208)

entire sample holder assembly, kept in a three zone furnace (Fig. 9.6), was evacuated to 1.3×10^{-5} atm and purged with the high purity argon (containing oxygen impurity less than 2 ppm) repeatedly four times. The tellurium vapors of desired pressure were generated by heating tellurium, kept in a silica boat at a predetermined temperature which was measured with a chromel/alumel thermocouple positioned in the protection tube (Fig. 9.6). This thermocouple was also used to observe temperatures of various zones of the furnace. The CdTe pellet was maintained at higher temperature as compared to that of the tellurium to prevent condensation of tellurium vapor on the silica reaction tube in between the tellurium reservoir and the CdTe sample pellet (cylindrical rod). The temperature of each zone of the furnace was independently controlled by three different temperature controllers each having an accuracy of ± 0.5 K. A slow stream of argon (10 c. c. per minute) was passed in the experimental assembly throughout the entire measurements.

The temperature of both sides of the sample rod and tellurium reservoir were measured by a digital multimeter (Keithley, USA; Model 195A). The temperature of the CdTe rod was taken as an average of these two temperatures. The platinum wires of the thermocouples were used as lead wires for the measurement of conductivity as well as the thermo-emf of CdTe sample. The conductivity of CdTe was measured as a function of temperature (675–975 K), and partial pressure of tellurium ($1.0 \times 10^{-15} - 2.0 \times 10^{-5}$ atm) by Impedance Analyser (Hewlett–Packard, Model 4192A LF) at a frequency of 1 kHz.

In order to check the reproducibility of the results, readings were taken during heating as well as cooling cycles of both tellurium reservoir and sample. In addition, results were obtained on pellets prepared by compaction of CdTe powders as well as cast rods of CdTe. The trend observed were identical in each case within the experimental error.

9.2.2 Role of Tellurium Pressure on the Mode of Conduction in Cadmium Telluride

The specific conductivity (σ) was calculated from the measured values of resistance (R), cross-sectional area (a), and length (l) of the sample (cylindrical pellet/rod) using the relationship ($\sigma = l/R.\ a$). The effect of temperature and ambient tellurium pressure on the specific conductivity of CdTe has been presented in Fig. 9.7. At all the temperatures the conductivity increases with the increase in tellurium pressure up to a critical pressure. After attaining the maximum, it started decreasing with further increase in tellurium pressure. The abrupt change in conductivity occurs in the pressure range of $1.3 \times 10^{-9} - 1.3 \times 10^{-6}$ atm with the minimum around 4.2×10^{-8} atm. However, the conductivity does not vary significantly beyond this pressure range. From Fig. 9.7, it is clearly evident that conductivity versus tellurium pressure plot is divided into two regions, namely, low tellurium pressure region: 1.0×10^{-15} – $\sim 4.2 \times 10^{-8}$ atm and high tellurium pressure region: $\sim 4.2 \times 10^{-8} - 2.0 \times 10^{-5}$ atm. From the thermo-emf measurements of CdTe it is concluded that the conduction in CdTe is p type in the low tellurium pressure region and n type in the high tellurium pressure region. However, doping with monovalent silver and trivalent indium has shown different effects. The conduction in all of the samples was found to be n type at the tellurium pressure of 1.7×10^{-6} whereas substitution of cadmium with indium in the low pressure region changes the conductivity of CdTe from p to n type.

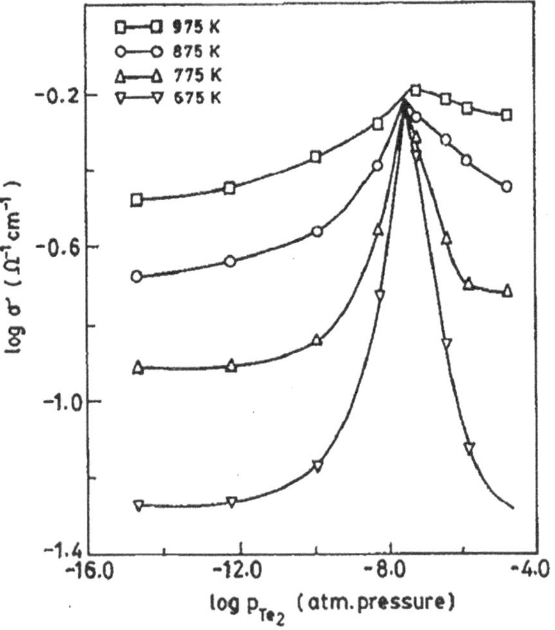

Fig. 9.7 Effect of ambient tellurium pressure on the electrical conductivity of cadmium telluride. (Reproduced from M. Shamsuddin et al. [27] J. Appl. Phys. **74** (1993) 6208)

The predominant *n* type conduction, noted in the tellurium pressure range ~$4.2 \times 10^{-8} - 2.0 \times 10^{-5}$ atm is due to the presence of either the excess of cadmium as interstitials or deficiency of tellurium as a vacancy at the tellurium site. In this region the electrical conductivity of CdTe increases with the decrease of tellurium pressure at all temperatures (Fig. 9.7). However, the change is more significant at lower temperatures. In this region, the increase in conductivity with a decrease in tellurium pressure is due to the increase of cadmium interstitials or tellurium vacancies. These electroactive point defects increase the concentration of electrons according to the following reactions:

$$Cd_i = Cd^{2+} + 2e^- \tag{9.7}$$

$$V_{Te} = V_{Te}^{2+} + 2e^- \tag{9.8}$$

The generation of electrons in the tellurium pressure range ~$4.2 \times 10^{-8} - 2.0 \times 10^{-5}$ atm confirms the existing hypothesis of n type conduction.

Fig. 9.7 further indicates that the conductivity of CdTe increases with the increase of tellurium pressure in the low tellurium pressure region: 1.0×10^{-15} – ~4.2×10^{-8}atm. Occurrence of *p*-type conduction in this region is due the replacement of tellurium (at low pressure) by cadmium (present at high pressure). This happens to be a case of anti-structural point defect and hence, the conduction is governed by the concentration of holes generated according to the following equilibrium:

$$Cd_{Te} = Cd_{Te}^{2-} + 2h^+ \tag{9.9}$$

Thus, any decrease in tellurium pressure results in increase in hole concentration and hence the conductivity should increase. But the conductivity has to decrease with the decrease in tellurium pressure. It may be due to precipitation of excess cadmium at very low tellurium pressure. This influences the total conduction which may attributed due to the presence of impurities. Nonstoichiometric defect concentration in CdTe is of the order of 10^{16} cm^{-3}. Since impurities less than 1 ppm exert substantial influence on the overall charge carriers the most likely impurity oxygen occupying the interstitial site may give rise to positive holes as per the following equilibrium:

$$\frac{1}{2}O_2 = O^{2-} + 2h^+ \tag{9.10}$$

Thus the above combination technique has been useful in studying the effect of vapor pressure on mode of conduction in polycrystalline cadmium telluride.

References

1. O.J. Kleppa, M.E. Melnichak, T.V. Charlu, J. Chem. Thermodyn. **5**, 595 (1973)
2. G. Boureau, O.J. Kleppa, J. Chem. Thermodyn. **9**, 543 (1977)
3. T. Graham, Philos. Trans. R. Soc. London **156**, 415 (1966)
4. E.A. Lewis, *The Palladium Hydrogen System* (Academic, London, 1967)
5. T. Skoskiewicz, Ber. Bunsenges. Phys. Chem **76**, 874 (1972).; Phys. Status Solidi. **K123** 6 (1972)
6. E. Veleckis, R.K. Edwards, J. Phys. Chem. **73**, 683 (1969)
7. G. Baureau, O.J. Kleppa, P. Dantzer, J. Chem. Phys. **64**, 5247 (1976)
8. G. Boureau, O.J. Kleppa, J. Chem. Phys. **65**, 3915 (1976)
9. C. Picard, O.J. Kleppa, G. Boureau, J. Chem. Phys. **70**, 2710 (1979)
10. O.J. Kleppa, M. Shamsuddin, C. Picard, J. Chem. Phys. **71**, 1656 (1979)
11. M. Shamsuddin, O.J. Kleppa, J. Chem. Phys. **71**, 5154 (1979)
12. R.C. Phutela, O.J. Kleppa, J. Chem. Phys. **75**, 4095 (1981)
13. R.C. Phutela, O.J. Kleppa, J. Chem. Phys. **76**, 1525 (1982)
14. M. Shamsuddin, O.J. Kleppa, J. Chem. Phys. **80**, 3760 (1984)
15. A. Sieverts, Z. Phys. Chem. **88**, 451 (1914)
16. C. Wagner, *Thermodynamics of Alloys* (Addison-Wesley, Reading, 1952)
17. M. Shamsuddin, J. Less Common Metals **154**, 285 (1989)
18. F.T.J. Smith, Metall. Trans. **1**, 617 (1970)
19. R.C. Whelan, D. Shaw, Phys. Status Solidi **29**, 145 (1968)
20. K. Zanio, Appl. Phys. Lett. **15**, 260 (1969)
21. K. Zanio, Appl. Phys. **41**, 1935 (1970)
22. Y.V. Rud, K.V. Sanin, Sov. Phys. Semicond. **5**, 244 (1971)
23. M.R. Lorenz, B. Segall, H.H. Woodbury, Phys. Rev. A **134**, 751 (1964)
24. O.M. Matveev, Y.V. Rud, K.V. Sanin, Sov. Phys. Semicond. **3**, 779 (1969)
25. L. Carlsson, C.N. Ahlquist, J. Appl. Phys. **43**, 2529 (1972)
26. F.A. Selim, F.A. Kroger, J. Electrochem. Sot. **124**, 401 (1977)
27. M. Shamsuddin, A. Nasar, V.B. Tare, J. Appl. Phys. **74**, 6208 (1993)

Chapter 10
Worked Out Examples

This chapter provides over 100 worked out problems related to classical as well as solution thermodynamics and measurement techniques. The problems cover both theoretical as well as practical and technical aspects. The wide range of problems include calculation of the adiabatic flame temperature, refining of lead bullion, and crude nickel, production of magnesium by Pidgeon process, solubility of gases in metals and deoxidation of steel.

Problem 1 Cadmium boils at 766 °C under 1 atm with a latent heat of evaporation of 23.87 kcal g atom^{-1}. Calculate the increment of the internal energy accompanying the volatilization of 1 mol of liquid cadmium at the boiling point.

Solution
For the calculation of change in internal energy (ΔU) the following equation applies to the reactions under constant pressure:

$$\Delta U = \Delta H - P\Delta V$$

$$\begin{aligned}
\Delta H &= \mathrm{L^e} = 23870 \text{ cal g atom}^{-1}, \text{atomic wt.of Cd} = 112.40 \\
\Delta V &= \text{volume of } 1 \text{ mol of Cd (g)} - \text{volume of } 1 \text{ mol of Cd (l)} \\
&= 22.4 \text{ (liter)} \times \frac{766 + 273}{273} - \frac{112.4}{8} = 85251 \text{ (liter)–14 cc} = 85.24 \text{ liters}
\end{aligned}$$

$$\begin{aligned}
P\Delta V &= 1 \text{ atm} \times 85.24 \text{ liter} = 85.24 \text{ l.atm} \qquad \left(1 \text{ lit atm} = 24.20 \text{ cal mol}^{-1}\right) \\
&= 85.24 \times 24.20 = 2063.2 \text{ cal mol}^{-1}
\end{aligned}$$

$$\therefore \Delta U = \Delta H - P\Delta V = 23870\text{–}2063 = 21807 \text{ cal mol}^{-1} \qquad \text{Ans.}$$

M. Shamsuddin, *Thermodynamic Measurement Techniques*, The Minerals, Metals & Materials Series, https://doi.org/10.1007/978-3-031-47118-6_10

Problem 2 Calculate the standard heat of formation of solid aluminum carbide from the following data:

$$\text{Heat of formation of CO (g)} = -26.40 \text{ kcal mol}^{-1}$$
$$CO_2(g) = -94.05 \text{ kcal mol}^{-1}$$
$$Al_2O_3(s) = -400.0 \text{ kcal mol}^{-1}$$

The heat of reaction between 1 mole of Al_4C_3 and 9 moles of CO_2 to produce alumina and CO at 25 °C is −223.7 kcal.

Solution
For the reaction: $Al_4C_3(s) + 9\,CO_2\,(g) = 2\,Al_2O_3(s) + 12CO\,(g)$, $\Delta H = -223.7$ kcal
From the given data and according to the Hess's law we can write:

$$\Delta H = 2\Delta H_{Al_2O_3} + 12\Delta H_{CO} - \Delta H_{Al_4C_3} - 9\Delta H_{CO_2}$$

$$\begin{aligned}\therefore\ \Delta H_{Al_4C_3} &= 2\Delta H_{Al_2O_3} + 12\Delta H_{CO} - 9\Delta H_{CO_2} - \Delta H\\ &= 2 \times (-400) + 12 \times (-26.40) - 9 \times (-94.05) - (-223.7)\\ &= -46.65 \text{ kcal}\end{aligned}$$ Ans.

Problem 3 Calculate the heat of reaction on dissolution of solid magnesium in a dilute acid consisting of HCl (100 H_2O) from the following data:

1. $Mg\ (s) + 2HCl(g) = MgCl_2(s) + H_2(g)$, $\Delta H_1 = -109.4$ kcal
2. $HCl\ (g) + 100H_2O(l) = HCl\ (100H_2O)\ (l)$, $\Delta H_2 = -17.4$ kcal
3. $MgCl_2\ (s) + 200\ H_2O(l) = MgCl_2(200H_2O)\ (l)$, $\Delta H_3 = -35.5$ kcal

Solution
The heat of reaction (ΔH) for the reaction under question:
$Mg\ (s) + 2\ [HCl\ (100H_2O)]\ (l) = MgCl_2\ (200H_2O)\ (l) + H_2\ (g)$ is obtained as:

$$\Delta H = \Delta H_1 + \Delta H_3 - 2\Delta H_2 = -109.4 - 35.5 - 2 \times (-17.4) = -110.1 \text{ kcal}$$ Ans.

Problem 4 Calculate the heat of reaction for the reduction of hematite (Fe_2O_3) in the blast furnace according to the reaction: $Fe_2O_3 + 3C + \frac{3}{2}\ O_2 = 2Fe + 3CO_2$ from the following data:

1. $\frac{3}{2}C + \frac{3}{2}O_2 = \frac{3}{2}CO_2$, $\Delta H_1 = -591.6$ kJ
2. $\frac{3}{2}CO_2 + \frac{3}{2}C = 3CO$, $\Delta H_2 = +258.6$ kJ
3. $Fe_2O_3 + \frac{1}{3}CO = \frac{2}{3}Fe_3O_4 + \frac{1}{3}CO_2$, $\Delta H_3 = -17.6$ kJ
4. $\frac{2}{3}Fe_3O_4 + \frac{8}{3}CO = 2Fe + \frac{8}{3}CO_2$, $\Delta H_4 = +10.5$ kJ

Solution
According to the Hess' law of constant heat summation, the heat of reaction (ΔH) for the reaction under question: $Fe_2O_3 + 3C + \frac{3}{2}\ O_2 = 2Fe + 3CO_2$ is equivalent to

$\Delta H = \Delta H_1 + \Delta H_2 + \Delta H_3 + \Delta H_4 = 258.6 - (591\,6 + 17.6 + 10.5) = -361.1$ kJ Ans.

Problem 4 The heat of evolution during oxidation of metallic iron to hematite and magnetite are given below:

(a) 2 Fe (s) $+ \frac{3}{2} O_2\ (g) = Fe_2O_3(s) + 196.3$ kcal mol^{-1} of hematite
(b) 3 Fe (s) + 2 $O_2\ (g) = Fe_3O_4(s) + 266.9$ kcal mol^{-1} of magnetite

Calculate the heat accompanying the following reaction.

$$\text{(c) } 2\ Fe_3O_4(s) + \frac{1}{2} O_2\ (g) = 3\ Fe_2O_3\,(s)$$

Solution
According to Hess' law the required reaction (c) is equivalent to 2 × (b) – 3 × (a)

$$\therefore \Delta H_c = -2 \times (-266.9) - 3 \times (-196.3) = -55.1 \text{ kcal} \qquad \text{Ans.}$$

Problem 5 The heat of combustion of H_2 (g), C (s) and that of C_2H_2 (g) are respectively, −68.317 kcal, −94.052 kcal and −310.62 kcal. Calculate the amount of heat required to synthesize C_2H_2 (g).

Solution
It is recommended to write each reaction explicitly, and then apply Hess' law of constant heat summation:

(a) $H_2\ (g) + \frac{1}{2} O_2\ (g) = H_2O\ (l)$ $\quad \Delta H_a = -68.317$ kcal
(b) C (s) + $O_2\ (g) = CO_2\,(g)$ $\quad \Delta H_b = -94.052$ kcal
(c) $C_2H_2\ (g) + \frac{5}{2} O_2\ (g) = 2CO_2\,(g) + H_2O\ (l)$ $\quad \Delta H_c = -310.62$ kcal

The reaction under question (d) 2C (s) + H_2 (g) = C_2H_2 (g) is equivalent to (a) + 2 × (b) – (c)

$$\therefore \Delta H_d = -68.317 - 2 \times 94.052 + 310.62 = 54.199 \text{ kcal}$$

Hence, 54.199 kcal is required to synthesize C_2H_2. Ans.

Problem 6 How many calories are required to heat 107.87 g of pure silver from 25 to 1127 °C? Silver melts at 960.8 °C with the latent heat of fusion, L^f of 2.65 kcal mol^{-1}.

Given:

$$C_{p(Ag,\ s)} = 5.09 + 2.04 \times 10^{-3}\ \text{T} - 0.36 \times 10^5 \text{T}^{-2} \quad \text{cal deg}^{-1}\text{mol}^{-1}$$

$$C_{p(Ag,\ l)} = 7.30 \text{ cal deg}^{-1}\text{mol}^{-1}, \text{Atomic weight of silver 107.87.}$$

Solution

1. Heat required to heat 1 mol (107.87 g) of siver from 298 to 1233.8 K

$$= \int_{298}^{1233.8} \left(5.09 + 2.04\ x\ 10^{-3} T - 0.36\ x\ 10^5 T^{-2}\right) dT$$
$$= \left[5.09\ T + \frac{1}{2} \times 2.04 \times T^2 - (-1)\ 0.36\ x\ 10^5 T^{-1}\right]_{298}^{1233.8}$$
$$= 4764.2 + 1462.6 + 254.5 = 6481.3\ \text{cal}$$

2. Heat of fusion, $L^f = 2650$ cal
3. Heat required to heat molten silver from 1233.8 to 1400 °C

$$= \int_{1233.8}^{1400} 7.30.\ dT = 7.30\ \times\ (1400–1233.8) = 1211.8\ \text{cal}$$

$$\therefore \text{Total heat required} = 6481.3 + 2650 + 1211.8 = 10343\ \text{cal}/107.87\ g$$
$$\left(\text{i.e.}10343\ \text{cal mol}^{-1}\right) \qquad \text{Ans.}$$

Problem 7 The enthalpy of combustion of 1 mol of CO (g) to CO_2 (g) at 25 °C and at a constant pressure of 1 atm is –67.6 kcal. What is the maximum temperature that may be reached by burning CO (g) in air (20% O_2 and rest N_2)? Neglect heat loss by radiation. Respective values of the molar heat capacity of N_2 (g) and CO_2 (g) may be assumed as 7 and 9 cal deg^{-1} mol^{-1}.

Solution

The reaction under question is

$$CO(g) + \frac{1}{2} O_2(g) = CO_2(g)\Big(1\ \text{mol of CO needs 0.5 mol of}\ O_2, \text{which is 1}$$
$$0.5\ O_2 + 2\ N_2 \quad 1 \quad \text{accompanied by 2 mol}\ N_2\Big)$$

Let ΔT °C is the temperature after combustion.

$$67600\ \text{cal} = \Delta T\ \big(1\ \text{mol} \times 9\ \text{cal deg}^{-1}\ \text{mol}^{-1} + 2\ \text{mol} \times 7\ \text{cal deg}^{-1}\ \text{mol}^{-1}$$

Or

$$67600 = \Delta T\,(9 + 14)$$
$$\therefore\ \Delta T = 67600/23 = 2939\ \text{deg}$$

Since 67600 cal has been measured at 25 °C, the maximum temperature attained would be

$$2939 + 25 = 2964°\text{C} \qquad \text{Ans.}$$

Problem 8 The mean molar heat capacities at constant pressure, of hydrogen, oxygen and water vapor in the temperature range from 25 to 100 °C are as follows: H_2 (g) 6.92, O_2 (g) 7.04, H_2O (g) 8.03 cal deg^{-1} mol^{-1}. Assuming that the heat of formation of water at 25 °C is −57.80 kcal mol^{-1}. Calculate the standard heat of formation of water vapor at 100 °C.

Solution
The reaction under question is

$$H_2\ (g) + \frac{1}{2}O_2\,(g) = H_2O\ \ (g)$$

Since the heat of formation of water at 25 °C is known, it can be calculated at 100 °C making use of the Kirchhoff's equation: $\Delta H_{373} = \Delta H_{298} + \int_{298}^{373} \Delta C_p.dT$

$$\begin{aligned}\Delta C_p &= C_p(\text{product}) - C_p(\text{reactant})\\ &= C_{p,H_2O} - \left[C_{p,H_2} + \frac{1}{2}C_{p,}O_2\right] = 8.03 - 6.92 - 3.52 = -2.41\end{aligned}$$

$$\begin{aligned}\therefore\ \Delta H_{373} &= -57800 - \int_{298}^{373} 2.41.\ dT\\ &= -57800 - 2.41\ (373 - 298)\\ &= -57980\ \text{cal mol}^{-1} \qquad \text{Ans.}\end{aligned}$$

Problem 9 Calculate the heat of the following reaction at 1000 K
Fe_2O_3(s) + 3C (s) = 2Fe_α(s) + 3CO (g) from the following data:

$$C_{p(\alpha,\ Fe)} = 4.18 + 5.92 \times 10^{-3}\ \text{T cal deg}^{-1}\text{mol}^{-1}$$

$$C_{p(CO)} = 6.79 + 0.98 \times 10^{-3}\text{T} - 0.11 \times 10^5\ \text{T}^{-2}\ \text{cal deg}^{-1}\text{mol}^{-1}$$

$$C_{p(Fe_2O_3)} = 23.39 + 18.6 \times 10^{-3}T - 3.55 \times 10^5 T^{-2}\ \text{cal deg}^{-1}\text{mol}^{-1}$$

$$C_{p(C)} = 4.10 + 1.02 \times 10^{-3}\ \text{T} - 2.10 \times 10^5\text{T}^{-2}\ \ \text{cal deg}^{-1}\text{mol}^{-1}$$

The heats of formation of Fe_2O_3 and CO at 298 K are −19700 – 26400 cal mol^{-1}, respectively.

Solution

$$\Delta H_{298} = 3 \times (-26400) - (-19700) = 117800\ \text{cal}$$

$$\begin{aligned}\Delta C_p &= C_p(\text{product}) - C_p(\text{reactant})\\ &= \left[2C_{p(\alpha,Fe)} + 3C_{p(CO)}\right] - \left[C_{p(Fe_2O_3)} - 3C_{p(C)}\right]\end{aligned}$$

$= [28.73 + 14.78 \times 10^{-3}T - 0.33 \times 10^5\ T^{-2}] - [35.79 + 21.66 \times 10^{-3}T - 8.85 \times 10^5\ T^{-2}]$

$= -7.06\ -6.88 \times 10^{-3}T + 9.52 \times 10^5\ T^{-2}$

$$\Delta H_{1000} = \Delta H_{298} + \int_{298}^{1000} \Delta C_p .\ dT$$

$$= 117800 + \int_{298}^{1000} \left(-7.06 - 6.88\ x\ 10^{-3}T + 9.52\ x 10^5 T^{-2}\right) dT$$

$= 117800 - 7.06\,(1000 - 298) - 6.88.\ \frac{1}{2}\left(1000^2 - 298^2\right) \times 10^{-3} - 9.52 \times 10^5\left(1000^{-1} - 298^{-1}\right)$

$= 117800 - 5850 = 111950$ cal Ans.

Problem 10 The standard heat of formation of ammonia gas is −11.03 kcal mol^{-1} at 25 °C. Calculate ΔH^o_{1000} from following data given below:

$$C_{p(NH_3)} = 6.189 + 7.787\ x 10^{-3}T - 0.728\ x 10^{-6}T^2 \quad \text{cal deg}^{-1}\text{mol}^{-1}$$

$$C_{p(N_2)} = 6.450 + 1.414\ x\ 10^{-3}T - 0.0808\ x 10^{-6}T^2 \quad \text{cal deg}^{-1}\text{mol}^{-1}$$

$$C_{p(H_2)} = 6.947 + 0.2\ x 10^{-3}T - 0.4808\ x 10^{-6}T^2 \quad \text{cal deg}^{-1}\text{mol}^{-1}$$

Develop a general expression for the heat of formation applicable in the temperature range: 273 K – T K. Use this equation to check its accuracy in the calculation of the heat of formation of ammonia gas at 1000 K.

Solution

The reaction under question is:

$$\frac{1}{2}N_2(g) + \frac{3}{2}H_2(g) = NH_3\,(g),\quad \Delta H^o_{298} = -11030 \ \text{cal mol}^{-1}$$

$$\Delta C_p = C_p(NH_3, g) - \left[\frac{1}{2}\ C_p\,(N_2, g) + \frac{3}{2}C_p\ \ (H_2,\ g)\right]$$

$$= -7.457 + 7.38 \times 10^{-3}T - 1.409 \times 10^{-6}T^2 \ \text{cal deg}^{-1}$$

$$\Delta H^o_{1000} = \Delta H^o_{298} + \int_{298}^{1000} \Delta C_p .\ dT$$

$$= -11030 - \int_{273}^{298} \left(-7.457 + 7.38\ x\ 10^{-3}T - 1.409\ x\ 10^{-6}T^2\right) dT$$

$= -11030 - 7.457\ (1000 - 298) + 3.69 \times 10^{-3}\ \left(1000^2 - 298^2\right) - 0.47 \times 10^{-6}\ \left(1000^3 - 298^3\right)$

$= -11030 - 5234.8 + 3362.3 - 457.6 = -13360$ cal mol^{-1} (1)

In order to develop the required expression in the temperature range: 273 – T K, ΔH^o_{273}should be calculated by making use of the Kirchhoff's equation:

$$\Delta H^o_{298} = \Delta H^o_{273} + \int_{273}^{298} \Delta C_p.\ dT$$

$$\therefore\ \Delta H^o_{273} = \Delta H^o_{298} - \int_{273}^{298} \Delta C_p.\ dT$$

$$= -11030 - \int_{273}^{298} \left(-7.457 + 7.38\ x\ 10^{-3}T - 1.409\ x\ 10^{-6}T^2\right)\ dT$$

$$= -11030 + 7.457\ (298\text{–}273) - \frac{1}{2}\ x7.38\ x\ 10^{-3}\left(298^2 - 273^2\right)$$

$$+ \frac{1}{3}x1.409\ \left(298^3 - 273^3\right)$$

$$= -11030 + 186.4\text{–}52.7 + 2.9 = -10893\ \text{cal mol}^{-1}$$

Hence, the general expression valid in the temperature range 273 – T K is presented as:

$$\Delta H^o_T = \Delta H^o_{273} + \int_{273}^{T} \left(-7.457 + 7.38\ x\ 10^{-3}T - 1.409\ x\ 10^{-6}T^2\right)\ dT$$

$$= -10893\text{–}7.457\ \text{T} + 3.69\ \text{x}\ 10^{-3}\text{T}^2 - 0.47 \times 10^{-6}\text{T}^3$$

$$-(-7.457 \times 273 + 3.69 \times 10^{-3} \times 273^2 - 0.47 \times 10^{-6} \times 273^3)$$

$$= -10893 + 2036\text{–}275 + 10\text{–}7.457\ \text{T} + 3.69 \times 10^{-3}\text{T}^2 - 0.47 \times 10^{-6}\text{T}^3$$

$$= -9122\text{–}7.457\ \text{T} + 3.69 \times 10^{-3}\text{T}^2 - 0.47 \times 10^{-6}\text{T}^3\ \ \text{cal mol}^{-1}$$

The standard heat of formation of ammonia gas at 1000 K can be obtained by substituting T = 1000 in the derived equation:

$$\Delta H^o_{1000} = -9122\text{–}7.457 \times 1000 + 3.69 \times 10^{-3} \times 1000^2\text{–}0.47 \times 10^{-6} \times 1000^3\ \ \text{cal mol}^{-1}$$

$$= -9122\text{–}7457 + 3690\text{–}470 = -13359\ \text{cal mol}^{-1} \quad (2)$$

Since the value of $\Delta H^o_{1000} = -13359$ cal mol^{-1} (2) obtained from the derived expression is almost the same as calculated above (-13360 cal mol^{-1}), the expression is correct.

Problem 11 What do you understand by adiabatic flame temperature? Find the adiabatic flame temperature for the combustion of CH_4 (g) with stoichiometric air (20 % O_2 and 80 % N_2).

Given: The heats of formation of CH_4 (g), CO_2 (g) and H_2O (g) are –17.89, –94.05, and –57.80 kcal mol^{-1}, respectively.

$$C_{p(CO_2)} = 10.55 + 2.16\ x\ 10^{-3}\text{T} - 2.04\ x10^5\text{T}^{-2}\quad \text{cal deg}^{-1}\text{mol}^{-1}$$
$$C_{p\ (H_2O)} = 7.17 + 2.56 \times\ 10^{-3}\text{T} - 0.08\ \times 10^5\text{T}^{-2}\quad \text{cal deg}^{-1}\text{mol}^{-1}$$
$$C_{p(N_2)} = 6.66 + 1.02\ x\ 10^{-3}\text{T}\quad \text{cal deg}^{-1}\text{mol}^{-1}$$

Solution

The short and non-luminous flame produced on burning of a gaseous fuel under such conditions that the heat evolved during oxidation is completely absorbed by the products of combustion. The theoretical maximum temperature achieved under these conditions is called the adiabatic flame temperature. On the basis of the assumption that the combustion within flame is adiabatic, the temperature can be calculated from values of the heats of formation and heat capacities of the combustion products.

The heat generated during combustion of CH_4 is calculated as: in air containing 20 % O_2 and 80 % N_2:

$$\text{CH}_4(\text{g}) + 2\ O_2\ (g) = CO_2\ (g) + 2\ H_2O\ (l)$$
$$\Delta H^o_{298} = -94.05 + 2 \times (-57.80) - (-17.89) = -191.76\ \text{kcal}$$

This heat raises the temperature of combustion products [CO_2 (g) and H_2O (g)] and also the N_2present in the air used for combustion. The amount of nitrogen is 4 times that of oxygen (i.e. 8 moles).

Let T be the maximum temperature (adiabatic flame temperature) achieved on complete combustion. We can frame the following equation making use of the capacities of CO_2 (g) and H_2O (g) and N_2:

$$\Delta H^o_{298} = \int_{298}^{T} C_{p\ (CO2)}\,dT + \int_{298}^{T} C_{p\ (H_2O)}\,dT + \int_{298}^{T} C_{p\ (N2)}\,dT$$
$$191760 = \int_{298}^{T} \left(10.55\ T + 2.16\ x\ 10^{-3}T - 2.04\ x\ 10^5 T^{-2}\right) dT$$
$$+2 \int_{298}^{T} \left(+7.17\ T + 2.56\ x\ 10^{-3}T + 0.08\ x10^5 T^{-2}\right) dT$$
$$+8 \int_{298}^{T} \left(6.66 + 1.02\ x\ 10^{-3}T\right)\ dT$$

Following the procedure adapted in Problem 10, we arrive on the general expression:

$$78.17\ (\text{T}-298) + 7.72 \times 10^{-3}\left(T^2 - 298^2\right) + 1.88 \times 10^5\ \left(\frac{1}{T} - \frac{1}{298}\right) = 191760$$

Solving by trial and approximation:

$$\text{T} = 2243\ \text{K}\ \ (= 1970^\circ\text{C})\qquad \text{Ans.}$$

Problem 12 Calculate the heat of formation of FeO at 850 °C from the following data:

$$Fe\ (\alpha\ ferromagnetic) + ½\ O_2\ (g) = FeO\ (s),\ \Delta H^o_{298} = -63.2\ kcal\ mol^{-1}$$

1. C_p (Fe, α-mag) = 4.18 + 5.92 x 10^{-3}T cal deg^1 mol^{-1}
2. C_p (FeO) = 11.66 + 2.00 × 100^{-3}T – 0.67 × 10^5 T^{-2} cal deg^1 mol^{-1}
3. C_p ($O_{2,}$ g) = 7.16 + 1.00 × 10^{-3}T – 0.40 × 10^5 T^{-2} cal deg^1 mol^{-1}
4. C_p (Fe, α-paramagnetic) = 9.0 cal deg^1 mol^{-1}
5. Fe (α-ferromagnetic) = Fe (α-paramagnetic)
   L^t = 660 cal

Solution

The heat of reaction at temperature T K is given as:

$$\Delta H^o_T = \Delta H^o_{298} - \int_{273}^{T} \Delta C_p .\ dT$$

At 760 °C (1033 K) α-Fe changes from the ferromagnetic to the paramagnetic state with a latent heat of transformation, L^t. Hence, the heat of reaction has to be known first at 760 °C (1033 K):

$$\Delta H^o_{1033} = \Delta H^o_{298} + \int_{298}^{1033} \Delta C_p .\ dT$$

$$\Delta C_p = C_p(FeO) - \left[C_p\ (\alpha - Fe\ mag) + \frac{1}{2} C_p\ (O_2,\ g)\right]$$

$$= 3.90–4.42 \times 10^{-3}T - 0.47 \times\ 10^5T^{-2}\ cal\ deg^{-1}$$

$$\therefore\ \Delta H^o_{1033} = -63200 + \int_{298}^{1033} (3.90–4.42\ x\ 10^{-3}T - 0.47\ x\ 10^5T^{-2})\,dT$$

$$= -63200 + 3.90\ (1033 - 298)–2.21\ x\ 10^{-3}\ (1033^2 - 298^2)$$

$$+0.47\ x\ 10^5 \left(\frac{1}{1033} - \frac{1}{298}\right)$$

$$= -\ 63200 + 2867 - 2152 - 112 = -\ 62600\ cal$$

Apply Hess Law at 1033 to the reaction: $Fe_{(\alpha\text{-paramag})} + \frac{1}{2}\ O_2\ (g) = FeO\ (s)$
Simultaneous reactions at 1033 K

1. $Fe_{(\alpha\text{-ferro})} + \frac{1}{2}\ O_2 = FeO\ (s),\ \Delta H^o_1$
2. $Fe_{(\alpha\text{-ferro})} = Fe\,(\alpha\text{-para}),\ \Delta H^o_2\ (= L^t)$
3. $Fe_{(\alpha\text{-para})} + \frac{1}{2} O_2 = FeO\ (s),\ \Delta H^o_3$

Equation (3) ≅ Eqs. 1 – 2

$$\therefore \Delta H^o_{1033} \left(= \Delta H^o_3\right) = \Delta H^o_1 - \Delta H^o_2$$
$$= -62600 - 660 = -63260 \text{ cal}$$

As Fe (α-paramagnetic) has a different heat capacity

$$\Delta C_p(1033 - 1123) = C_p(FeO) - \left[C_p\,(\alpha - Fe\;para) + \frac{1}{2} C_p\,(O_2,\ g)\right]$$
$$= -\ 0.92–1.50 \times 10^{-3}\mathrm{T} - 0.47 \times 10^5\mathrm{T}^{-2}\ \text{cal deg}^{-1}$$

$$\Delta H^o_{1123} = \Delta H^o_{1033} + \int_{1033}^{1123} \Delta C_p.dT$$

$$\Delta H^o_{1123} = -63260 + \int_{1033}^{1123} (0.92–1.50\ x\ 10^{-3}T - 0.47\ x\ 10^5 T^{-2})\,dT$$
$$= -63260 + 0.92\ (1123 - 1033) – \frac{1}{2} x1.50\ (1123^2 - 1033^2)$$
$$+0.47\ x\ 10^5\ \left(\frac{1}{1033} - \frac{1}{298}\right)$$
$$= -63157 \text{ cal}$$
Ans.

Problem 13 A Carnot engine where low temperature reservoir is at 7 °C has an efficiency of 50%. It is desired to increase the efficiency to 70%. By how many degrees should be temperature of the high temperature reservoir be raised?

Solution

Efficiency = 0.5, $T_2 = 280$ K, T_1?
Efficiency $= \frac{T_1 - T_2}{T_1} = 0.5$

$\therefore T_1 = 560$ K

Efficiency = 0.7, T_2 =280, $T_1{}' = 840$ K
∴ Increase in temperature = 840–560 = 280 K Ans.

Problem 14 A carnot engine is operated between two reservoirs at temperatures 450 K and 350 K. If the engine receives 1000 cal of heat from the source in each cycle, calculate the amount of heat rejected to the sink in each cycle. Calculate the efficiency of the engine and the work done in each cycle.

Solution

$$\mathrm{T}_1\ 450\ \mathrm{K};\ \ \mathrm{T}_2 = 350\ \ \mathrm{K},\ \ q_1 = 1000\ \ \text{cal}\ \ q_2 = ?$$

$$\frac{q_1}{T_1} = \frac{q_2}{T_2}$$

$$\therefore q_2 = \frac{T_2 q_1}{T_1} = \frac{350 \times 1000}{450} = 777.77 \ \text{cal}$$

$$\text{Efficiency} = \frac{T_1 - T_2}{T_1} \times 100 = \frac{450 - 350}{450} \times 100 = 22.22 \ \%$$

$$\text{Work done} = q_1 - q_2 = 1000 - 777.77 = 222.23 \ \text{cal}$$

Problem 15 A Carnot refrigerator takes heat from water at 0 °C and discards it to a room at 27 °C. 1 Kg of water at 0 °C is to be changed into ice. How many cal of heat are discarded to the room? What is the work done by the refrigerator in this process? Calculate the efficiency and the temperature to be raised to increase the efficiency further by 10%.

Solution

$$q_1 = ?, \ q_2 = 80 \times 1000 = 80000 \ \text{cal}$$

$$\mathrm{T_1} = 300 \ \mathrm{K}, \mathrm{T_2} = 273\mathrm{K}$$

$$\frac{q_1}{T_1} = \frac{q_2}{T_2},$$

1. $q_1 = \frac{T_1 q_2}{T_2} = \frac{300 \times 80000}{273} = 87900 \ \text{cal}$
2. work done, $\mathrm{w} = q_1 - q_2 = 4.2 \ (87900 - 80000) = 4.2 \times 7900 \ = 3.183 \times 10^4 \ \mathrm{J}$
3. Efficiency $= \frac{T_1 - T_2}{T_1} \times 100 = \frac{300 - 273}{300} \times 100 = 9 \ \%$
4. Efficiency = 9 +10 = 19%, $\mathrm{T_1}$?

$$19 = \frac{T_1 - T_2}{T_1} \times 100$$

When $\mathrm{T_2} = 273$, $\mathrm{T_1} = 337$

$$\therefore \Delta \mathrm{T_1} = 337\text{–}300 = 37^{\circ}$$

Therefore, the temperature has to be increased by 37 degrees.

Problem 16 Assuming that vapors of iron at one atmospheric pressure obeys the perfect gas equation calculate the entropy of evaporation, ΔS^e for 1 g atom of liquid iron using the following data:

Density of liquid iron 6.88 g cc^{-1}, at wt 55.85, boiling point 3070 °C.

Solution

$$\Delta S^{e} = R \ln \frac{v_{vap}}{v_l} \qquad v_{vap} \ at \ 3343 \ K = \frac{22400 \times 3343}{273}$$

$$= 1.987 \ \ln \frac{22400 \times 3343}{273 \times 55.85/6.88} = 20.72 \ \text{cal deg}^{-1} \text{ g atom}^{-1} \qquad \text{Ans.}$$

Problem 17 The heat capacity at 1 atm pressure of solid magnesium in the temperature range from 0 to 560 °C, is given by the expression. $C_p = 6.20 + 1.33 \times 10^{-3}\ T + 6.78 \times 10^4\ T^{-2}$ cal deg^{-1} g atom^{-1}. Determine the increase of entropy, per g atom for an increase of temperature from 300 to 800 K at 1 atm pressure.

Solution

$$\Delta S = \frac{C_p}{T} = \frac{6.20}{T} + 1.33 \times 10^{-3} + 6.78 \times 10^4 \mathrm{T}^{-3}$$

$$\therefore \ \mathrm{S}_{800}\text{–}\mathrm{S}_{300} = \int_{300}^{800} \frac{6.20}{T} + 1.33 \times 10^{-3} + 6.78 \times 10^4 \mathrm{T}^{-3}) \ \mathrm{dT}$$

$$= 6.20 \ \ln \frac{800}{300} + 1.33\mathrm{x}10^{-3} \ (800 - 300)$$

$$- 1/2 \times 6.78 \times 10^4 \ \left(800^{-2} - 300^{-2}\right)$$

$$= 6.083 + 0.665 + 0.324$$

$$= 7.07 \text{ cal deg}^{-1} \text{ g atom}^{-1} \qquad \text{Ans.}$$

Problem 17 Calculate the entropy change of a perfect gas which is allowed to expand isothermally so that its volume increases two folds.

Solution

Isothermal process $\Delta S_T = R \ \ln \frac{v_2}{v_1} = 1.987 \times \ln \ 2 = 1.38$ cal deg^{-1} mol^{-1} Ans.

Problem 18 Calculate the entropy change for pure iron at 2000 K. The entropy of iron at 298 K is 6.49 cal deg^{-1} mol^{-1}

$$\mathrm{Cp}\ (\alpha \ \mathrm{Fe}) = 3.04 + 7.58 \times 10^{-3}\mathrm{T} + 0.60 \times 10^{-5}\ \mathrm{T}^2 \text{ cal deg}^{-1} \text{ g atom}^{-1} \ (298\text{–}1033 \text{ K})$$

$$\mathrm{L}^{t}_{\alpha \to \beta} = 0$$

$$\mathrm{C_p}\ (\beta \ \mathrm{Fe}) = 11.13 \text{ cal deg}^{-1} \text{ g atom}^{-1} \ (1033\text{–}1183 \text{ K})$$

$$\mathrm{L}^{t}_{\beta \to \gamma} = 215 \text{ cal g atom}^{-1} \text{ at } 1183\mathrm{K}$$

$$\mathrm{Cp}\ (\gamma \ \mathrm{Fe}) = 5.80 + 1.98 \times 10^{-3}\mathrm{T} \text{ cal deg}^{-1} \text{ g atom}^{-1} \ (1183\text{–}1673 \text{ K})$$

$$L^t(\gamma \rightarrow \delta) = 165 \text{ cal g atom}^{-1} \text{at } 1673 \text{ K}$$

$$\text{Cp } (\delta \text{ Fe}) = 6.74 + 1.60 \times 10^{-3}\text{T cal deg}^{-1} \text{ g atom}^{-1} \text{ (1673–1812 K)}$$

$$\text{L}^\text{f} = 3670 \text{ cal g atom}^{-1}$$

$$\text{Cp (Fe, l)} = 9.77 + 0.40 \times 10^{-3}\text{T cal deg}^{-1} \text{ g atom}^{-1}$$

Solution

$$\text{Fe}_{(\alpha\text{-mag})} \; {}^{1033K} \rightarrow \text{Fe}_{(\beta,\text{nonmag})} \; {}^{1183K} \rightarrow \text{Fe } (\gamma) \; {}^{1673K} \rightarrow \text{Fe } (\delta) \; {}^{1812K} \rightarrow \text{Fe } (\text{l})$$

$$S_{2000} = S_{298} + \int_{298}^{1033} \frac{C_{p,\alpha}}{T} dT + \int_{1033}^{1183} \frac{C_{p,\beta}}{T} dT + \frac{215}{1183} + \int_{1183}^{1673} \frac{C_{p,\gamma}}{T} dT$$
$$+ \frac{165}{1673} + \int_{1673}^{1812} \frac{C_{p,\delta}}{T} dT + \frac{3670}{1812} + \int_{1812}^{2000} \frac{C_{p,l}}{T} dT$$
$$= 6.49 + \int_{298}^{1033} \left[\frac{(3.04 + 7.58 \ x \ 10^{-3}T + 0.60 \ x \ 10^{-5}T2)}{T} \right] dT$$
$$+ \int_{1033}^{1183} \left[\frac{11.13}{T} \right] dT + \frac{215}{1183} + \int_{1183}^{1673} \left[\frac{(5.80 + 1.98 \ x \ 10^{-3}T)}{T} \right] dT$$
$$+ \frac{165}{1673} + \int_{1673}^{1812} \left[\frac{(6.74 + 1.60 \ x \ 10^{-3}T)}{T} \right] dT + \frac{3670}{1812} +$$
$$\int_{1812}^{2000} \left[\frac{(9.77 + 0.4 \ x \ 10^{-3}T)}{T} \right] dT$$
$$= 24.98 \text{ cal deg}^{-1} \text{ g atom}^{-1}$$

Ans.

Problem 19 Differntiate between Trouton's and Rechard' rules.

Solution

Trouton's Rule: Entropy of evaporation, $\Delta S^e = L^e / T^e$ (latent heat of evaporation/temperature of evaporation) = 22.4, is almost constant for metallic substances. This has been very clearly demonstrated in Table 3.1.

Trouton's Rule fails for associated liquid e.g., water, alcohols etc. Based on the value of ΔS^e we can conversely estimate if there is any association or polymerization in the liquid. In other we can estimate the liquid structure from the entropy value.

Richard's Rule: The entropy of fusion, $\Delta S^f = L^f / T^f$ (latent heat of fusion/temperature of fusion) ~2.4, is is not as constant as that of evaporation. The change in state of order as melting is not as great as on evaporation and the variation in the state of order of a solid due to variety of the possible binding forces, has therefore a proportionately large effect on the entropy of fusion.

Problem 20 Suggest a simple method to write Maxwell's equations. With the aid these equations and the third law of thermodynamics (i) evaluate numerically the isothermal change of entropy with pressure for mercury vapors at the boiling point of

357 °C and the atmospheric pressure (ii) show that coefficient of thermal expansion α approaches zero at absolute zero but there is no evidence that coefficient of isothermal compressibility β does likewise.

Solution

(i) Maxwell's thermodynamic relations can be written by arranging S, P. T, V (a word-sportive as SPorTiVe) where S, P, T and V, respectively represent entropy, pressure, temperature and volume. The first two equations: $\left(\frac{\partial S}{\partial V}\right)_T = \left(\frac{\partial P}{\partial T}\right)_V$ and $\left(\frac{\partial S}{\partial P}\right)_T = -\left(\frac{\partial V}{\partial T}\right)_P$ can be obtained by writing ∂S in the numerator of the left hand side of the equation and ∂P, ∂T *and* ∂V are written in the clock wise direction and in the anticlock wise for Eqs. 1 and 2 respectively with positive and negative signs.

$$\left(\frac{\partial S}{\partial V}\right)_T = \left(\frac{\partial P}{\partial T}\right)_V \tag{1}$$

$$\left(\frac{\partial S}{\partial P}\right)_T = -\left(\frac{\partial V}{\partial T}\right)_P \tag{2}$$

Similarly for other Eqs. 3 and 4, ∂S is written in the denominator of the right hand side of the equation and the other parameters are substituted in the clock wise and in anti-clock wise direction, respectively.

$$\left(\frac{\partial T}{\partial P}\right)_S = \left(\frac{\partial V}{\partial S}\right)_P \tag{3}$$

$$\left(\frac{\partial T}{\partial V}\right)_S = -\left(\frac{\partial P}{\partial S}\right)_V \tag{4}$$

According to Eq. (2) we have

$$\begin{aligned}\left(\frac{\partial S}{\partial P}\right)_T &= -\left(\frac{\partial V}{\partial T}\right)_P \\ &= -\left(\frac{R}{P}\right) \\ &= -\frac{1.987}{1} = -1.987\ \text{cal deg}^{-1}\text{mol}^{-1}\text{atm}^{-1} \quad \text{Ans.}\end{aligned}$$

$$PV = RT$$

$$\frac{V}{T} = \frac{R}{P}$$

$$\therefore \frac{\partial V}{\partial T} = \frac{R}{P}$$

(ii) According to the third law of thermodynamics S = 0 as $T \rightarrow 0$, thus the entropy of a perfectly crystalline substance must be independent of pressure as well as volume.

Thus *as*

$$\mathrm{T} \rightarrow 0 \left(\frac{\partial S}{\partial P} \right)_T = 0 \tag{1}$$

$$\text{and} \left(\frac{\partial S}{\partial V} \right)_T = 0 \tag{2}$$

From Maxwell's relationship we have

$$\left(\frac{\partial S}{\partial P} \right)_T = - \left(\frac{\partial V}{\partial T} \right)_P \tag{3}$$

$$\text{and} \left(\frac{\partial S}{\partial V} \right)_T = \left(\frac{\partial P}{\partial T} \right)_V \tag{4}$$

Combining Eqs. 1 and 3 we can write:

$$\left(\frac{\partial V}{\partial T} \right)_P = 0 \qquad \text{as } T \rightarrow 0$$

Similarly, from Eqs. 2 and 4:

$$\left(\frac{\partial P}{\partial T} \right)_V = 0 \qquad \text{as } T \rightarrow 0$$

In other words, the temperature gradient of the pressure and volume vanish as absolute zero is approached.

Coefficient of thermal expansion, $\alpha = \frac{1}{V} \left(\frac{\partial V}{\partial T} \right)_P$

From Maxwell's equation and according to the third law of thermodynamics:

$$\left(\frac{\partial S}{\partial P} \right)_T = - \left(\frac{\partial V}{\partial T} \right)_P = 0$$

$$\therefore \alpha = 0, \ \text{when } T \rightarrow 0$$

The coefficient of isothermal compressibility, β

$$\beta = -\frac{1}{V}\left(\frac{\partial V}{\partial P}\right)_T$$

From Maxwell's equation and the third law of thermodynamics, we have

$$\left(\frac{\partial S}{\partial V}\right)_T = \left(\frac{\partial P}{\partial T}\right)_V = -\left(\frac{\partial P}{\partial V}\right)_T \left(\frac{\partial V}{\partial T}\right)_P = -\alpha V \left(\frac{\partial P}{\partial V}\right)_T = \frac{\propto}{\beta}$$

Thus α approaches zero at absolute zero, but there no evidence that β does likewise.

Problem 21 Correlate between van't Hoff isochore and Le Chatelier principle.

Correlation: See Sect. 2.11.3

Problem 22 Arrange Al_2O_3, BeO, Cu_2O and UO_2 in increasing order of stability from the data given below.

(a) $Al_2O_3(s) = 2\ Al\ (l) + \frac{3}{2}O_2(g)$, $\Delta G_a^o = 405760 + 3.75\ T\ log\ T - 92.22T$ cal per mole of oxide
(b) BeO (s) = Be (s) + $\frac{1}{2}O_2(g)$, $\Delta G_b^o = 143450 + 1.61\ T\ log\ T$ cal
(c) $Cu_2O(l) = 2\ Cu\ (l) + \frac{1}{2}O_2(g)$, $\Delta G_c^o = 46700 + 3.92\ T\ log\ T - 34.1T$ cal
(d) $UO_2(s) = U\ (l) + O_2(g)$, $\Delta G_d^o = 269700 + 15.4\ T\ log\ T - 97.0T$

Solution

The most stable oxide will have the least free energy of formation. Hence, for comparison of stability of different oxides we should calculate the free energy of formation (ΔG_f^o) of all the oxides formed from one mole of oxygen 1000 K.

$$\Delta G_a^o(1000\ K) = 405760 + 3.75 \times 1000 \times log(1000) - 92.22 \times 1000 = 324790\ \text{cal}$$

$$\frac{4}{3}Al(l) + O_2(g) = \frac{2}{3}Al_2O_3(s), \Delta G_f^o = -\frac{2}{3}\Delta G_a^o = -\frac{2}{3} \times (324790)$$
$$= -216526\ \text{cal per mole of}\ O_2$$

$$\Delta G_b^o(1000\,K) = 143450 + 1.61 \times 1000 \times log\ (1000) = 120230\ \text{cal}$$
$$2\ Be\ (s) + O_2(g) = 2\ BeO\ (s),\ \Delta G_f^o = -2\ \Delta G_b^o = -240460\ \text{cal per mole of}\ O_2$$

$$\Delta G_c^o(1000\,K) = 46700 + 3.92 \times 1000 \times log(1000) - 34.1 \times 1000 = 24360$$
$$2\ Cu\ (l) + O_2(g) = 2\ Cu_2O(l)\ \ \Delta G_f^o = -2\ \Delta G_c^o = -48720\ \text{cal per mole of}\ O_2$$

$$\Delta G_d^o(1000\,K) = 269700 + 15.4 \times 1000 \times log(1000) - 97.0 \times 1000 = 218900$$
$$U\,(l) + O_2(g) = UO_2(s),\quad \Delta G_f^o = -\Delta G_d^o = -218900\ \text{cal per mole of}\ O_2$$

Hence, the increasing order of stability is $Cu_2O < Al_2O_3 < UO_2 <$ BeO Ans.

Problem 23 Determine the temperatures at which nickel oxide can dissociate under (i) standard conditions (ii) a pressure of 10^{-4} mm Hg. Given that

$$Ni(s) + \frac{1}{2}\ O_2(g) = NiO(s), \quad \Delta G^o = -62650 + 25.98T \text{ cal}$$

Solution

(i) Under standard conditions, $p_{O_2} = 1$

For the dissociation reaction:

$$2NiO\ (s) = 2\ Ni(s) + O_2\ (g), \quad \Delta G^o = 125300 - 51.96\ T, \text{ cal}$$

Since Ni and NiO are pure solids, $a_{Ni} = 1 = a_{NiO}$, $K = p_{O_2} = 1$

$$\Delta G^o = -RT \ln K = 0$$

$$\therefore \quad 125300 - 51.96\ T = 0$$

$$T = \frac{125300}{51.96} = 2411\ K = 2138\,°C \quad \text{Ans}$$

(ii) $p_{O_2} = 10^{-4}$ mm Hg $= \frac{10^{-4}}{760}$ atm $= K$

$$\Delta G^o = -RT \ln K = -1.987 \times T \times ln\ \left(\frac{10^{-4}}{760}\right) = 31.48T$$

$$\therefore \quad 31.48T = 125300 - 51.96\ T$$

$$T = 1502\ K = 1229\,°C \quad \text{Ans.}$$

Problem 24 The standard free energy change for the reaction: NiO (s) + CO (g) = Ni (s) + CO_2 (g) at 1125 K is −5147 cal. Would an atmosphere of 15% CO_2, 5% CO and 80% N_2 oxidize nickel at 1125 K?

Solution

$$NiO(s) + CO(g) = Ni(s) + CO_2(g), \quad \text{at } 1125\,K, \Delta G^o = -\ \ 5147 \text{ cal}$$

$$\Delta G^o = -RT \ln K = -5147 \text{ cal}$$

$$\ln K = \frac{5147}{1.987 \times 1125} = 2.303$$

$$K = 10$$

For the above reaction, K $= \frac{p_{CO_2}.\ a_{Ni}}{p_{CO}\cdot\ a_{NiO}} = \frac{p_{CO_2}}{p_{CO}} = 10\ (a_{Ni} = 1 = a_{NiO})$

$$\therefore p_{CO} = 0.1.p_{CO_2}$$

In the given atmosphere, $p_{CO} = \frac{5}{15}\cdot p_{CO_2} = 0.33.p_{CO_2}$

This means the atmosphere under question is more reducing, and hence will not oxidize nickel at 1125 K.

Problem 25 What is the maximum partial pressure of moisture which can be tolerated in H_2-H_2O mixture at 1 atm total pressure without oxidation of nickel at 750 °C? Given that:

$$\text{Ni (s)} + \tfrac{1}{2}\ O_2(g) = \text{NiO (s)},\quad \Delta G_1^o = -58450 + 23.55T \text{ cal}$$

$$H_2(g) + \tfrac{1}{2}\ O_2(g) = H_2O\ (g),\quad \Delta G_2^o = -58900 + 13.1T \text{ cal}$$

Solution
Reaction under question: $Ni(s) + H_2O(g) = NiO(s) + H_2(g)$

$$\Delta G_R^o = \Delta G_1^0 - \Delta G_2^o = 450 + 10.45\ \text{T cal}$$

$$T = 750 + 273 = 1023\ \text{K}$$

$$\therefore\ \Delta G_{1023}^0 = 450 + 10.45 \times 1023 = 11140$$

$$= -RT\ \ln K$$

$$\ln K = -\frac{11140}{1.987 \times 1023}$$

$$= -5.4804$$

$$K = 0.0041677 = \frac{p_{H_2}}{p_{H_2O}} \quad (a_{Ni} = 1 = a_{NiO})$$

$$p_{H_2} = 0.0041677.p_{H_2O} \tag{1}$$

and

$$p_{H_2} + p_{H_2O} = 1 \tag{2}$$

from 1 and 2 we get:

$$p_{H_2O} = 0.99585\ \text{atm}$$

$$p_{H_2} = 0.00415\ \text{atm}$$

Hence, maximum tolerable partial pressure of moisture is 0.99585 atm i.e., 99.6 %

Problem 26 Calculate the maximum moisture content which can be tolerated in the gaseous mixture of H_2-H_2O for reduction of WO_3 by hydrogen gas in a system maintained at a total pressure of one atmosphere and 400, 700 and 1000 K? Assuming that WO_3 and W are present as pure solids, comment on the utilization efficiency of hydrogen.

(a) Does the efficiency improve with increasing temperature?
(b) What will be effect of increasing the total pressure in the system on this efficiency?

Given that:

1. $WO_3\,(s) = W\ (s) + 3/2O_2\,(g),\ \ \Delta G_1^o = 201500 + 10.2T\ log\ T - 91.7T\ \text{cal}$
2. $H_2\,(g) + ½\ O_2\,(g) = H_2O\ (g),\ \ \Delta G_2^o = -58900 + 13.1\text{T cal}$

Solution
The free energy change, ΔG^o for the required reaction: $WO_3(s) + 3H_2(g) = W(s) + 3H_2O(g)$,

$$\Delta G^o = \Delta G_1^o + 3\ \Delta G_2^o = 24800 + 10.2\ T\ \log T - 52.4\ T\ \text{cal}$$

(i) ΔG^o at T = 400 K, $\Delta G^o_{400} = 14460$ cal
(ii) ΔG^o at T = 700 K, $\Delta G^o_{700} = 8440$ cal
(iii) ΔG^o at T = 1000 K, $\Delta G^o_{1000} = 2800$ cal

$$\Delta G^o = -RT\ln K$$

at T = 400 K, $K_{400} = 1.253 \times 10^{-8}$

$$T = 700\ K,\ \ K_{700} = 2.317 \times 10^{-3}$$

$$T = 1000\ K,\ \ K_{1000} = 0.244$$

$$K = \left\{\frac{p_{H_2O}}{p_{H_2}}\right\}^3 \text{or} \left\{\frac{p_{H_2O}}{p_{H_2}}\right\} = (K)^{1/3}$$

$$\left\{\frac{p_{H_2O}}{p_{H_2}}\right\} = K_{400}^{1/3} = \left(1.253 \times 10^{-8}\right)^{1/3} = 2.323 \times 10^{-3} \quad (1)$$

$$p_{H_2} + p_{H_2O} = 1 \quad (2)$$

From Eqs. 1 and 2 we get:

$$p_{H_2} = 0.9981\ \text{atm and}\ p_{H_2O} = 0.0019\ \text{atm}$$

∴ The maximum permissible moisture content in the gaseous mixture of H_2-H_2O at 400 K is 0.19%.

By similar calculations we find that the maximum permissible moisture content in the gaseous mixture at 700 K is 11.66% and at 1000 K is 38.5%.

(a) Thus, we find that the maximum permissible moisture content in the gaseous mixture of H_2-H_2O in reduction of WO_3 increases with increase of the reduction temperature. This demonstrates that the efficiency of utilization of hydrogen increases with increase of temperature. Hence, reduction should be carried out at higher temperature.

(b) Since the number of moles of gases, H_2 on the reactant side and H_2O vapors on the product side are equal, the efficiency of hydrogen utilization will remain unaffected with the increase of total pressure of the system.

Problem 27 Using the data given below, calculate the precise temperature at which wustite decomposes into metallic iron and magnetite: $4FeO(s) \rightarrow Fe(s) + Fe_3O_4(s)$ where 'FeO' refers to wustite saturated with metallic iron and magnetite. Which of the two oxides will be formed first at 500 °C, if pure iron is oxidized by gradual increase of partial pressure of oxygen from $p_{O_2} = 0$ to higher? What is the triple point in this system?

$$\text{Given that } Fe(s) + \frac{1}{2}O_2(g) = FeO(s), \quad \Delta G_1^o = -62952 + 15.493 \ \text{cal}$$

$$3FeO\ (s) + \frac{1}{2}O_2(g) = Fe_3O_4(s), \quad \Delta G_2^o = -74538 + 29.373 \ \text{cal}$$

Solution

Wustite is a nonstoichiometric solid solution of metallic iron and magnetite (Fe_3O_4). The stoichiometric FeO does not exist in the phase diagram. Wustite can exist stably only above 560 °C, below which it decomposes to metallic iron and magnetite. At the decomposition temperature three phases of metallic iron, wustite and magnetite can co-exist. This temperature is called the triple point. At the decomposition temperature (triple point):

$$\Delta G_1^o = \Delta G_2^o, \quad \therefore T = \frac{11586}{13.88} = 834.7\text{K} = 561.7\,°\text{C} \ \text{Ans.}$$

Thus, wustite decomposes into metallic iron and magnetite at 561.7 °C. It means below this temperature wustite does not exist stably.

In order to know whether magnetite (Fe_3O_4) or wustite (FeO) is formed first at 500 °C (773 K) with gradual increase of p_{O_2}, we must have the values of standard free energy of formation of the two oxides with one mole of oxygen. These may be obtained from the given data:

$$\frac{3}{2}Fe\ (s) + O_2(g) = \frac{1}{2}Fe_3O_4\ (s) \tag{3}$$

$$\Delta G_3^o = \frac{1}{2}(3\ \Delta G_1^o + \Delta G_2^o) = -131697 + 37.926\ T$$

$$2Fe\ (s) + O_2(g) = 2FeO(s) \tag{4}$$

$$\Delta G_4^o = 2\ \Delta G_1^o = -125904 + 30.986\ T$$

For both the reactions (3 and 4) $K = \frac{1}{p_{O_2}}$, hence $\Delta G^o = -RT\ln K = RT\ln p_{O_2}$ (T = 773 K)

$$\therefore p_{O_2} = 1.13 \times 10^{-29} \text{atm for reaction (3) : magnetite formation}$$

$$\text{and } p_{O_2} = 1.49 \times 10^{-29} \text{atm for reaction (4) : wustite formation}$$

Thus, reaction (3) has a lower p_{O_2} than reaction (4). In other words, when we start the oxidation of pure iron by increasing gradually the partial pressure from $p_{O_2} = 0$ to higher, the first iron oxide to form at 500 °C would be magnetite, and not wustite.

Problem 28 Some investigators based on their microscopic studies reported that Cu_2O underwent thermal decomposition into CuO and Cu at 375 °C. Making use of the following thermodynamic data, show that the old microscopic observation was in error. In reality, Cu_2O does not decompose.

$$2Cu(s) + \frac{1}{2}\ O_2(g) = Cu_2O(s), \quad \Delta G_1^o = -39855 + 17.041\ T\ \text{cal}$$

$$Cu_2O(s) + \frac{1}{2}\ O_2(g) = 2CuO(s), \quad \Delta G_2^o = -31347 + 22.643\ T\ \text{cal}$$

Solution

Consider the free energy change for the formation of CuO(s) from pure elements:

$$Cu(s) + \frac{1}{2}\ O_2(g) = CuO(s) \tag{3}$$

$$\Delta G_3^o = \frac{1}{2}\left(\Delta G_1^o + \Delta G_2^o\right) = -35601 + 19.842\ T\ \text{cal}$$

Decomposition of Cu_2O into CuO and Cu can be represented by the reaction:

$$Cu_2O(s) = CuO(s) + Cu(s) \tag{4}$$

$$\Delta G_4^o = \Delta G_3^o - \Delta G_1^o = -RT\ln K_4$$

$$\text{and} \quad K_4 = \frac{a_{CuO} \cdot a_{Cu}}{a_{Cu_2O}}$$

At the decomposition temperature, three solid phases (Cu, Cu_2O and CuO each of at unit activity) coexist, hence

$$K_4 = \frac{a_{CuO} \cdot a_{Cu}}{a_{Cu_2O}} = \frac{1 \times 1}{1} = 1$$

$$\therefore \quad \Delta G_4^o = -RT\ln K_4 = -RT\ln(1) = 0$$

$$\text{hence} \quad \Delta G_3^o - \Delta G_1^o = 0 \quad \text{or} \quad \therefore \Delta G_3^o = \Delta G_1^o$$

$$\therefore \; -35601 + 19.842 \;\; T = -39855 + 17.041 \;\; T$$

$$\text{or } T = -1519 \;\; K.$$

As a temperature of –1519 K is not feasible, $Cu_2O(s)$ does not decompose into metallic Cu and CuO. Thus, the microscopic observation was in erroneous.

Problem 29 The partial pressure of oxygen in equilibrium with Cu_2O and CuO has been found to be 0.0208 atm at 900 °C and 0.1303 atm at 1000 °C, respectively. Calculate heat of reaction for the reaction: $2Cu_2O$ (s) + O_2 (g) = 4CuO (s).

Solution

$$2Cu_2O \;\; (s) + O_2\,(g) = 4CuO \;\; (s), \qquad K = \frac{1}{p_{O_2}}$$

$$K_{900} = \frac{1}{0.0208} \;\; \text{and} \;\; K_{1000} = \frac{1}{0.1306}$$

Considering the unit of K in atmosphere, reaction under question would be

$$2Cu_2O \;\; (s) = 4Cu \;\; (s) + O_2\,(g), \quad K_1 = p_{O_2} \;\; \text{at } 900\,°C$$

$$\therefore \frac{1}{K_1} = \frac{1}{p_{O_2}} \;\; i.e. \;\; K_{900} = \frac{1}{K_1} = \frac{1}{0.0208}$$

$$\text{And} \quad 2CuO \;\; (s) = 2Cu\,(s) + O_2\,(g), \quad K_2 = p_{O_2} \, \text{at } 1000\,°C$$

$$\therefore \frac{1}{K_2} = \frac{1}{p_{O_2}} \;\; i.e. \;\; K_{1000} = \frac{1}{K_2} = \frac{1}{0.1306}$$

For mean heat of reaction use van’t Hoff equation:

$$\ln\left(\frac{K_2}{K_1}\right) = \frac{\Delta H}{R}\left(\frac{T_2 - T_1}{T_1 T_2}\right)$$

$$\therefore \Delta H = \frac{R \ln \;\; (K_{1000}/K_{900}) \times 1173 \times 1273}{(1273 - 1173)}$$

$$= \frac{1.987 \ln \;\; (0.0208/0.1306) \times 1173 \times 1273}{100}$$

$$= -54510 \;\; \text{cal} \qquad \text{Ans.}$$

Problem 30 The values of equilibrium constants for the reaction Fe (s) + H_2O (g) → FeO(s) + H_2 (g) are 2.38 and 1.67 at 900° and 1100 °C, respectively. Find the mean heat of reaction.

Solution

$$\int_{K_1}^{K_2} d\ln K = \left(\frac{\Delta H}{R}\right)\int_{T_1}^{T_2}\frac{dT}{T^2}$$

$$K_1 = 2.38 \qquad T_1 = 1173K, \qquad K_2 = 1.67 \qquad T_2 = 1373K$$

$$\ln K_2 - \ln K_1 = \frac{\Delta H}{R}\left[-\frac{1}{T}\right]_{T_1}^{T_2} = \frac{\Delta H}{R}\left[-\frac{1}{T_2}+\frac{1}{T_1}\right] = \frac{\Delta H}{R}\left[\frac{T_2-T_1}{T_1T_2}\right]$$

$$\ln\left(\frac{K_2}{K_1}\right) = \ln\left(\frac{1.67}{2.38}\right) = \frac{\Delta H}{1.987}\left(\frac{1373-1173}{1373\times 1173}\right)$$

or $$\Delta H = \frac{\ln(1.67/2.38)\times 1.987\times 1373\times 1173}{200} = -5670 \text{ cal} \qquad \text{Ans.}$$

Problem 31 Calculate the value of heat of reaction at 1000 K for the reaction:

$$WO_3(s) + 3H_2(g) = W(s) + 3H_2O(g), \quad \Delta G^0 = 24800 + 10.2T \log T - 52.4T \text{ cal}$$

Solution
We know that

$$\frac{\partial(\Delta G^0/T)}{\partial T} = -\frac{\Delta H^0}{T^2}$$

$$\frac{\partial}{\partial T}\left(\frac{24800}{T} - 52.4 + 10.2\log T\right) = -\frac{\Delta H^0}{T^2}$$

or

$$-\frac{24800}{T^2} + \frac{10.2}{T} = -\frac{\Delta H^0}{T^2}$$

or

$$\Delta H^0 = 24800 - 10.2T$$

$$\therefore \Delta H^0_{1000} = 24800 - 10.2\times 1000 = 14600 \text{ cal} \qquad \text{Ans.}$$

Problem 32 Determine the temperatures at which nickel sulfide can dissociate under (1) standard conditions (2) a pressure of 10^{-2} mm Hg and (3) a pressure of 10^{-5} mm Hg. Given

$$Ni(s) + ½\, S_2\,(g) = NiS\,(l), \quad \Delta G^o = -26730 + 10.5\,T \text{ cal}$$

Solution

Dissociation reaction: $2NiS(s) = 2\ Ni(s) + S_2\ (g)$, $K = p_{S_2}$

$$K = \frac{p_{S_2} . a_{Ni}^2}{a_{NiS}^2} = p_{S_2} \quad (a_{Ni} = 1 = a_{NiS})$$

(1) under standard conditions, $p_{S_2} = 1$ atm

$$\Delta G^o = -R\ T\ \ln K = 0\ \left(K = p_{S_2} = 1\right)$$

$$\therefore 0 = 53460 - 21.0T$$

$$T = 53460/21 = 2546\,K = 2273\,°C$$ Ans.

(2) $p_{S_2} = 10^{-2}$ mm Hg $= 10^{-2}/760$ atm

$$\Delta G^0 = -4.575T\log\left(\frac{10^{-2}}{760}\right) = 22.32T$$

$$\therefore 22.32T = 53460 - 21.0T$$

or $$T = \frac{53460}{43.32} = 1234\,K = 961\,°C$$ Ans.

(3) $p_{S_2} = 10^{-2}$ mm Hg $= 10^{-5}/760$ atm

$$\Delta G^0 = -4.575T\log\left(\frac{10^{-5}}{760}\right) = 36.08T$$

$$\therefore 36.08T = 53460 - 21.0T.$$

$$T = \frac{53460}{57.08} = 936\,K = 336\,°C$$ Ans.

Problem 33 Find the temperature at which reduction of NiO by carbon becomes feasible at (i) 1 atm and (ii) 10^{-1}atm. Given that

$$2C\ (s) + O_2\,(g) = 2\ CO\ (g) \quad \Delta G^o = -53400\text{–}41.9T\ \text{cal} \quad (1)$$

and

$$2Ni + O_2(g) = 2NiO\ (s) \quad \Delta G^o = -116900 + 47.1T\ \text{cal} \quad (2)$$

Solution

for the reaction: 2NiO (s) + 2C (s) = 2Ni (s) + 2CO (g)

$$\begin{aligned}\Delta G^o &= \Delta G_1^o - \Delta G_2^o\\ &= -53400–41.9T - (-116900 + 47.1T)\\ &= 116900 - 53400 - 41.9T - 47.1T\\ &= 63500 - 89.0T \text{ cal}\end{aligned}$$

$$\therefore K = \frac{a_{Ni}^2 p_{CO}^2}{a_{NiO}^2 a_C^2} = p_{CO}^2$$

$$(a_{N\ i} = 1 = a_{NiO} = a_C)$$

$$\text{When} \quad p_{CO} = 1, K = 1$$

$$\Delta G^o = -RT \ln K = 0 \quad (\text{when } p_{CO} = 1)$$

$$\therefore 63500 - 89.0T = 0$$

$$T = 713.5 \text{ K } (440\,°C)$$

$$\text{when } p_{CO} = 0.1$$

$$K = p_{CO}^2 = \left(\frac{0.1}{760}\right)^2 = \left(\frac{1}{7600}\right)^2$$

$$\therefore \Delta G^o = -4.575T \ln\left(\frac{1}{7600}\right)^2 = 2 \times 4.575T \log 7600 = 34.51T$$

$$\therefore 34.51T = 63500 - 89.0\ T$$

$$\therefore T = 63500/124.5 = 510 \text{ K } (237\,°C) \qquad \text{Ans.}$$

Problem 34 Calculate the free energy change at 1000 °C for the reaction CO (g) + H_2O (g) = CO_2 (g) + H_2 (g). The equilibrium constants K_1 and K_2 at 1000 °C for the following reactions 1 and 2 are given as log K_1 = −0.402 and log K_2 = −0.175

$$\text{Reaction 1}: FeO(s) + CO\ (g) = Fe\ (s) + CO_2\ (g)$$

$$\text{Reaction 2}: FeO\ (s) + H_2\ (g) = Fe\ (s) + H_2O\ (g)$$

Solution

For the reaction under question, $K = \dfrac{p_{CO_2} \cdot p_{H_2}}{p_{CO} \cdot p_{H2O}}$

$$K_1 = \frac{p_{CO_2}}{p_{CO}}, \qquad K_2 = \frac{p_{H2O}}{p_{H_2}}$$

$$\therefore K = \frac{K_1}{K_2}$$

$$\begin{aligned}\log K &= \log K_1 - \log K_2\\ &= -\ 0.402\text{–}(-0.175)\\ &= -0.227\end{aligned}$$

$$\begin{aligned}\Delta G &= -\text{RT In K} = -4.575\text{ T log K}\\ &= -4.575\times 1273\times(-0.227)\\ &= 1322\text{ cal} \qquad \text{Ans.}\end{aligned}$$

Problem 35 For the reaction NiO (s) + H_2 (g) = Ni (s) + H_2O (g), calculate the equilibrium constant at 750 °C from the following data:

$$\text{Ni (s)} + \tfrac{1}{2}\ O_2\text{(g)} = \text{NiO (s)},\quad \Delta G_1^o = -58450 + 23.55\text{T cal}$$

$$H_2\text{(g)} + \tfrac{1}{2}\ O_2\text{(g)} = H_2O\text{ (g)},\quad \Delta G_2^o = -58900 + 13.1\text{T cal}$$

Could pure nickel be annealed at 750 °C in an atmosphere containing 95% H_2O and 5% H_2 by volume without oxidation?

Solution
Reaction under question: Ni (s) + H_2O (g) = NiO (s) +H_2 (g)

$$\Delta G_R^o = \Delta G_2^0 - \Delta G_1^o = -450 - 10.45\ \text{T cal}$$

$$\text{T} = 750 + 273 = 1023\ \text{K}$$

$$\begin{aligned}\therefore\ \Delta G_{1023}^0 &= -450 - 10.45\times 1023 = -11140\\ &= -\text{RT ln K}\end{aligned}$$

$$\ln\text{K} = -\frac{-11140}{1.987\times 1023} = 2.38$$

$$\text{K} = 240 = \frac{p_{H_2O}}{p_{H_2}} \qquad (a_{\text{Ni}} = 1 = a_{\text{NiO}})$$

$$p_{H_2O} = 240.p_{H_2} \tag{1}$$

$$p_{H_2O} + p_{H_2} = 1 \tag{2}$$

or

$$240.p_{H_2} + p_{H_2} = 1$$

$$p_{H_2} = \frac{1}{241} = 0.004$$

$$p_{H_2O} = 0.996$$

Thus the atmosphere containing 5% H_2 under question is more reducing than the equilibrium requirement of only 0.4% H_2.

Hence the given atmosphere can be used for annealing nickel without oxidation.

Problem 36 From the following data comment on the feasibility of production of titanium (i) by carbothermic reduction of TiO_2 at 1200 °C and (ii) via chlorination of TiO_2 at 1200 °C under reducing conditions and subsequent reduction of $TiCl_4$ with magnesium at 750 °C. Given that:

$$\mathrm{Ti}\ (s) + \mathrm{O_2}\ (g) = \mathrm{TiO_2}\ (s), \quad \Delta G_1^o = -223500 + 41.55\ T,\ \text{cal} \tag{1}$$

$$2\ \mathrm{C}\ (s) + \mathrm{O_2}\ (g) = 2\mathrm{CO}\ (g), \quad \Delta G_2^o = -53400 - 41.90\ T,\ \text{cal} \tag{2}$$

$$\mathrm{Ti}\ (s) + 2\mathrm{Cl_2}\ (g) = \mathrm{TiCl_4}\ (g), \quad \Delta G_3^o = -180700 - 1.8\ T \log T + 34.65\ T,\ \text{cal} \tag{3}$$

$$\mathrm{Mg}\ (l) + \mathrm{Cl_2}\ (g) = \mathrm{MgCl_2}\ (l), \quad \Delta G_4^o = -147850 - 13.58\ T \log T + 72.77\ T,\ \text{cal} \tag{4}$$

Solution

Route I: Carbothermic reduction of TiO_2 at 1200 °C:

$$\mathrm{TiO_2}\ (s) + 2\mathrm{C}\ (s) = \mathrm{Ti}\ (s) + 2\mathrm{CO}\ (g) \tag{5}$$

$$\Delta G_5^o = \Delta G_2^o - \Delta G_1^o = -53400 - 41.90T - (-223500 + 41.55\ T) = 170100 - 83.45\ T$$

At 1473 K, $\Delta G_5^o = 170100 - 83.45\ T = 170100 - 83.45 \times 1473 = +47178$ cal

Since the free energy change for the carbothermic reduction of TiO_2 at 1200 °C is positive reaction will not be feasible.

Route II: Chlorination of TiO_2 at 1200 °C in the presence of carbon:

$$\mathrm{TiO_2}\ (s) + 2\mathrm{C}\ (s) + 2\mathrm{Cl_2}\ (g) = \mathrm{TiCl_4}\ (g) + 2\mathrm{CO}\ (g) \tag{6}$$

$$\Delta G_6^o = \Delta G_2^o + \Delta G_3^o - \Delta G_1^o = -10600 - 1.8\ T \log T - 48.80\ T$$

$$\text{At}\ \ 1473\ \text{K},\ \Delta G_6^o = -90883\ \text{cal}$$

Since the free energy change for chlorination of TiO_2 at 1200 °C under reducing conditions is negative, reaction is feasible.

Reduction of $TiCl_4$ (*g*) with Mg (*l*) at 750 °C:

$$\mathrm{TiCl_4}\ (g) + 2\mathrm{Mg}\ (l) = \mathrm{Ti}\ (s) + \mathrm{MgCl_2}\ (l) \tag{7}$$

$$\Delta G_7^o = 2\Delta G_4^o - \Delta G_3^o = -115000 - 25.36\ T \log T + 110.89\ T$$

$$\text{At}\ \ 1023\ \text{K},\ \Delta G_7^o = -79646\ \text{cal}$$

Since the free energy change for reduction of $TiCl_4$ (g) with Mg (l) at 750 °C is negative, reaction is feasible.

Problem 37 What is the permissible limit of HCl gas in the gaseous mixture: H_2-HCl for reduction of chromium chloride by hydrogen at 1100 K? Given that:

$$H_2\,(g) + Cl_2\,(g) = 2HCl\,(g),\quad \Delta G_1^o = -43540 + 1.98\,T\log T - 10.44\,T\ \text{cal} \quad (1)$$

$$Cr\ (s) + Cl_2\,(g) = CrCl_2\ (s),\quad \Delta G_2^o = -93900 - 8.7\ T\log T + 55.0\ T\ \text{cal} \quad (2)$$

Solution

The reaction under consideration:

$$\text{CrCl}_2\ (\text{s}) + \text{H}_2\,(\text{g}) = \text{Cr}\ (\text{s}) + 2\text{HCl}\ (\text{g}), \quad (3)$$

$$\Delta\text{G}_3^\text{o} = \Delta\text{G}_1^\text{o} - \Delta\text{G}_2^\text{o}$$

$$= 50360 + 10.68\ \text{T}\log\text{T} - 65.44\ \text{T}\ \text{cal}$$

$$\text{At}\quad 1100\ \text{K},\ \Delta\text{G}_3^\text{o} = 14106\ \text{cal}$$

$$\therefore\ -4.575\ \text{T}\log\text{K} = 14106,\quad \text{K} = 0.001574$$

for reaction (3), $\text{K} = \frac{p_{\text{HCl}}^2}{p_{\text{H}_2}}$

$$\therefore\ \frac{p_{\text{HCl}}^2}{p_{\text{H}_2}} = 0.001574 \quad \text{or}\ p_{\text{H}_2} = \frac{1}{0.001574}p_{\text{HCl}}^2 = 635.32.p_{\text{HCl}}^2 \quad (4)$$

$$\text{and}\quad p_{\text{H}_2} + p_{\text{HCl}} = 1 \quad (5)$$

From (4) and (5) we get: $635.32.p_{\text{HCl}}^2 + p_{\text{HCl}} - 1 = 0$

$$p_{\text{HCl}} = 0.039\ \text{atm}$$

$$p_{\text{H}_2} = 0.961\ \text{atm}$$

In the light of the above calculations it is evident that reduction of $CrCl_2$ by hydrogen will proceed in the forward direction only if the concentration of HCl gas in the gaseous mixture: H_2-HCl does not exceed beyond 3.9% by volume. Hence, HCl gas has to be removed by inserting fresh hydrogen gas into the reduction chamber.

Problem 38 In the Mond carbonyl process crude nickel is purified by volatilization and subsequent decomposition of $Ni(CO)_4$ according to the reactions: Ni(s) + 4CO (g) $\rightleftharpoons$ $Ni(CO)_4$(g). What are the effects of temperature and pressure on this reaction? Discuss these aspects by calculating percent volumes of CO and $Ni(CO)_4$ in CO-Ni $(CO)_4$ gaseous mixture at 100, 150 and 200 °C, at total pressures of 20, 50, and 100 atm. At what temperature can equal volume fractions of both the gases be obtained? The free energy change for the reaction is given as: $\Delta G^\circ = -3500 + 95\,T$ cal.

Solution

For the reaction: Ni (s) + 4 CO (g) $\rightleftharpoons$ $Ni(CO)_4$ (g), $\Delta G^o = -33500 + 95\,T$ cal

$$\text{At } 100\,°\text{C } (373\text{ K}), \Delta G^o_{373} = -33500 + 95 \times 373 = +1935 \text{ cal}$$

$$K_{373} = 0.0735$$

$$K = \frac{p_{(NiCO)_4}}{p^4_{CO}} = 0.0735$$

$$p_{(NiCO)_4} = p^4_{CO} \cdot 0.0735 \qquad (1)$$

$$p_{(NiCO)_4} + p_{CO} = 20 \qquad (2)$$

From (1) and (2) we get:

$$p^4_{CO} \cdot 0.0735 + p_{CO} - 20 = 0 \qquad (3)$$

On solving by trial and error we get: $p_{CO} = 3.85$ atm and $p_{(NiCO)_4} = 16.15$ atm.

Assuming ideal behavior of gases: % CO = 19.25 and % $Ni(CO)_4$ = 80.75 (by volume).

Following the above procedure at different temperatures and pressures values of *K*, partial pressures and % volume composition of CO and $Ni(CO)_4$ gases in the CO-$Ni(CO)_4$ gaseous mixture, calculated, are listed in table below:

		Pressure (atm)					
		20		50		100	
Temperature (K)	*K*	%CO	%$Ni(CO)_4$	%CO	%$Ni(CO)_4$	%CO	%$Ni(CO)_4$
100	0.0735	19.25	80.75	9.95	90.05	5.98	94.02
150	0.00035	61.0	39.0	34.92	65.08	21.74	78.26
200	0.0000052	96.4	3.6	77.06	22.94	54.39	45.41

Since the carbonyl reaction is only slightly exothermic it proceeds in reverse direction even at 100 °C due to large contribution of entropy which makes ΔG^o_{373} positive (= + 1935 cal). From the above Table it is quite evident that the equilibrium constant for the reaction: Ni (s) + 4CO (g) = $Ni(CO)_4$ decreases drastically from 0.0735 to 0.0000052 with the corresponding increase of temperature from 100 to 200 °C. Hence, the formation of $Ni(CO)_4$ decreases with increase of temperature.

The table clearly shows that $Ni(CO)_4$ content in the gaseous mixture decreases with increase of temperature at all pressures. According to the Le Chatelier's principle, formation of $Ni(CO)_4$ will be favored with increase of pressure because there is severe reduction in volume on formation of one mol of $Ni(CO)_4$ by consuming 4 mol of CO gas. Table clearly demonstrates that the carbonyl content in the gaseous mixture increases with increase of pressure at all temperatures.

Assuming ideal behavior, gaseous mixture having equal volume fractions of CO and $Ni(CO)_4$ at a total pressure of 1 atm, the partial pressure of both the gases will be equal, hence $p_{CO} = p_{(NiCO)_4} = 0.5$ atm

$$\text{Ni (s)} + 4\ \text{CO (g)} = \text{Ni(CO)}_4, \quad \Delta G^o = -33500 + 95\ \text{T cal} \tag{1}$$

$$K = \frac{p_{(NiCO)_4}}{p_{CO}^4} = \frac{0.5}{(0.5)^4} = 8$$

$$\Delta G^o = -RT \ln K = -4.575\ \text{T} \log(8) = -4.132\ \text{T} \tag{2}$$

From Equations 1 and 2:

$$\Delta G^o = -33500 + 95\ T = -4.132\ \text{T}$$

$$\text{or } 95\ \text{T} + 4.132\ \text{T} = 99.132\ \text{T} = 33500$$

$$\text{T} = 338\ \text{K} = 65\ °\text{C} \qquad \text{Ans.}$$

Problem 39 In the Pidgeon's process for the manufacture of magnesium, calcined dolomite is reduced with ferrosilicon according to the equation:

$$\text{2MgO.CaO } (s) + \text{Si } (s) = \text{2Mg } (g) + \text{2CaO.SiO}_2\,(s)$$

Given that:

$$\text{2MgO } (s) + \text{Si } (s) = \text{2Mg } (g) + \text{SiO}_2\ (s)$$

$$\Delta G_1^o = 152600 + 11.37\ \text{T} \log\ \text{T} - 99.18\text{T cal}$$

$$\text{2CaO } (s) + \text{SiO}_2\ (s) = 2\ \text{CaO.SiO}_2\,(s)$$

$$\Delta G_2^o = -30200 - 1.2\ \text{T cal}$$

Evaluate the partial pressure of magnesium vapor at 1200 °C and discuss the feasibility of the process.

Solution

For the reduction reaction under question: 2MgO.CaO (s) + Si (s) = 2Mg (g) + 2CaO.SiO_2 (s) the standard free change is obtained as:

$$\Delta G^o = \Delta G_1^o + \Delta G_2^o = 122400 + 11.37\ \text{T} \log\ \text{T} - 100.38\ \text{T} \quad (\text{T} = 1473\ \text{K})$$

$$\begin{aligned}\Delta G_{1473}^o &= 122400 + 11.37 \times 1473\ \log\ 1473 - 100.38 \times 1473 \\ &= 122400 + 53061 - 147860 \\ &= +27601\ \text{cal}\end{aligned}$$

For the above reaction, K = p_{Mg}^2 (activities of solid components being unity)

$$\Delta G^o = -\text{RT lnK} = -\text{RT } \ln p_{Mg}^2 = -4.575 \times \text{T} \times 2 \ \log p_{Mg}$$

$$p_{Mg} = 0.00896 \ \text{atm}$$

since $p_{Mg} < 1$ atm and $a_{SiO_2} = 0.001$ (in $2CaO.SiO_2$) slag, consider van't Hoff isotherm:

$$\Delta G = \Delta G^o + \text{RT ln K} = 27601 + \text{RT ln}\left[\frac{p_{Mg}^2 . a_{SiO_2}}{a_{MgO} . a_{Si}}\right]$$

In the Pidgeon process, reduction is carried out at 1200 °C under a vacuum of the order of 10^{-4} atm.

$$\therefore p_{Mg} = 10^{-4} \ \text{atm}$$

$$\text{in } 2CaO.SiO_2, \ a_{SiO_2} = 0.001 \ \text{and}$$

$$\text{in } 2MgO.CaO, \ a_{MgO} = 1$$

$$a_{Si} = 1$$

$$\therefore \Delta G = 27601 + 1.987x2.303x1473 \log\left[\frac{(10^{-4})^2 10^{-3}}{1 \times 1}\right]$$

$$= 27600 - 74145 = -46545 \ \text{cal}$$

Since ΔG is negative, reaction is feasible under the reduced pressure of 10^{-4} atm at 1200 °C in the presence of a basic oxide (CaO). Thus it is important to note that the van't Hoff isotherm is more useful in judging the thermodynamic feasibility of reactions whenever the activities of reactants and products differ significantly from unity. The silicothermic reduction of MgO (the Pidgeon process) sets a good example for understanding the problem of deviation from unit activity.

Problem 40 Calculate (i) the elevation of the boiling point of zinc when the external pressure is 2 atm and (ii) the depression of the freezing point when the external pressure is 50 atm. The latent heat of vaporization of zinc is 27.3 kcal g atom^{-1} and the normal temperature of boiling is 907 °C. Zinc melts at 419.5 °C with the heat of fusion is 1.74 kcal g atom^{-1}. The density of solid zinc is 7.0 g cc^{-1} and that of liquid zinc is 6.48 g cc^{-1} at 1 atm pressure.

Solution

In case of vaporization of a liquid into its vapors, according to the Clausius–Clapeyron equation we have:

$$\frac{dP}{dT} = \frac{L^e}{T^e (v_g - v_l)} \tag{1}$$

Where v_g is the volume of 1 g atom of gas and v_l is that of liquid.

The volume occupied by 1 g atom of a perfect gas at STP = 22400 cc
∴ The volume occupied by 1 g atom of zinc vapors at 907 °C (1180 K),

$$v_g = \frac{22400 \times 1180}{273}$$

Volume occupied by 1 g atom of liquid zinc, $v_l = \frac{63.38}{6.48} = 10.09$ cc

Substituting the values in Eq. 1, $dP = 1$ atm

$$\begin{aligned} dT &= \frac{dP.T^e(v_g - v_l)}{L^e} = \frac{dP.T^e.v_g}{L^e} && \text{since } v_g \gg v_l,\; v_g - v_l = v_g \\ &= \frac{1 \times 1180 \times 22400 \times 1180 \text{ cc atm } K}{273 \times 27300 \text{ cal}} && 1 \text{ cc atm} = 0.024212 \text{ cal} \\ &= \frac{1 \times 1180 \times 22400 \times 1180 \times 0.024212 \text{ cal } K}{273 \times 27300 \text{ cal}} \\ &= 101.3 \text{ K} \end{aligned}$$

Let us repeat the calculation for the of zinc vapors at the external pressure of 2 atm.

$$\frac{P_1 V_1}{T_1} = \frac{P_2 V_2}{T_2} \qquad P_1 = 1 \text{ atm}, \; T_1 = 273 \; K, P_2 = 2 \text{ atm}, \; T_2 = 1180 \; K$$

$$\begin{aligned} V_2 &= \frac{P_1 V_1 T_2}{P_2 T_1} = v_g \\ &= \frac{1 \times 22400 \times 1180}{2 \times 273} = \frac{22400 \times 1180}{2 \times 273} \\ dT &= \frac{dP.\; T^e.\; v_g}{L^e} \\ &= \frac{1 \times 1180 \times 22400 \times 1180 \text{ cc atm } K}{2 \times 273 \times 27300 \text{ cal}} \\ &= 50.66 \text{ K} \end{aligned}$$

$$\text{Average } dT = \frac{101.32 + 50.66}{2} = 75.99 \text{ deg}$$

Integral method

$$\ln \left[\frac{p_2}{p_1}\right] = \frac{L}{R} \left[\frac{T_2 - T_1}{T_1 T_2}\right], \qquad P_1 = 1 \text{ atm}, T_1 = 1180 \; K, \; P_2 = 2 \text{ atm}, \; T_2?$$

$$ln \left[\frac{2}{1}\right] = \frac{27300}{1.987} \left[\frac{T_2 - 1180}{T_2 \times 1180}\right]$$

$$\frac{T_2 - 1180}{T_2} = 1180 \times \left(0.6931471 \times \frac{1.987}{27300}\right)$$

$$T_2 = 1254.7 \;\; K$$

$$T_1 = 1180$$

$$\therefore \Delta T = T_2 - T_1 = 1254.7 - 1180 = 74.7 \;\; \text{deg}$$

Thus integral method gives the average value of ΔT.

(ii) For solid-liquid transformation, we have

$$\frac{dP}{dT} = \frac{L^f}{T^f(v_l - v_s)} \qquad v_l = \frac{65.38}{6.48} \qquad \text{and } v_s = \frac{65.38}{7}$$

$$dT = \frac{dP.T^f(v_l - v_s)}{L^f}$$

$$= \frac{49 \times 692.5\left(\frac{65.38}{6.48} - \frac{65.38}{7}\right) \text{ cc atm } K}{1740 \text{ cal}} \qquad 1 \text{ cc atm} = 0.024212 \text{ cal}$$

$$= \frac{49 \times 692.5(10.089 - 9.34) \times 0.024212}{1740}$$

$$= 0.354 \;\; \text{K} \qquad \text{Ans.}$$

Problem 41 The latent heat of vaporization of zinc is 27.3 kcal mol^{-1} at the boiling point of 907 °C. Find the vapor pressure over pure zinc at 850 °C.

Solution

For liquid → gas transformation, we have

$$\frac{dP}{dT} = \frac{L^e}{T^e(v_g - v_l)} \qquad (v_g \gg v_l, \qquad v_g - v_l = v_g)$$

$$\frac{dP}{dT} = \frac{L^e}{T^e.v_g} \qquad \left(\text{metallic vapors behave ideally, } v = \frac{RT}{P}\right)$$

$$\frac{dP}{dT} = \frac{LP}{RT^2} \qquad P_1 = 1 \text{ atm}, \;\; T_1 = 1180 \;\; K, P_2 = ?, \;\; T_2 = 1123 \;\; K$$

$$\int \frac{dP}{P} = \frac{L}{R} \int \frac{1}{T^2}$$

or

$$\ln \frac{P_2}{P_1} = \frac{L}{R}\left[\frac{T_2 - T_1}{T_1 T_2}\right]$$

or

$$\ln P_2 - 0 = \frac{27300}{1.987}\left[\frac{1123-1180}{1180\times 1123}\right] = -0.5909869$$

$$P_2 = 0.55378 \ \text{atm} = 420.9 \ \text{mm Hg} \qquad \text{Ans.}$$

Problem 42 At one atmospheric pressure sodium melts at 97.8 °C with a latent heat of fusion of 630 cal mol^{-1} and an increase in specific volume of 0.0279 cc g^{-1}. Calculate the melting point of sodium at 10 atm pressure.

Solution
The effect of pressure on a phase change of a component system is given by the Clausius–Clapeyron equation:

$$\frac{dP}{dT} = \frac{L^f}{T^f \cdot \Delta v}$$

Care must be taken in selecting the units of the quantities. For consistency the latent heat of fusion must be expressed in cal g^{-1}.

At wt of sodium = 23, L^f = 630 cal mol^{-1} = $\frac{630}{23}$ = 27.4 cal g^{-1}

$$\frac{dT}{dP} = \frac{T^f.\Delta v}{L^f} = \frac{370.8\times 0.0279}{27.4}\ \frac{ccg^{-1}K}{\text{cal g}^{-1}} \text{ i.e } \frac{\text{cc.K}}{\text{cal}}$$

$$= 0.378 \ \text{K cc cal}^{-1}$$

$$\therefore \frac{dT}{dP} = \frac{0.378}{41.293}\ \frac{\text{ccK}}{\text{cc atm}}, \qquad 1\ \text{cal} = 41.293\ \text{cc atm},$$

$$\text{cal}^{-1} = \frac{1}{41.293}\text{cc}^{-1}\ \text{atm}^{-1}$$

$$= 0.00915 \ \text{K atm}^{-1}$$

for a change of 9 atm $dT = dP \times 0.00915$

$$= 9 \times 0.00915 \ \text{atm K atm}^{-1}$$

$$= 0.08235 \ \text{K}$$

∴ Melting point of sodium at 10 atm = 97.8 + 0.08 = 98.6 °C.

There is no significant change. In fact it is within experimental error limit.

Problem 43 The vapor pressure of liquid carbon tetrachloride is given as a function of temperature as: log p = −2400T^{-1}–5.30 log T + 23.60, where p is in mm Hg. Calculate the latent heat of vaporization of CCl_4 at its boiling point, 77 °C.

Solution

$$\log\ p = -\frac{2400}{T} - 5.30\ \log\ T + 23.60$$

$$\therefore \frac{d\ logP}{dT} = \frac{2400}{T^2} - \frac{5.3}{T}$$

For l → g transformation we have

$$\frac{dP}{dT} = \frac{LP}{RT^2}$$

or
$$\frac{dP/P}{dT} = \frac{L}{RT^2}$$

or
$$\frac{dlnP}{dT} = \frac{L}{RT^2}$$

or
$$\frac{d\ logP}{dT} = \frac{L}{2,303 \times RT^2}$$

or
$$\frac{d\,logP}{dT} = \frac{L}{4.575\,T^2}$$

or
$$L = 4.575\,T^2\ \left[\frac{d\ logP}{dT}\right]$$

$$\therefore L = 4.575\,T^2\ \left[\frac{2400}{T^2} - \frac{5.3}{T}\right]$$
$$= 2400 \times 4.575 – 5.3 \times 4.575 \times T \qquad (T = 350\ K)$$
$$= 10980 – 8487$$
$$= 2493\ \text{cal mol}^{-1} \qquad \text{Ans.}$$

Problem 44 Calculate the depression in the freezing point of cadmium when the external pressure is 50 atm. The latent heat of fusion of cadmium is 1.53 kcal g atom^{-1} and the normal temperature of fusion is 321 °C. The density of solid and liquid cadmium are 8.65 and 7.35 g cc^{-1} respectively, at one atm pressure. Atomic weight of Cd 112.4.

Solution

$$\mathrm{Cd\,(s)} \rightarrow \mathrm{Cd\,(l)}$$

$$\frac{dP}{dT} = \frac{L^f}{T^f(v_l - v_s)} \quad v_l = \frac{112.4}{7.35}, v_s = \frac{112.4}{8.65}$$

$$\begin{aligned}
\mathrm{dT} &= \frac{dP\,T^f}{L^f}(v_l - v_s) \\
&= \frac{49 \times 594}{1530}\left(\frac{112.4}{7.35} - \frac{112.4}{8.65}\right) \\
&= \frac{49 \times 594(15.293 - 12.944) \times 0.024212}{1530} \\
&= \frac{49 \times 594 \times 2.349 \times 0.024212}{1530} = \frac{1620.138}{1530} \\
&= 1.06 \text{ K} \qquad \text{Ans.}
\end{aligned}$$

Problem 45 Zinc melts at 419.5 °C and boils at 907 °C with a latent heat of vaporization of 27300 cal g atom^{-1}. (a) Express the vapor pressure of liquid zinc as $\ln p_{Zn} = \frac{A}{T} + B$. (b) Calculate the free energy change of the evaporation of liquid zinc and express it in the form of ΔG^o = a + bT for the reaction Zn (l) = Zn(g). (c) Calculate ΔG^o and the partial pressure of zinc vapor over zinc bath at 800 °C. (d) If we apply a vacuum of 1 mm Hg over the molten bath of pure zinc, at what temperature will we observe the boiling evaporation?

Solution

According to van't Hoff equation:

$$\frac{dln\,K}{dT} = \frac{\Delta H}{RT^2}$$

In case of vaporization:

$$Zn\ (l) \rightarrow Zn(g)$$

$$K = p^o_{Zn}\ and\ \Delta H = L^e(\text{latent heat of evaporation})$$

$$\therefore \frac{dlnp^o_{Zn}}{dT} = \frac{L^e}{RT^2}$$

On integration, $\int d\ ln\,p^o_{Zn} = \frac{L^e}{R}\int\frac{dT}{T^2}$

or $\ln p_{Zn} = -\frac{L^e}{RT} + B$ (1) substitute $-\frac{L^e}{R} = A$

or $\ln p_{Zn} = \frac{A}{T} + B$ (2) $\therefore A = -\frac{27300}{1.987} = -13739$

or $\ln p_{Zn} = -\frac{13739}{T} + B$ when p = 1 atm, T = 1180 K

$$\ln(1) = -\frac{13739}{T} + B$$

$$\therefore B = -\frac{13739}{1180} = 11.64$$

(a) $\ln p_{Zn} = -\frac{13739}{T} + 11.64$ Ans.

(b) $\Delta G^o = -RT \ln K = -RT \ln p_{Zn}$

$$= -RT\left[-\frac{13739}{T} + 11.64\right]$$
$$= 13739\ R - 11.64\ \mathrm{RT}$$
$$= 27300 - 23.14\ \mathrm{T\ cal} \qquad \text{Ans.}$$

(c) $\frac{dP}{dT} = \frac{L\,P}{RT^2}$

Integration of the above equation gives:

$$\therefore \int_{p_1}^{p_2}\frac{dP}{P} = \frac{L}{R}\int_{T_1}^{T_2}\frac{dT}{T^2}$$

Where p_1 and p_2 are the initial and final pressures at temperatures T_1 and T_2, respectively.

On integration we get:

$$ln\left[\frac{p_2}{p_1}\right] = \frac{L}{R}\left[\frac{T_2 - T_1}{T_1 T_2}\right]$$

$$ln\ p_2 - ln\,p_1 = \frac{L}{R}\left[\frac{T_2 - T_1}{T_1 T_2}\right] \qquad p_1 = 1\ \text{atm},\ T_1 = 1180\ K$$

$$ln\ p_2 - 0 = \frac{27300}{1.987}\left[\frac{1073 - 1180}{1180 \times 1073}\right] = -1.161 \qquad p_2\ ? \quad T_2 = 1073\ K$$

$$p_2 = 0.312\ \text{atm} = 2.38\ \text{mm Hg} \qquad \text{Ans.}$$

$$\Delta G^o = -RT\ ln p_{Zn} = -1.987 \times 1073 \times ln p_2$$
$$= -1.987 \times 1073 \times (-1.161) = 2475\ \text{cal} \qquad \text{Ans.}$$

(d) $ln\ p_2 - ln\,p_1 = \frac{L}{R}\left[\frac{T_2 - T_1}{T_1 T_2}\right]$ $\qquad p_1 = 1$ atm, $T_1 = 1180\ K$,

$$p_2 = 1\ \text{mm Hg} = \frac{1}{760}\text{atm},\ T_2?$$

$$\Delta G^o = -RT\ ln p_{Zn}$$

$$= -1.987\ T \times ln\left[\frac{1}{760}\right]$$

$$= 13.18\ \mathrm{T} \tag{1}$$

$$\text{From (b)} \quad \Delta G^o = 27300 - 23.14 \ \text{T cal} \tag{2}$$

$$\therefore 13.18 \ \text{T} = 27300 - 23.14 \ T$$

$$T = \frac{27300}{36.32} = 751.65 \ K = 478.35\,°\text{C} \qquad \text{Ans.}$$

At 1 mm Hg zinc will boil at 478.35 °C.

Problem 46 The vapor pressure of liquid chlorine in cm of mercury is expressed as:

$$\log p = -\frac{1414.8}{T} + 9.91635 - 1.206 \times 10^{-3} T + 1.34 \times 10^{-6} T^2$$

The specific volume of chlorine gas at its boiling point is 269.1 cc g^{-1} and that of the liquid is 0.7 cc g^{-1}. Calculate heat of vaporization of liquid chlorine in cal cc g^{-1} at its boiling point, 269.05 K.

Solution

$$\log p = -\frac{1414.8}{T} + 9.91635 - 1.206 \times 10^{-3} T + 1.34 \times 10^{-6} T^2$$

$$\frac{d \log p}{dT} = \frac{1414.8}{T^2} - 1.206 \times 10^{-3} + 2.68 \times 10^{-6} T \tag{1}$$

Further, since

$$\frac{d \log p}{dT} = \frac{1}{2.303} \cdot \frac{d \ln p}{dT} = \frac{1}{2.303\,p} \cdot \frac{dp}{dT} \tag{2}$$

From Eqs. 1 and 2, we get:

$$\boldsymbol{\frac{dp}{dT}} = \left[\boldsymbol{\frac{d \log p}{dT}}\right] \times (2.303p) \qquad \text{boiling point} = 269.05 \ \text{K}, \ p = 76 \ cm$$

$$= \left[\frac{1414.8}{T^2} - 1.206 \times 10^{-3} + 2.68 \times 10^{-6} T\right] \times (2.303p)$$

$$= 3.343 \ \text{cm Hg deg}^{-1}$$

$$= \frac{3.343}{76.00} = 0.04398 \ \text{atm deg}^{-1}$$

For liquid → gas transformation, we have

$$\frac{dP}{dT} = \frac{l^e}{T^e (v_g - v_l)}$$

$$L^e = T^e\left(v_g - v_l\right)\frac{dP}{dT}$$

$$v_g = 269.1 \text{ cc g}^{-1}, \quad v_l = 0.7 \text{ cc g}^{-1}$$

$$\begin{aligned}\therefore L^e &= 269.05 \ (269.1 - 0.7) \times 0.04398 \text{ cc atm g}^{-1} \\ &= 269.05 \times 268.4 \times 0.04398 \times 0.0242 \qquad (1 \text{ cc atm} = 0.0242 \text{ cal}) \\ &= 76.9 \text{ cal g}^{-1} \qquad \text{Ans.}\end{aligned}$$

Problem 47 The vapor of mercury, cadmium and zinc are given as function of temperature by expressions.

$$\log p_{Hg} = -\frac{3305}{T} - 0.795 \log T + 10.36$$

$$\log p_{Cd} = -\frac{5707}{T} - 1.086 \ \log \ T + 8.776$$

$$\log p_{Zn} = -\frac{6620}{T} - 1.255 \ \log \ T + 12.34$$

where p is in mm Hg. Calculate the latent heat of evaporation of mercury, cadmium and zinc at their boiling points.

Solution

From Problem 37, for a liquid-gas transformation, we know:

$$\frac{d\,logP}{dT} = \frac{L}{4.575\,T^2}$$

or

$$L = 4.575\,T^2\left[\frac{d\,logP}{dT}\right]$$

From the expression given for Hg:

$$\frac{d \log p}{dT} = \frac{3305}{T^2} - \frac{0.795}{2.303\,T}$$

$$\therefore L^e_{Hg} = 4.575\ T^2 \left[\frac{3305}{T^2} - \frac{0.795}{2.303T}\right]$$

$$= 4.575 \times 3305 - \frac{4.575 \times 0.795 \times 630}{2.303} \qquad \left(T^e_{Hg} = 357 + 273 = 630\ \text{K}\right)$$

$$= 15120 - 995$$

$$= 14125\ \text{cal g atom}^{-1}$$

Expression given for Cd:

$$\frac{d \log p}{dT} = \frac{5707}{T^2} - \frac{1.086}{2.303T}$$

$$\therefore L^e = 4.575 T^2 \left[\frac{5707}{T^2} - \frac{1.086}{2.303\,T}\right]$$

$$= 4.575 \times 5707 - \frac{4.575 \times 1.086 \times 1036}{2.303} \qquad \left(T^e_{Cd} = 763 + 273 = 1036\ \text{K}\right)$$

$$= 26109.5 - 2235.57$$

$$= 23874\ \text{cal g atom}^{-1}$$

Similarly, for Zn;

$$L^e_{Zn} = 27345\ \text{cal g atom}^{-1}$$

Problem 48 Develop a general expression for liquid → gas transformation relating vapor pressure with latent heat of vaporization free energy change during evaporation.

Solution
For liquid → gas transformation, we have

$$\frac{dP}{dT} = \frac{L^e}{T^e\left(v_g - v_l\right)} \qquad \left(v_g \gg v_l,\ \ v_g - v_l = v_g\right)$$

$$\frac{dP}{dT} = \frac{L^e}{T^e . v_g} \qquad \left(\text{if vapors behave ideally},\ \ v = \frac{RT}{P}\right)$$

$$\frac{dP}{dT} = \frac{L\,P}{RT^2}$$

$$\int \frac{dP}{P} = \frac{L}{R} \int \frac{1}{T^2}$$

$$ln\ p = -\frac{L}{RT} + B = \frac{A}{T} + B \qquad \left(\text{Where } A = -\frac{L}{R}\right)$$

Substituting $p = 1$ atm at the boiling point, we get the value of B as:

$$ln\,(1) = -\frac{L}{RT} + B$$

$$or\ \ 0 = -\frac{L}{RT} + B \qquad \text{(since } A \text{ and } T \text{ are known, } B \text{ can be calculated)}$$

$$\text{For} \quad l \rightarrow g, \quad K = p$$

$$\therefore \Delta G = -RT \ln p = -RT\left[\frac{A}{T} + B\right] = -RA - RBT = a + bT$$

Problem 49 Differenciate between ideal and regular solutions.

Answer

Ideal solution	Regular solution
1. obeys Raoult's law i.e. (activity,a_i) = (mole fraction, x_i) $(A \leftrightarrow B) = (A \leftrightarrow A) = (B \leftrightarrow B)$ $= \frac{1}{2}\{(A \leftrightarrow A) + (B \leftrightarrow B)\}$	1. small deviation from Raoult's ideal behavior. $(A \leftrightarrow B) <$ or $> \frac{1}{2}\ \{(A \leftrightarrow A)$ $+(B \leftrightarrow B)\}$
2. $\Delta H^M = 0,\ \Delta V^M = 0$	2. $\Delta H^M + or - ve,\ \Delta H^M = \Delta G^{xs}$ $\Delta H^M = -RT(x_A\ \ ln\gamma_A + x_B\ ln\gamma_B)$ $= RT\alpha x_A x_B$
3. $\Delta S^{M,\,id} = -R(x_A\,lnx_A + x_B lnx_B)$	3. $\Delta S^M = \Delta S^{M,\,id}$
4. $\Delta G^M = RT(x_A lnx_A + x_B lnx_B)$	4. $\Delta G^M = RT(x_A\,lna_A + x_B lna_B)$

Problem 50 Show that volume of mixing and heat of mixing are zero for ideal solutions.

Solution

To show $\Delta V^{M,\,id} = 0$ or $\overline{V}_i^{M,id} = 0$

We know that $\left[\frac{\partial \overline{G}_i}{\partial P}\right]_{T,x} = \overline{V}_i$ and $\left[\frac{\partial G_i^o}{\partial P}\right]_{T,x} = V_i^o$

$$\therefore \left[\frac{\partial\left(\overline{G}_i - G_i^o\right)}{\partial P}\right]_x = \overline{V}_i - V_i^o$$

$$\text{or} \qquad \frac{\partial \Delta \overline{G}_i^M}{\partial P} = \Delta \overline{V}_i^M$$

For an ideal solution: $\Delta \overline{G}_i^M = RT\ \ ln\,x_i$

$$\Delta \overline{G}_i^M = \overline{G}_i - G_i^o = RT\ \ ln\,x_i$$

$$\therefore \left[\frac{d(RT\ \ ln\,x_i)}{dP}\right] = \overline{V}_i - V_i^o = \Delta \overline{V}_i^{M,id}$$

Since x_i is not a function of pressure, $\Delta \overline{V}_i^{M,id} = 0$.

$$\begin{aligned}\Delta V'^{M} &= \left(n_A.\overline{V}_A + n_B.\overline{V}_B\right) - \left(n_A.V_A^o + n_B.V_B^o\right)\\ &= n_A\left(\overline{V}_A - V_A^o\right) + n_B\left(\overline{V}_B - V_B^o\right)\\ &= n_A\Delta\overline{V}_A^M + n_B\Delta\overline{V}_B^M = 0\end{aligned}$$

$\therefore\ \Delta V^{M,id} = x_A\Delta\overline{V}_A^{M,id} + x_B\Delta\overline{V}_B^{M,id} = 0$ for ideal solutions

To show that $\Delta H^{M,\ id} = 0$

Gibbs–Helmholtz equation: $\frac{\partial(G/T)}{\partial T} = -\frac{H}{T^2}$

$$\therefore\ \frac{\partial\left[\overline{G}_i/T\right]}{\partial T} = -\frac{\overline{H}_i}{T^2} \tag{1}$$

For pure components:

$$\left[\frac{\partial\left(G_i^o/T\right)}{\partial T}\right]_{P,x} = -\frac{H_i^o}{T^2} \tag{2}$$

substracting Eqs. 2 from 1, we get:

$$\left[\partial\frac{\left(\overline{G}_i - G_i^o\right)/T}{\partial T}\right]_{P,x} = -\frac{\left(\overline{H}_i - H_i^o\right)}{T^2}$$

$$\left[\partial\frac{\left(\Delta\overline{G}_i^M\right)/T}{\partial T}\right]_{P,x} = -\frac{\Delta\overline{H}_i^M}{T^2} \tag{3}$$

Where $\Delta\overline{H}_i^M$ is the partial molar heat of mixing of i in the solution

$$\text{Since}\Delta\overline{G}_i^M = RT\ \ ln\,x_i,\ \frac{d[R\ \ ln\,x_i]}{dT} = -\frac{\Delta\overline{H}_i^{M,id}}{T^2}$$

As x_i is independent of T, $\Delta\overline{H}_i^{M,id} = 0$

$$\text{Therefore, }\Delta H_i^{M,id} = x_A\Delta\overline{H}_A^{M,id} + x_B\Delta\overline{H}_B^{M,id} = 0$$

Problem 51 In the formation of liquid brass: $(1-x)$ Cu(l) + x Zn(l) = Cu-Zn (l), the molar heat of formation is given by ΔH = $-7100\,x\,(1-x)$ cal, where x is the atom fraction of zinc. Derive expressions for partial molar heat of mixing of Cu and Zn in the liquid brass as a function of composition.

Solution

Molar heat of mixing, $\Delta H = -7100\,x\,(1-x)$

The partial molar heat of mixing of Zn:

$$\Delta \overline{H}_{Zn} = \Delta H + (1 - x_{Zn}).\frac{\partial \Delta H}{\partial x_{Zn}}$$

$$\frac{\partial \Delta H}{\partial x_{Zn}} = \frac{\partial \Delta H}{\partial x} = \frac{d}{dx}[-7100x(1-x)]$$

$$= \frac{d}{dx}\left[-7100\left(x - x^2\right)\right] = -7100(1-2x)$$

$$\therefore\ \Delta \overline{H}_{Zn} = -7100x(1-x) + (1-x)[-7100(1-2x)]$$

$$= (1-x)\,[-7100x + (-7100)(1-2x)]$$

$$= -7100(1-x)[x+1-2x] = -7100(1-x)(1-x) = -7100(1-x)^2 \quad \text{Ans.}$$

$$\Delta \overline{H}_{Cu} = \Delta H + (1 - x_{Cu})\left(\frac{\partial \Delta H}{\partial x_{Cu}}\right) = \Delta H + x_{Zn}\left(\frac{\partial \Delta H}{\partial x_{Cu}}\right) = \Delta H + x_{Zn}\left(\frac{\partial \Delta H}{-\partial x_{Zn}}\right)$$

$$= \Delta H - x.\frac{\partial \Delta H}{\partial x}$$

$$= -7100\ x\ (1-x)-x\ [-7100\ (1-2x)]$$

$$= -7100x\ [1-x-(1-2x)]$$

$$= -7100\ x\ [1 - \mathrm{x} - 1 + 2\mathrm{x}] = -7100\ x\ (x) = -7100\ x^2 \quad \text{Ans.}$$

Problem 52 The enthalpies of mixing of Cd-Sn alloys at 500 °C are given below:

x_{Cd}	0	0.1	0.3	0.5	0.7	0.9	1.0
ΔH^M cal mol^{-1}	0	298.2	652.4	800.0	620.5	251.5	0

Calculate the values of the partial molar enthalpies of mixing of cadmium and tin in a Cd-Sn alloy containing 60 at % cadmium.

Solution

To find the value of partial molar enthalpies of mixing of cadmium, $\Delta \overline{H}^M_{Cd}$ and that of tin i.e., $\Delta \overline{H}^M_{Sn}$, plot ΔH^M vs x_{Cd} (as illustrated in Fig. 3.4). Draw a tangent at $x_{Cd} = 0.6$. The ordinates of this tangent at $x_{Cd} = 1$ and $x_{Sn} = 1$ are equal to $\Delta \overline{H}^M_{Cd}$ and $\Delta \overline{H}^M_{Sn}$, respectively. (Refer Sect. 3.3.1, Graphical method).

$$\Delta \overline{H}^M_{Cd} = 320\ \text{cal mol}^{-1} \text{ and } \Delta \overline{H}^M_{Sn} = 1360\ \text{cal mol}^{-1} \quad \text{Ans.}$$

Problem 53 At 746 K the activity coefficient of lead in liquid Pb-Bi alloys is expressed as $ln\gamma_{Pb} = -0.74\ (1 - x_{Pb})^2$. Calculate the integral molar free energy of the alloy containing 40 at % lead at 746 K.

Solution

The expression: $ln\ \gamma_{Pb} = -0.74\ (1 - x_{Pb})^2$, indicates that α is constant and Pb-Bi solutions obeys regular solution model.

$$\alpha = \frac{ln\,\gamma_{Pb}}{(1 - x_{Pb})^2} = -0.74$$

$$\therefore\ \alpha_{Pb} = \alpha_{Bi} = \alpha = -0.74$$

$$\begin{aligned} G^{xs} &= RT\alpha x_{Pb} x_{Bi} \\ &= 1.987 \times 746\ \ (-0.74)\ \ 0.4 \times 0.6 \\ &= -263.3\ \ \text{kcal mol}^{-1} \end{aligned}$$

$$\begin{aligned} \Delta G^M &= \Delta G^M_{id} + G^{XS} \\ &= RT(x_{Pb}\ \ln x_{Pb} + x_{Bi}\ \ln x_{Bi}) + G^{XS} \\ &= 1.987 \times 746\ \ (0.4\ \ \ln\ \ 0.4 + 0.6\ \ \ln\ \ 0.6) - 263.3 \\ &= 1.987 \times 746\ \ (-0.6725) - 263.3 \\ &= -996.9 - 263.3 = -1260.2\ \ \text{cal mol}^{-1} \end{aligned}$$ Ans.

or

$$\alpha_{Pb} = -0.74,\ \ x_{Pb} = 0.4,\ \ x_{Bi} = 0.6,$$

$$ln\,\gamma_{Pb} = -0.74 \times (1 - 0.4)^2 = -0.74 \times 0.36 = -0.2664$$

$$\gamma_{Pb} = 0.766$$

$$a_{Pb} = 0.766 \times 0.4 = 0.3065$$

$$ln\,\gamma_{Bi} = -0.74 \times (1 - x_{Bi})^2 = -0.74 \times 0.16 = -0.1184$$

$$\gamma_{Bi} = 0.8901$$

$$a_{Bi} = 0.8901 \times 0.6 = 0.5341$$

$$\begin{aligned} \Delta G^M &= RT\ (x_{Pb}\ \ln a_{Pb} + 0.x_{Bi}\ \ln a_{Bi}) \\ &= 1.987 \times 746\ \ (0.4\ \ \ln\ \ 0.3065 + 0.6\ \ \ln\ \ 0.5341) \\ &= 1.987 \times 746\ \ (-0.8493) \\ &= -1259.0\ \ \text{cal mol}^{-1} \end{aligned}$$ Ans.

Problem 54 The variation of $G^{xs}$with composition for liquid Fe-Mn alloys at 1863 K is given below:

x_{Mn}	0.1	0.2	0.3	0.4	0.5	0.6	0.7	0.8	0.9
G^{xs} J mol^{-1}	395	703	925	1054	1100	1054	925	703	395

(a) Does the system obey regular solution behavior at 1863 K?
(b) Calculate $\overline{G}^{xs}_{Fe}$ and $\overline{G}^{xs}_{Mn}$ at $x_{Mn} = 0.6$.
(c) Calculate ΔG^M at $x_{Mn} = 0.4$
(d) Calculate the partial pressure of Fe and Mn exerted by the alloy of $x_{Mn} = 0.2$

$$\ln p_{Fe}\ (\text{atm}) = -45390/\text{T} - 1.2\ \ln \text{T} + 23.93$$

$$\ln p_{Mn}\ (\text{atm}) = -33440/\text{T} - 3.02\ \ln \text{T} + 37.68$$

Solution

$$G^{xs} = \Delta H^M = RT\alpha x_{Mn} x_{Fe} = \alpha' x_{Mn} x_{Fe}$$

$$\alpha' = \frac{G^{xs}}{x_{Mn} x_{Fe}}$$

$$x_{Mn} = 0.1, \quad \alpha' = \frac{395}{0.1 \times 0.9} = 4389$$

$$x_{Mn} = 0.2, \quad \alpha' = \frac{703}{0.2 \times 0.8} = 4393$$

$$x_{Mn} = 0.3, \quad \alpha' = \frac{925}{0.21} = 4404.76$$

$$x_{Mn} = 0.4, \quad \alpha' = \frac{1054}{0.4 \times 0.6} = 4391.66$$

$$x_{Mn} = 0.5, \quad \alpha' = \frac{1100}{0.5 \times 0.5} = 4400$$

The following table based on the above calculations shows that α′ is almost constant i.e. independent of composition. Hence, the system obeys regular solution behavior at 1863 K.

x_{Mn}	0.1	0.2	0.3	0.4	0.5	0.6	0.7	0.8	0.9
G^{xs} J mol^{-1}	395	703	925	1054	1100	1054	925	703	395
α′	4389	4393	4405	4392	4400	4392	4405	4393	4389

(b) $$\overline{G}^{XS}_{Fe} = RT \ln \gamma_{Fe} \alpha_{Mn} = \frac{G^{XS}}{RT\, x_{Mn} \cdot x_{Fe}} = \frac{1054}{8.314 \times 1863 \times 0.6 \times 0.4} = 0.2835 = \alpha_{Fe}$$

$$\alpha_{Fe} = \frac{\ln \gamma_{Fe}}{x^2_{Mn}}; \quad \ln \gamma_{Fe} = \alpha\ x^2_{Mn} = 0.2835 \times 0.6^2 = .2835 \times 0.36 = 0.10207$$

$$\gamma_{Fe} = 1.1075, \quad \ln \gamma_{Mn} = \alpha x^2_{Fe} = 0.2835 \times 0.16 = 0.04536$$

$$\therefore \overline{G}^{XS}_{Fe} = RT \ln \gamma_{Fe} = 1863 \times 8.314 \times 0.10207 = 1581\,\text{J} \quad \text{Ans.}$$
$$\overline{G}^{XS}_{Mn} = RT \ln \gamma_{Mn} = 1863 \times 8.314 \times 0.04536 = 703\,\text{J} \quad \text{Ans.}$$

(c) $\Delta G^M = \Delta G^M_{id} + G^{XS}$

$$\Delta G^M_{id} = RT(0.4 \ln 0.4 + 0.6 \ln 0.6)$$
$$= 8.314 \times 1863\,(-0.6730) = -1044 \text{ J mol}^{-1}$$
$$\therefore \Delta G^M = \Delta G^M_{id} + G^{XS}$$
$$= -10424 + 1054 = -9370 \text{ J mol}^{-1} \quad \text{Ans}$$

(d)
$$\ln p_{Mn}(\text{atm}) = -33440/\text{T}–3.02 \ln \text{T} + 37.68$$
$$= -33440/1863–3.02 \times \ln 1863 + 37.68$$
$$= -17.94954–22.74043 + 37.68 = 3.09972$$
$$\therefore p^0_{Mn}\ (1863) = 0.04929 \text{ atm}$$

$$\ln p_{Fe}(\text{atm}) = -\frac{45390}{1863} - 1.27 \ln 1863 + 23.93$$
$$= -\frac{45390}{1863} - 1.27 \times 7.5299 + 23.93$$
$$= -24.364–9.963 + 23.93 = -10.3973$$

$$\therefore p^0_{Fe}\,(1863) = 0.000030515 \text{ atm} = 3.0515 \times 10^{-5} \text{ atm}$$

$$\alpha = \frac{703}{8.314 \times 1863 \times 0.2 \times 0.8} = 0.283669$$

$$\alpha_{Mn} = \frac{\ln \gamma_{Mn}}{x^2 Fe}; \qquad \ln \gamma_{Mn} = \alpha . x^2_{Fe} = 0.283669 \times 0.64 = 0.181548$$

$$\gamma_{Mn} = 1.19907, \quad a_{Mn} = 1.199 \times 0.2 = 0.2398$$

$$p_{Mn} = p^o_{Mn} . a_{Mn} = 0.049279 \times 0.2398 = 0.0118 \text{ atm} \quad \text{Ans.}$$

$$\ln \gamma_{Fe} = \alpha . x^2_{Mn} = 0.283669 \times .04 = 0.011346$$

$$\gamma_{Fe} = 1.0114, \quad a_{Fe} = 1.0114 \times 0.8 = 0.80912$$

$$p_{Fe} = p^0_{Fe} . a_{Fe} = 3.0515 \times 10^{-5} \times 0.80912 = 2.45 \times 10^{-5} \text{ atm} \quad \text{Ans.}$$

Problem 55 The activity coefficient of copper in liquid brass in the temperature range 1000–1500 K is expressed as RT ln $\gamma_{\text{Cu}} = -4600\ (1-x_{\text{Cu}})^2$. Calculate the partial pressure of zinc in brass containing 40 at % copper at 1100 K. The latent heat of vaporization of zinc at its boiling point of 907 K is 27.3 kcal g atom^{-1}.

Solution

$$p_{Zn} = p^0_{Zn}.a_{Zn} = p^0_{Zn}.\gamma_{Zn}x_{Zn}, \quad x_{Cu} = 0.4$$

$$\ln \gamma_{Cu} = -\frac{4600}{RT}(1 - x_{Cu})^2$$

$$\alpha_{Cu} = \frac{\ln \gamma_{Cu}}{(1 - x_{Cu})^2} = -\frac{4600}{RT}$$

This means α_{Cu} is constant which indicates that Cu-Zn solution follows regular model.

$$\therefore \alpha_{Cu} = \alpha_{Zn} = \alpha$$

Hence, $$\ln \gamma_{Zn} = -\frac{4600}{RT}(1 - x_{Zn})^2$$

$$= -\frac{4600}{1.987 \times 1100}(1 - 0.6)^2 = -\frac{46 \times 0.16}{1.987 x 11} = -0.3367$$

$$\therefore \gamma_{Zn} = 0.714$$

$$\therefore p_{Zn} = p^0_{Zn}.\gamma_{Zn}.x_{Zn} = 325.88 \times 0.714 \times 0.6 = 135.25\text{mm Hg}$$ Ans.

In order to calculate the value of p^o_{Zn} at 1100 K make use of the Clausius–Clapeyron equation.

$$\ln\left(\frac{p_2}{p_1}\right) = \frac{L^e}{R}\left(\frac{T_2 - T_1}{T_1 T_2}\right) \qquad T_2 = 1180K, \; p_2 = 1 \text{ atm}, \; T_1 = 1100K, p_1 = ?$$

$$\ln\left(\frac{1}{p_1}\right) = -\ln p_1 = \frac{27300}{1.987}\left(\frac{1180 - 1100}{1180 \times 1100}\right)$$

$$or \ln p_1 = -\frac{27300 \times 80}{1.987 \times 1180 \times 1100} = -0.8467$$

$$\therefore p_1 = 0.4288 \text{ atm}$$

$$\therefore p^o_{Zn} = 0.4288 \times 760 = 325.88 \text{ mm Hg}$$ Ans.

Problem 56 Solid Au-Cu alloys are regular in their thermodynamic behavior. ΔH^M at 773 K is listed below as a function of composition:

x_{Cu}	0.1	0.2	0.3	0.4	0.5	0.6	0.7	0.8	0.9
ΔH^M, cal mol^{-1}	−355	−655	−910	−1120	−1230	−1240	−1130	−860	−460

(a) Find $\Delta\overline{H}_{Cu}$ and $\Delta\overline{H}_{Au}$ at $x_{Cu} = 0.3$
(b) Find ΔG^M at at $x_{Cu} = 0.3$
(c) Calculate the partial pressure of Cu for an alloy containing 30 at % Cu. Given that:

$$\log p_{Cu} = \frac{-17770}{T} - 0.86 \log T + 12.39\,(\text{mm})$$

Solution

(a) To find the value of partial molar enthalpies of mixing of cadmium, $\Delta\overline{H}^M_{Cu}$ and that of tin i.e., $\Delta\overline{H}^M_{Au}$, plot ΔH^M vs x_{Cu} (as illustrated in Fig. 3.4). Draw a tangent at $x_{Cu} = 0.3$. The ordinates of this tangent at $x_{Cu} = 1$ and $x_{Au} = 1$ are equal to $\Delta\overline{H}^M_{Cu}$ and $\Delta\overline{H}^M_{Au}$, respectively. (Refer Sect. 3.3.1, Graphical method).

$$\Delta\overline{H}^M_{Cu} = -2540\ \text{cal mol}^{-1} \text{ and } \Delta\overline{H}^M_{Au} = -210\ \text{cal mol}^{-1} \qquad \text{Ans.}$$

(b) $\Delta G^M = G^{XS} + \Delta G^{id} = \Delta H^M + \Delta G^{id}$ (for regular solution $G^{XS} = \Delta H^M$)

$$= -910 + RT\ (x_{Cu}\ln x_{Cu} + x_{Au}\ln x_{Au})$$
$$= -910 + 1.987 \times 773\ (0.3\ \ln\ 0.3 + 0.7\ \ln\ 0.7) = -1848\ \text{cal mol}^{-1}$$

or

$$\begin{aligned}\Delta\overline{G}^M_{Cu} &= \overline{G}^{XS}_{Cu} + \Delta G^{M,id}_{Cu} \\ &= \Delta\overline{H}^M_{Cu} + RT\ln x_{Cu} \\ &= -2540 + 4.575 \times 773\ \log\ 0.3 \\ &= -2540 - 1849 = -4389\ \text{cal mol}^{-1}\end{aligned}$$

$$\begin{aligned}\Delta\overline{G}^M_{Au} &= \overline{G}^{XS}_{Au} + \Delta G^{M,id}_{Au} \\ &= \Delta\overline{H}^M_{Au} + RT\ln x_{Au} \\ &= -210 + 4.575 \times 773 \log 0.7 \\ &= -210 - 548 = -758 \\ \Delta G^M &= x_{Cu}\Delta\overline{G}^M_{Cu} + x_{Au}\Delta\overline{G}^M_{Au} \\ &= 0.3(-4389) + 0.7(-758) \\ &= -1317 - 529 \\ &= -1846\ \text{cal mol}^{-1}\end{aligned}$$

(c) $\log a_{Cu} = \dfrac{\Delta\overline{G}^M_{Cu}}{2.303RT} = \dfrac{-4389}{4.575 \times 773} = -1.242$

$$a_{Cu} = 0.05728$$

$$\Delta \overline{H}^M_{Cu} = RT \ln \gamma_{Cu} = -2540$$

$$\log \gamma_{Cu} = -\frac{2540}{4.575 \times 773} = 0.7183, \ \gamma_{Cu} = 0.1713$$

$$a_{Cu} = \gamma_{Cu} . x_{Cu} = 0.05139$$

$$p_{Cu} = p^o_{Cu} . a_{Cu} = 1.5 \times 10^{-13} \times 0.05139 = 7.7 \times 10^{-15} \text{ mm Hg}$$

$$\log p_{Cu} = \frac{-17770}{T} - 0.86 \ \log T + 12.39 \ (\text{mm})$$

$$\log p^o_{Cu}(at773) = -13.178$$

$$p^o_{Cu} = 1.5 \times 10^{-13} \text{ mmHg}$$

∴ Partial pressure of Cu over Au-Cu alloy (at 500 °C and x_{Cu}= 0.3) = 7.7×10^{-15} mm Hg. Ans.

Problem 57 Alloys in the system A–B obey regular solution model. If the enthalpy of mixing of an alloy containing 60 at % B at 500 °C is 2030 J mol^{-1}, calculate the activities of A and B in this alloy at 600 °C.

Solution

In case of regular solution: $G^{xs} = \Delta H^M = RT\alpha \, x_A x_B$ (R = 8.314 J deg^{-1} mol^{-1})

$$\therefore \ \Delta H^M = 2030 = 8.314 \times 773 \times 0.4 \times 0.6 \ \alpha$$

for regular solution:

$$\alpha_A = \alpha_B = \alpha = \frac{2030}{8.314 \times 773 \times 0.24} = 1.3162$$

$$\alpha_A = \frac{\ln \gamma_A}{x_B^2} \quad \text{and} \quad \alpha_B = \frac{\ln \gamma_B}{x_A^2}$$

$$\ln \gamma_B = \alpha_B . x_A^2 = \alpha x_A^2$$

$$\frac{\ln \gamma_B(T_2)}{\ln \gamma_B(T_1)} = \frac{T_1}{T_2}$$

$$\therefore \ln\gamma_B(T_2) = \ln\gamma_B(T_1).\frac{T_1}{T_2}$$
$$= \alpha.x_A^2.\frac{T_1}{T_2}(T_1 = 773K, T_2 = 873\ \text{K},\ \ x_A = 0.4, x_B = 0.6)$$
$$= 1.3162 \times (0.4)^2 \times \frac{773}{873}$$
$$= 0.4662$$
$$\therefore \gamma_B(873K) = 1.594$$
$$a_B = \gamma_B.x_B = 1.594 \times 0.6 = 0.956 \qquad \text{Ans.}$$
$$\ln\gamma_A(T_2) = \frac{T_1}{T_2}\alpha x_B^2 = \frac{773}{873}1.3162 \times 0.6^2 = 0.419$$
$$\gamma_A(873K) = 1.521$$
$$a_A = \gamma_A.x_A = 1.521 \times 0.4 = 0.608 \qquad \text{Ans.}$$

Problem 58 From electrochemical measurements the partial molar free energy of mixing of silver in a liquid Au-Ag alloy containing 80 at % Ag has been reported as -700 cal mol^{-1} at 1085 °C, relative to liquid silver as standard state. Assuming that Au-Ag alloys behave regularly at this temperature calculate the excess integral molar free energy of the alloy and the integral molar free energy of mixing. Check your answer by two different methods.

Solution

$$\Delta\overline{G}_{Ag}^{M} = -700\ \text{cal mol}^{-1} = RT\ln a_{Ag}$$
$$\ln a_{Ag} = -\frac{700}{1.987 \times 1358}$$
$$a_{Ag} = 0.7715$$
$$\gamma_{Ag} = \frac{a_{Ag}}{x_{Ag}} = \frac{0.7715}{0.8} = 0.964375$$
$$\ln\gamma_{Ag} = -0.0363$$

Since the Au-Ag solution is regular in behavior, we have

$$\frac{\ln\gamma_{Ag}}{x^2Au} = \frac{\ln\gamma_{Au}}{x^2Ag}$$
$$\therefore \ln\gamma_{Au} = \frac{\ln\gamma_{Ag}.x_{Ag}{}^2}{x_{Au}{}^2} = \frac{-0.0363 \times 0.8^2}{0.2^2} = -0.5804$$

$$\begin{aligned} G^{xs} &= RT\left(x_{Ag}\ln\gamma_{Ag} + x_{Au}\ln\gamma_{Au}\right) \\ &= x_{Ag}\overline{G}_{Ag}^{XS} + x_{Au}\overline{G}_{Au}^{XS} = 1.987 \times 1358 \ \ [0.8(-0.0363) + 0.2(-0.5804)] \\ &= -1.987 \times 1358 \ \ (0.02904 + 0.11608) \\ &= -1.987 \times 1358 \times 0.14512 = -391.6 \text{ cal mol}^{-1} \end{aligned}$$

Ans (i).

$$\text{or } \alpha_{Ag} = \alpha_{Au} = \frac{\ln\gamma_{Ag}}{x^2 Au} = -\frac{0.0363}{0.04} = 0.9075 = \alpha$$

$$\begin{aligned} G^{XS} &= RT\alpha x_{Ag} x_{Au} \\ &= 1.987 \times 1358 \times (-0.9075) 0.8 \times 0.2 = -391.8 \text{ cal mol}^{-1} \end{aligned}$$

Ans (ii)

Integral molar free energy of mixing: $\Delta G^M = RT(x_{Ag} \ln a_{Ag} + x_{Au} \ln a_{Au})$

$$x_{Ag} = 0.8, \ \ a_{Ag} = 0.7715$$

$$x_{Au} = 0.2, \ \ a_{Au} = 0.1119$$

$$\ln\gamma_{Au} = -0.5804, \gamma_{Au} = 0.5597$$

$$a_{Au} = \gamma . x = 0.5597 \times 0.2 = 0.1119$$

$$\begin{aligned} \Delta G^M &= RT(0.8\ln 0.7715 + 0.2\ln 0.1119) \\ &= 1.987 \times 1358[0.8(-0.2594) + 0.2(-2.1898)] \\ &= -1741.9 \text{ cal mol}^{-1} \end{aligned}$$

Since the solution is regular, we can calculate given value G^{xs} by the following expression: making use of the above calculated value of ΔG^M

$$G^{xs} = \Delta G^M - \Delta G^M_{id}$$

$$\begin{aligned} \Delta G^M_{id} &= RT\left(x_{Ag}\ln x_{Ag} + x_{Au}\ln x_{Au}\right) = RT(0.8\ln 0.8 + 0.2\ln 0.2) \\ &= 1.987 \times 1358 \ \ (-0.5003) = -1350.0 \text{ cal mol}^{-1} \end{aligned}$$

$$\therefore G^{XS} = -1741.9 - (-1350.0) = -391.9 \text{ cal mol}^{-1}$$

Ans (ii)

The value of G^{xs}calculated by two methods are the same, hence correct.

Problem 59 Zn-Cd alloys form a regular solution. Molar enthalpy of mixing, ΔH^M at 773K at various compositions is given below.

x_{Zn}	0.0901	0.214	0.6086	0.8602	0.9286
ΔH^M, J mol^{-1}	714	1458	2030	1060	592

Calculate
(a) Activities of Zn and Cd in each alloys at 873 K.
(b) Excess molar free energy of the alloys at 773 K.

Solution

$$\Delta H^M = RT\,\alpha\, x_{Zn}.x_{Cd}$$

$$x_{Zn} = 0.0901, \quad x_{Cd} = 0.9099$$

$$\alpha = \frac{\Delta H^M = G^{XS}}{RT\, x_{A_{Zn}} x_{B_{Cd}}} = \frac{714}{8.314 \times 773 \times 0.0901 \times .9099} = 1.448876733$$

$$\alpha = \alpha_{Zn} = \frac{\ln \gamma_{Zn}}{(1 - x_{Zn})^2} = \frac{\ln \gamma_{Zn}}{x_{Cd}^2} \quad \therefore \ \ln \gamma_{Zn} = \alpha . x_{Cd}^2 = 1.19955 \ \ (\text{at } 773K)$$

$$\gamma_{Zn} = 3.3186$$

$$\frac{\alpha(T_2)}{\alpha(T_1)} = \frac{T_1}{T_2}; \qquad T_1 = 773K, \qquad T_2 = 873K$$

$$\therefore \frac{\frac{\ln \gamma_{Zn}^{873}}{x_{Cd}^2}}{\frac{\ln \gamma_{Zn}^{773}}{x_{Cd}^2}} = \frac{773}{873}$$

$$\text{or} \quad \ln \gamma_{Zn}^{873} = \frac{773}{873} . \ln \gamma_{Zn}^{773} = \frac{773}{873} \alpha x_{Cd}^2$$

$$\ln \gamma_{Zn} = \frac{773}{873} \times 1.19955 = 0.99344$$

$$\gamma_{Zn}(873) = 2.7005$$

$$a_{Zn} = \gamma_{Zn}.x_{Zn} = 2.7005 \times .0901 = 0.2433$$

$$\alpha_{Cd} = \alpha = \frac{\ln \gamma_{Cd}}{x_{Zn}^2}$$

$$\therefore \ \ln \gamma_{Cd} = \alpha x_{Zn}^2 = 1.448876733 \times .0901^2 = 0.0117619$$

$$\gamma_{Cd} = 1.01183 \ \ (\text{at } 773K)$$

$$\frac{\alpha_2(T_2)}{\alpha_1(T_1)} = \frac{T_1}{T_2}$$

$$\therefore \frac{\frac{\ln \gamma_{Cd}^{873}}{x_{Zn}^2}}{\frac{\ln \gamma_{Cd}^{773}}{x_{Zn}^2}} = \frac{773}{873}$$

$$\text{or} \quad \ln \gamma_{Cd}^{873} = \frac{773}{873} . \ln \gamma_{Cd}^{773} = \frac{773}{873} \alpha x_{Zn}^2$$

$$\therefore \ \ln \gamma_{Cd}^{873} = \frac{773}{873} \times 0.0117619 = 0.00974$$

$$\gamma_{Cd}^{873} = 1.0098$$

$$a_{Cd} = 1.0098 \times 0.9099 = 0.9188$$

$$x_{Zn} = 0.6086 \quad x_{Cd} = 0.3914$$

$$\Delta H^M = G^{XS} = RT\alpha x_{Zn} x_{Cd}$$

since it is regular solution, $\alpha_{Zn} = \alpha_{Cd} = \alpha$

$$\therefore \alpha_{773} = \frac{\Delta H^M}{RTx_{Zn}x_{Cd}} = \frac{2030}{8.314 \times 773 \times 0.6086 \times 0.3914} = 1.4177$$

$$\alpha_{Zn} = \frac{\ln \gamma_{Zn}}{x_{Cd}^2}; \quad \ln \gamma_{Zn}(773) = \alpha x_{Cd}^2$$

For regular solution:

$$\alpha_2 T_2 = \alpha_1 T_1; \quad \frac{\alpha_2(T_2)}{\alpha_1(T_1)} = \frac{T_1}{T_2}$$

$$\therefore \frac{\ln \gamma_{Zn}^{873}}{\ln \gamma_{Zn}^{773}} = \frac{773}{873}$$

$$\therefore \ln \gamma_{Zn}^{873} = \frac{773}{873} . \ln \gamma_{Zn}^{773} = \frac{773}{873} \alpha x_{Cd}^2$$

$$= \frac{773}{873} .1.4177 \times (0.3914)^2 = 0.17987$$

$$\gamma_{Zn}^{873} = 1.1971$$

$$a_{Zn} = 1.1971 \times 0.6086 = 0.7285 \qquad \text{Ans.}$$

$$\alpha_{Cd} = \alpha_{Zn} = \alpha = \frac{\ln \gamma}{x_{Cd}^2} \therefore \ln \lambda_{Cd} = \alpha x_{Zn}^2$$

$$\frac{\ln \gamma_{Cd}^{873}}{\ln \gamma_{Cd}^{773}} = \frac{773}{873}$$

$$\therefore \ln \gamma_{Cd}^{873} = \frac{773}{873} \ln \gamma_{Cd}^{773}$$

$$= \frac{773}{873} . \alpha x_{Zn}^2 = \frac{773}{873} .1.4177\,(.6086)^2$$

$$= 0.434882$$

$$\ln \gamma_{Cd}^{873} = 1.5448$$

$$a_{Cd} = 1.5448 \times 0.3914 = 0.6046$$

Similarly activities of Zn and Cd for each alloy were calculated and the results are presented in the table:

x_{Zn}	0.0901	0.214	0.6086	0.8602	0.9286
x_{Cd}	0.9099	0.786	0.3914	0.1348	0.0714
$\Delta H^M = G^{XS}$ (J mol^{-1})	714	1458	2030	1060	592
α	1.44888	1.4420	1.4177	1.4664	1.4854
$\gamma_{Zn}(873)$	2.7005	2.0913	1.1971	1.0240	1.0063
a_{Zn} (873)	0.2433	0.4475	0.7285	0.8808	0.9344
$\gamma_{Cd}(873)$	1.0098	1.0568	1.5448	2.5462	1.8886
a_{Cd} (873)	0.9188	0.8302	0.6046	0.3434	0.2062

(b) for regular solutions we have $G^{xs} = \Delta H^M = RT\alpha\, x_A x_B$

$$\text{and} \quad \alpha_1 T_1 = \alpha_2 T_2$$

$$\therefore\ G^{xs}\,(\text{at } T_1) = G^{xs}\,(\text{at } T_2) = \Delta H^M$$

The excess molar free energy of the alloy at 773 and 873 will be the same and that is equal to ΔH^M at 773 K as given in the problem.

Problem 60 If the activity coefficients of a binary solution (1–2) can be expressed as:

$$\ln\gamma_1 = \alpha_1 x_2 + \frac{1}{2}\alpha_2 x_2^2 + \frac{1}{3}\alpha_3 x_2^3 + - - - -$$

$$\ln\gamma_2 = \beta_1 x_2 + \frac{1}{2}\beta_2 x_2^2 + \frac{1}{3}\beta_3 x_2^3 + - - - -$$

Show that for the above expressions to be valid for the entire composition range, $\alpha_1 = \beta_1 = 0$. Also prove that if the variation of the activity coefficients with composition is represented by the quadratic terms alone, $\alpha_2 = \beta_2$.

Solution

$$\ln\gamma_1 = \alpha_1 x_2 + \frac{1}{2}\alpha_2 x_2^2 + \frac{1}{3}\alpha_3 x_2^3 + - - - - \tag{1}$$

$$\ln\gamma_2 = \beta_1 x_1 + \frac{1}{2}\beta_2 x_1^2 + \frac{1}{3}\beta_3 x_1^3 + - - - - \tag{2}$$

We make use of the Gibbs–Duhem equation to show, $\alpha_1 = \beta_1 = 0$, if these equations hold over the entire composition range. A suitable Gibbs–Duhem equation for a binary system (1–2) is expressed as:

$$x_1 dln\gamma_1 + x_2 dln\gamma_2 = 0$$

or

$$x_1 dln\gamma_1 = -x_2 dln\gamma_2 \tag{3}$$

On differentiation of the given Eqs. (1) and (2) we get:

$$dln\gamma_1 = \alpha_1 dx_2 + \alpha_2 x_2 dx_2 + \alpha_3 x_2^2 dx_2 + ----\quad (4)$$

$$dln\gamma_2 = \beta_1 dx_1 + \beta_2 x_1 dx_1 + \beta_3 x_1^2 dx_1 + ----\quad (5)$$

For a binary solution, $x_2 = 1 - x_1 \quad or\ dx_2 = -dx_1$, on substituting these in Eq. (4), we get:

$$\begin{aligned} dln\gamma_1 &= -\alpha_1 dx_1 - \alpha_2(1 - x_1)dx_1 - \alpha_3(1 - x_1)^2 dx_1 + ---- \\ &= -\alpha_1 dx_1 - \alpha_2 dx_1 + \alpha_2 x_1 dx_1 - \alpha_3 dx_1 + 2\ \alpha_3 x_1 dx_1 - \alpha_3 x_1^2 dx_1 + ---- \end{aligned}\quad (6)$$

Substituting Eqs. (6) and (5) in Eq. (3), we get:

$$\begin{aligned} &-\ \alpha_1 x_1 dx_1 - \alpha_2 x_1 dx_1 + \alpha_2 x_1^2 dx_1 - \alpha_3 x_1 dx_1 + 2\alpha_3 x_1^2 dx_1 - \alpha_3 x_1^3 dx_1 + ---- \\ &= -(1 - x_1)\beta_1 dx_1 - \beta_2(1 - x_1)x_1 dx_1 - \beta_3(1 - x_1)x_1^2 dx_1 + ---- \end{aligned}$$

or $\alpha_1 x_1 + \alpha_2 x_1 - \alpha_2 x_1^2 + \alpha_3 x_1 - 2\alpha_3 x_1^2 + \alpha_3 x_1^3 + --- = \beta_1 - \beta_1 x_1 + \beta_2 x_1 - \beta_2 x_1^2 + \beta_3 x_1^2 - \beta_3 x_1^3 + ---$

or $(\alpha_1 + \alpha_2 + \alpha_3)\ x_1 + (-\alpha_2 - 2\alpha_3)x_1^2 + \alpha_3 x_1^3 + --- = \beta_1 + (\beta_2 - \beta_1)x_1 + (\beta_3 - \beta_2)x_1^2 - \beta_3 x_1^3 + ---$

On equating the coefficients of both sides of the equation, we get:

$$\beta_1 = 0$$

$$\alpha_1 + \alpha_2 + \alpha_3 = \beta_2 - \beta_1 \qquad (\text{as}\ \beta_1 = 0)$$

$$\beta_2 = \alpha_1 + \alpha_2 + \alpha_3 \quad (7)$$

$$\text{and} \quad -\alpha_2 - 2\alpha_3 = \beta_3 - \beta_2$$

$$\therefore \beta_2 = \alpha_2 + 2\alpha_3 + \beta_3 \quad (8)$$

$$\text{and} \quad \alpha_3 = -\beta_3 \quad (9)$$

From Eqs. (7) and (8), we get: $\alpha_1 + \alpha_2 + \alpha_3 = \alpha_2 + 2\alpha_3 + \beta_3$. On substituting Eq. (9) into this we get: $\alpha_1 = 0$.

Thus for the given equations to be valid for the entire composition range: $\alpha_1 = \beta_1 = 0$.

Similarly if the variation of the activity coefficients with composition has to be represented by the quadratic terms according to the equations:

$$\ln\gamma_1 = \frac{1}{2}\alpha_2 x_2^2, \text{ on differentiation: } dln\gamma_1 = \alpha_2 x_2 dx_2 \quad (10)$$

$$\ln \gamma_2 = \frac{1}{2}\beta_2 x_1^2, \quad \text{on differentiation: } dln\gamma_2 = \ \beta_2 x_1 dx_1 \tag{11}$$

Substituting Eqs. (10) and (11) in the Gibbs–Duhem equation of the form:

$$x_1\, dln\gamma_1 + x_2 dln\gamma_2 = 0$$

$$x_1 \ \ \alpha_2 x_2 dx_2 + x_2 \ \ \beta_2 x_1 dx_1 = 0 \qquad (\text{since } dx_2 = -\ \ dx_1)$$

$$\therefore \ \alpha_2 x_1 x_2 dx_2 - \beta_2 x_1 x_2 dx_2 = 0$$

$$\text{Hence} \quad \alpha_2 = \beta_2 \qquad (\text{the required condition})$$

Problem 61 Using Quasi-Chemical model calculate the integral molar enthalpy of an iron-chromium alloy containing 25 at % Cr at 1600 °C, if activity of Cr in this alloy at 1600 °C is 0.31.

Solution

$$\gamma_{Cr} = \frac{a_{Cr}}{x_{Cr}} = \frac{0.31}{0.25} = 1.25$$

According to Q-C model $\Delta \overline{H}_{Cr}^{M} = \Omega x_{Cr}^2$ and Integral molar enthalpy $\Delta H^M = \Omega x_{Cr} x_{Fe}$

$$\alpha = \frac{\Omega}{RT}$$

$$\begin{aligned}
\Delta H^M &= \Omega x_{Fe} x_{Cr} = \Omega x_{Fe}(1 - x_{Fe}) \\
&= \Omega x_{Cr}(1 - x_{Cr}) = \Omega x_{Cr} - \Omega x_{Cr}^2 \\
\frac{\partial \Delta H^M}{\partial x_{Cr}} &= \Omega - 2\Omega x_{Cr} \\
&= \Omega(1 - 2x_{Cr}) = \Omega(x_{Fe} - x_{Cr}) \\
\Delta H^M &= G^{XS} = \Omega x_{Fe} x_{Cr} = RT\alpha x_{Fe} x_{Cr} \\
\Delta \overline{H}_{Cr}^{M} &= \Delta H^M + x_{Fe} \frac{\partial \Delta H^M}{\partial x_{Cr}} \\
&= \Omega x_{Fe} x_{Cr} + \Omega x_{Fe}(x_{Fe} - x_{Cr}) \\
\Delta \overline{H}_{Cr}^{M} &= \Omega x_{Fe}^2 \\
\because \Delta \overline{H}_{Cr}^{M} &= RT \ln \gamma_{Cr} \\
\therefore \ \Omega x_{Fe}^2 &= RT \ln \gamma_{Cr} \\
\Omega &= \frac{RT \ln \gamma_{Cr}}{x_{Fe}^2} \\
&= \frac{8.31 \times 1600 \times \ln 1.25}{0.75^2} = 5275 \\
\Delta H^M &= \Omega x_{Fe} x_{Cr} = 5275 \times 0.75 \times 0.25 \\
&= 990 \text{ J g atom}^{-1}. \qquad \text{Ans.}
\end{aligned}$$

Problem 62 The activity coefficient of zinc in liquid Cd-Zn alloys at 435 °C has been expressed as $ln\,\gamma_{Zn} = 0.87\,x_{Cd}^2 - 0.30\,x_{Cd}^3$

(a) Calculate the activity of cadmium in a 30 at % Cd at 435 °C.
(b) Develop a corresponding equation for the activity coefficient of cadmium in the alloy system at this temperature.

Solution
From Gibbs–Duhem equation we have

$$x_{\mathrm{Zn}}\mathrm{d}\ln\gamma_{\mathrm{Zn}} + x_{\mathrm{Cd}}\mathrm{d}\ln\gamma_{\mathrm{Cd}} = 0$$

on integration we get:

$$\begin{aligned}
\ln\gamma_{Cd} &= -\int_{x_{Cd}=1}^{x_{Cd}=x_{Cd}} \frac{x_{Zn}}{x_{Cd}}\, d\ln\gamma_{Zn} \\
&= -\int_{x_{Cd}=1}^{x_{Cd}=0.3} \left(1.74 - 2.64x_{Cd} + 0.90x_{Cd}^2\right)dx_{Cd} \\
&= -\left[1.74x_{Cd} - 1.32x_{Cd}^2 + 0.3x_{Cd}^3\right]_1^{0.3} \\
&= \left[-1.74x_{Cd} + 1.32x_{Cd}^2 - 0.3\,x_{Cd}^3\right]_1^{0.3} \\
&= -1.74\ (0.3-1) + 1.32\ (0.09-1) - 0.3\ (0.027-1) \\
&= 1.218 - 1.201 + 0.292 = 0.309
\end{aligned}$$

$$\therefore \gamma_{Cd} = 1.362$$

and $a_{Cd} = \gamma_{Cd} \cdot x_{Cd} = 1.362 \times 0.3 = 0.4086$ Ans.

(b) $\ln\gamma_{Cd} = -\int \frac{x_{Zn}}{x_{Cd}} d\ln\ \gamma_{Zn}$

$$\begin{aligned}
&= -\int_{x_{Cd}=1}^{x_{Cd}=x_{Cd}} \frac{1-x_{Cd}}{x_{Cd}}\left(1.74x_{Cd} - 0.9x_{Cd}^2\right)dx_{Cd} \\
&= -\int (1-x_{Cd})(1.74 - 0.9x_{Cd})dx_{Cd} \\
&= -\int \left(1.74 - 2.64x_{Cd} + 0.9x_{Cd}^2\right)dx_{Cd} \\
&= \int_{x_{Cd}=1}^{x_{Cd}=x_{Cd}} \left(-1.74 + 2.64x_{Cd} - 0.90x_{Cd}^2\right)dx_{Cd} \\
&= \left[-1.74x_{Cd} + 1.32x_{Cd}^2 - 0.3\,x_{Cd}^3\right]_{x_{Cd}=1}^{x_{Cd}=x_{Cd}} \\
&= -1.74x_{Cd} + 1.32x_{Cd}^2 - 0.3\,x_{Cd}^3 - (-1.74 + 1.32 - 0.3) \\
&= 0.72 - 1.74x_{Cd} + 1.32x_{Cd}^2 - 0.3\,x_{Cd}^3
\end{aligned}$$

In order to get the expression in the desired, substitute: $x_{Cd} = 1 - x_{Zn}$

$$=0.72-1.74(1-x_{Zn})+1.32(1-x_{Zn})^2-0.3(1-x_{Zn})^3$$
$$=0.72-1.74+1.74x_{Zn}+1.32-2.64x_{Zn}+1.32x_{Zn}^2-0.3(1-3x_{Zn}+3x_{Zn}^2-x_{Zn}^3)$$
$$=0.3x_{Zn}^3+x_{Zn}^2(-0.9+1.32)+x_{Zn}(0.9-2.64+1.74)-0.3+1.32-1.74+0.72$$
$$\therefore\ ln\ \gamma_{Cd}=0.3x_{Zn}^3+0.42x_{Zn}^2 \qquad \text{Ans.}$$

Alternatively, we can also solve as

$$\ln\gamma_{Cd} = \int -\frac{x_{Zn}}{x_{Cd}}(1.74x_{Cd}-0.9x_{Cd}^2)dx_{Cd} = +\int x_{Zn}(1.74-0.9x_{Cd})(-dx_{Cd})$$
$$= \int_{x_{Cd}=1}^{x_{Cd}=x_{Cd}} x_{Zn}(1.74-0.9x_{Cd}).dx_{Zn} = \int_{x_{Zn}=0}^{x_{Zn}=x_{Zn}} [1.74x_{Zn}dx_{Zn}-0.9x_{Zn}(1-x_{Zn})]dx_{Zn}$$
$$= \int 1.74x_{Zn}dx_{Zn} - \int 0.9x_{Zn}dx_{Zn} + \int 0.9x_{Zn}^2dx_{Zn}$$
$$= \int_0^{x_{Zn}} 0.84\ x_{Zn}dx_{Zn} + \int_0^{x_{Zn}} 0.90\ x_{Zn}^2dx_{Zn}$$
$$ln\gamma_{Cd} = 0.42\ x_{Zn}^2 + 0.30\ x_{Zn}^3$$

Problem 63 The activity coefficient of copper in Cu-Fe alloys has been expressed as

$$\log\gamma_{Cu} = 1.45x_{Fe}^2 - 1.86x_{Fe}^3 + 1.41x_{Fe}^4$$

(a) Using Gibbs–Duhem equation show that activity coefficient of iron can be expressed as

$$\log\gamma_{Fe} = 1 - 2.90x_{Fe} + 4.24x_{Fe}^2 - 3.74x_{Fe}^3 + 1.41x_{Fe}^4.$$

(b) Calculate the activity of iron in a 30 at % Fe alloy.

Solution

$$\log\gamma_{Cu} = 1.45x_{Fe}^2 - 1.86x_{Fe}^3 + 1.41x_{Fe}^4$$
$$\therefore\ dlog\gamma_{Cu} = 2.90x_{Fe} - 5.58x_{Fe}^2 + 5.64x_{Fe}^3$$

(a) Appropriate form of Gibbs–Duhem equation: $x_{Fe}d\ \log\ \gamma_{Fe} + x_{Cu}d\ \log\ \gamma_{Cu} = 0$

$$\text{or}\quad d\ \log\gamma_{Fe} = -\frac{x_{Cu}}{x_{Fe}}.d\ \log\gamma_{Cu}$$

$$\log\gamma_{Fe} = -\int_{x_{Fe}=1}^{x_{Fe}=x_{Fe}} \frac{x_{Cu}}{x_{Fe}} d\log\gamma_{Cu} = -\int_{x_{Fe}=1}^{x_{Fe}=x_{Fe}} \frac{1-x_{Fe}}{x_{Fe}}.(2.90x_{Fe} - 5.58x_{Fe}^2 + 5.64x_{Fe}^3)dx_{Fe}$$

$$= -\int_{x_{Fe}=1}^{x_{Fe}=x_{Fe}} (1-x_{Fe})(2.90 - 5.58x_{Fe} + 5.64x_{Fe}^2)dx_{Fe}$$

$$= -\int (2.90 - 8.48x_{Fe} + 11.22x_{Fe}^2 - 5.64x_{Fe}^3)dx_{Fe}$$

$$= \left[-2.90x_{Fe} + 4.24x_{Fe}^2 - 3.74x_{Fe}^3 + 1.41x_{Fe}^4\right]_{x_{Fe}=1}^{x_{Fe}=x_{Fe}}$$

$$= -2.90x_{Fe} + 4.24x_{Fe}^2 - 3.74x_{Fe}^3 + 1.41x_{Fe}^4 - (-2.90 + 4.24 - 3.74 + 1.41)$$

$$= -2.90x_{Fe} + 4.24x_{Fe}^2 - 3.74x_{Fe}^3 + 1.41x_{Fe}^4 - (-1)$$

$$\therefore\ log\gamma_{Fe} = 1 - 2.90\ x_{Fe} + 4.24x_{Fe}^2 - 3.74x_{Fe}^3 + 1.41x_{Fe}^4$$

(b) $\therefore\ \log\gamma_{Fe} = 1 - 2.90\times0.3 + 4.24\times0.3^2 - 3.74\times0.3^3 + 1.41\times0.3^4\ (x_{Fe} = 0.3)$

$= 1 - 0.87 + 0.3816 - 0.10098 + 0.01142$

$= 0.422$

$\gamma_{Fe} = 2.642$

$a_{Fe} = \gamma_{Fe}.x_{Fe} = 2.64\times.3 = 0.7926$ Ans.

Problem 64 At 746 K the activity coefficient of lead in liquid Pb-Bi alloy is expressed as:

$$ln\,\gamma_{Pb} = -0.74(1 - x_{Pb})^2$$

Making use of the Gibbs–Duhem equation develop the corresponding equation for the activity coefficient of bismuth in the alloy at 746 K.

(a) Calculate the activity of lead at 746 K and 100 K in the Pb-Bi alloy containing 50 at % lead
(b) Calculate the integral molar heat of mixing/excess free energy of the alloy containing 40 at % Pb at 746 K
(c) What is the integral molar free energy of mixing of the above alloy in (a) at 1000 K
(d) Calculate the difference in change in free energy when 1 g atom of lead dissolves in a very large amount of the above alloy at 746 and 1000 K

Solution

(a) For Pb-Bi system the Gibb–Duhem equation may be written as:

$$x_{Bi}\ln\gamma_{Bi} + x_{Pb}ln\gamma_{Pb} = 0$$

on integration, we get:

$$\ln \gamma_{Bi} = - \int_{ln\gamma_{Pb}\ at x_{Bi}=1}^{ln\gamma_{Pb}\ at x_{Bi}=x_{Bi}} \frac{x_{Pb}}{x_{Bi}} dln\gamma_{Pb} \tag{1}$$

$$\text{Given :}\quad ln\,\gamma_{Pb} = -0.74(1 - x_{Pb})^2$$

$$dln\gamma_{Pb} = 1.48\ (1 - x_{Pb})dx_{Pb} \tag{2}$$

From Eqs. 1 & 2:

$$\begin{aligned} \ln \gamma_{Bi} &= - \int \frac{x_{Pb}}{(1 - x_{Pb})} 1.48\ (1 - x_{Pb})\, dx_{Pb} \\ &= -1.48 \int_{x_{Bi}=1}^{x_{Bi}} x_{Pb} dx_{Pb} \\ &= -1.48 \left[\frac{x_{Pb}^2}{2} \right]_{x_{Pb}=0}^{x_{Pb}=(1-x_{Bi})} \\ &= -0.74(1 - x_{Bi})^2 \qquad \text{Ans.} \end{aligned}$$

$$\alpha_{Pb} = \frac{ln\,\gamma_{Pb}}{(1 - x_{Pb})^2} = -0.74$$

Alternatively, we may use the α-function:

$$\ln \gamma_A \Big|_{x_A = x_A} = -\alpha_B x_A x_B - \int_{\alpha_B\ at\ x_A=1}^{\alpha_B\ at\ x_A=x_A} \alpha_B dx_A$$

$$\begin{aligned} \ln \gamma_{Bi} &= -\alpha_{Pb} x_{Bi} x_{Pb} - \int_{x_{Bi}=1}^{x_B=x_{Bi}} \alpha_{Pb} dx_{Bi} \\ &= -[-0.74 x_{Bi}(1 - x_{Bi})] - \int_{x_{Bi}=1}^{x_{Bi}=x_{Bi}} -0.74 dx_{Bi} \\ &= 0.74\ x_{Bi}(1 - x_{Bi}) + 0.74(x_{Bi} - 1) \\ &= 0.74(1 - x_{Bi})(x_{Bi} - 1) \\ ln\,\gamma_{Bi} &= -0.74(1 - x_{Bi})^2 \qquad \text{Ans.} \end{aligned}$$

(a) Thus from the above calculations, we find:

$$\alpha_{Pb} = \frac{ln\,\gamma_{Pb}}{(1 - x_{Pb})^2} = -0.74 = \alpha_{Bi} = \frac{ln\,\gamma_{Bi}}{(1 - x_{Bi})^2}$$

Hence, (Pb-Bi) alloy system obeys regular solution model.

$$ln\,\gamma_{Pb} = -0.74(1 - x_{Pb})^2, \quad \gamma_{Pb} = 0.83$$

$$\gamma_{Pb} = 0.5 \times 0.83 = 0.415 \ \ (\text{at } 746\ K) \qquad \text{Ans.}$$

For regular solution: $\alpha_1 T_1 = \alpha_2 T_2$

$$\alpha_1 = -0.74\ (746 = T_1)$$

$$\alpha_2 = (1000\ \text{K}) = \frac{\alpha_1 T_1}{T_2} = -\frac{0.74 \times 746}{1000} = -0.55$$

From the equation: $ln\gamma_{Pb} = \alpha_2(1 - x_{Pb})^2$ at 1000 K

$$= -0.55\ (1 - 0.5)^2 = -0.138$$

$$\therefore \gamma_{Pb} = 0.87$$

$$a_{Pb} = 0.5 \times 0.87 = 0.435\ \ (\text{at } 1000\ \text{K}) \qquad \text{Ans.}$$

(b) $\alpha_{Pb} = \alpha_{Bi} = \frac{ln\,\gamma_{Pb}}{x_{Bi}^2} = \frac{ln\,\gamma_{Bi}}{x_{Pb}^2} = \alpha$

$$\text{at } 746\ \text{K}, \quad ln\,\gamma_{Pb} = -\alpha x_{Bi}^2 = -074 x_{Bi}^2$$

$$ln\,\gamma_{Bi} = -\alpha x_{Pb}^2 = -074 x_{Pb}^2$$

$$\begin{aligned} G^{xs} &= RT(x_{Pb} ln x_{Pb} + x_{Bi} ln x_{Bi}) = \Delta H^M \\ &= RT\alpha x_{Pb} x_{Bi} \qquad (\text{at } 746\ \text{K}, \ \alpha = -0.74) \\ &= -1.987 \times 746 \times 0.74 \times 0.4 \times .6 \\ &= -263.3\ \text{cal mol}^{-1} \qquad \text{Ans.} \end{aligned}$$

(c) From (a), $a_{Pb} = 0.435$ (at 1000 K,)

$$\text{At } 1000\ \text{K}, \ a_{Pb} = 0.435 = a_{Bi}\ \ (\text{when } x_{Pb} = 0.5 = x_{Bi})$$

$$\begin{aligned} G^M &= RT(x_{Pb} ln a_{Pb} + x_{Bi} ln a_{Bi}) \\ &= 1.987 \times 1000\ \ (0.5\ \ln\ 0.435 + 0.5\ \ln\ 0.435) \\ &= 1.987 \times 1000 \times \ln\ 0.435 \quad = -1654\ \text{cal mol}^{-1} \qquad \text{Ans.} \end{aligned}$$

(d) difference in free energy = change in partial molar free energy of lead

$$= \Delta\overline{G}_{Pb}(1000\ K) - \Delta\overline{G}_{Pb}(746)$$
$$= RT\ \ ln a_{Pb}(1000\ K) - RT\ \ ln a_{Pb}(746\ K)$$
$$= 1.987\ \ (1000 \times \ln\ \ 0.435\text{–}746\ \ \ln\ \ 0.415)$$
$$= -\ 350.5\ \text{cal mol}^{-1}$$

Ans.

Problem 65 Al-Zn alloys exhibit the following relationship at 477 °C:

RT ln γ_{Zn}= 1750 $(1-x_{Zn})^2$ where R and T in expressed in cal deg^{-1} mol^{-1} and K, respectively.

(i) Develop the corresponding expression ln γ_{Al}.
(ii) Calculate the heat of mixing of the alloy containing 40 at % Zn at 477 °C. What would be excess molar free energy of the alloy at this temperature?
(iii) Calculate the integral molar free energy of the above alloy at 507 °C.

Solution

(i) For Al-Zn system the Gibbs–Duhem equation may be written as:

$$x_{Al} d\ ln\gamma_{Al} + x_{Zn} d\ ln\gamma_{Zn} = 0$$

on integration, we get:

$$\ln\gamma_{Al} = -\int_{ln\gamma_{Zn}\ at x_{Al}=1}^{ln\gamma_{Zn}\ at x_{Al}=x_{Al}} \frac{x_{Zn}}{x_{Al}} dln\gamma_{Zn} \quad (1)$$

$$\text{Given}: \ ln\,\gamma_{Zn} = \frac{1750}{RT}(1 - x_{Zn})^2$$

$$dln\gamma_{Zn} = -\frac{3500}{RT}(1 - x_{Zn})dx_{Zn} \quad (2)$$

From Eqs. 1 & 2:

$$\ln\gamma_{Al} = -\int \frac{x_{Zn}}{(1-x_{Zn})}\left[-\frac{3500}{RT}\right](1-x_{Zn})\,dx_{Zn}$$
$$= \frac{3500}{RT}\int_{x_{Al}=1}^{x_{Al}} x_{Zn}dx_{Zn}$$
$$= \frac{3500}{RT}\left[\frac{x_{Zn}^2}{2}\right]_{x_{Zn}=0}^{x_{Zn}=(1-x_{Al})}$$
$$\therefore\ \ln\gamma_{Al} = \frac{1750}{RT}(1 - x_{Al})^2$$

Ans.

(ii) $\alpha_{Zn} = \frac{ln\,\gamma_{Zn}}{(1-x_{Zn})^2} = \frac{1750}{RT} = \alpha_{Al} = \frac{ln\,\gamma_{Al}}{(1-x_{Al})^2}$

The above equations show that α is independent of composition, and hence Al-Zn system follows regular solution model. For regular: $\Delta H^M = G^{xs}$

$$\therefore \Delta H^M = G^{xs} = RT\alpha x_{Zn}x_{Al} \qquad (x_{Zn} = 0.4,\ x_{Al} = 0.6)$$
$$= RT\frac{1750}{RT}0.4 \times 0.6$$
$$= 1750 \times 0.24$$
$$= 420 \text{ cal mol}^{-1} \qquad \text{Ans}$$

(iii) $\Delta G^M = RT(x_{Al}lna_{Al} + x_{Zn}lna_{Zn})$

for regular solutions, $\alpha_1 T_1 = \alpha_2 T_2$

$$T_1 = 477 + 273 = 750K,\ T_2 = 507 + 273 = 780K.$$

$$\alpha_{750} = \frac{\ln\gamma_{Zn}}{(1-x_{Zn})^2} = \frac{1750}{RT} = \frac{1750}{R750} = \frac{7}{3\times 1.987} = 1.174$$

$$\alpha_{780} = \frac{\alpha_{750}\times 750}{780} = \frac{1750}{R750}x\frac{750}{780} = \frac{175}{78x1.987} = 1.129$$

$$\alpha_{780} = \frac{\ln\gamma_{Zn}}{(1-x_{Zn})^2}$$

$$\therefore \ln\gamma_{Zn(780)} = \alpha_{780}\times(1-x_{Zn})^2 = 1.129\times 0.36 = 0.40644$$

$$\therefore \gamma_{Zn(780)} = 1.50146$$

$$\therefore a_{Zn} = \gamma_{Zn}.x_{Zn} = 1.50146\times 0.4 = 0.6005$$

$$\therefore \alpha_{780} = \frac{\ln\gamma_{Al}}{(1-x_{Al})^2}$$

$$\ln\gamma_{Al} = \alpha_{780}(1-0.6)^2 = 1.129\times 0.16 = 0.18664$$

$$\gamma_{Al} = 1.19798$$

$$a_{Al} = 1.19798\times 0.6 = 0.71879$$

$$\therefore \Delta G^M = RT(0.6\ln 0.718979 + 0.4\ln 0.6005)$$
$$= 1.987\times 780\ \ (0.6\times(-0.3301856) + 0.4\ \ (-0.5075)$$
$$= -622.8 \text{ cal mol}^{-1}. \qquad \text{Ans.}$$

Problem 66 At 1200 K the activity coefficient of zinc in liquid brass is expressed as

$$\ln\gamma_{Zn} = -1.929\,(1 - x_{Zn})^2$$

(a) Calculate the integral molar heat of mixing and the excess free energy of brass containing 40 at % copper at 1200 K.
(b) What is the integral molar free energy of mixing of the above alloy at 1300 K? Check your answer.
(c) Calculate the difference in change of free energy when one g atom of liquid zinc dissolves in a large amount of liquid brass at 1200 and 1300 K.

Solution

(a) The variation of activity coefficient with composition shows that $\alpha_{Zn} = -1.929$, is independent of composition. This indicates that brass follows regular solution model at 1200 K.

$$\therefore\ \alpha_{Zn} = \alpha_{Cu}$$

$$\alpha_{Zn} = \frac{\ln\gamma_{Zn}}{(1 - x_{Zn})^2} = \frac{\ln\gamma_{Cu}}{(1 - x_{Cu})^2} = \alpha_{Cu} = -1.929$$

For regular solution we have

$$\Delta H^M = G^{XS} = RT(x_A \ln\gamma_A + x_B \ln\gamma_B)$$

∴ Integral molar heat of mixing (or excess molar free energy) of brass,

$$\Delta H^M = G^{xs} = RT\alpha x_{Zn} x_{Cu}$$

$$\mathrm{T} = 1200\ \mathrm{K},\ x_{Cu} = 0.4,\ x_{Zn} = 0.6$$

$$\begin{aligned}\ln\gamma_{Zn} &= -1.929(1 - x_{Zn})^2 = -1.929x_{Cu}^2\\ &= -1.929(0.4)^2 = -0.30864\end{aligned}$$

$$\therefore\ \gamma_{Zn} = 0.7344 \quad \text{and} \quad a_{Zn} = 0.7344 \times 0.6 = 0.4406$$

$$\begin{aligned}\ln\gamma_{Cu} &= -1.929(1 - x_{Cu})^2 = -1.929x_{Zn}^2\\ &= -1.929(0.6)^2 = -0.6944\end{aligned}$$

$$\therefore\ \gamma_{Cu} 0.4993 \quad \text{and} \quad a_{Cu} = 0.4993 \times 0.4 = 0.1997$$

$$\begin{aligned}\therefore\ \Delta H^M &= G^{XS} = 1.987 \times 1200(0.6\ln\gamma_{Zn} + 0.4\ln\gamma_{Cu})\\ &= -2384.4(0.6 \times 0.30864 + 0.4 \times 0.6944)\\ &= -2384.4 \times 0.4629 = -1103.8\ \text{cal mol}^{-1}\end{aligned}$$

or $G^{xs} = RT\alpha x_{Zn} x_{Cu}$

$$= 1.987 \times 1200 \times (-1.929) 0.4 \times 0.6 = -1104 \text{ cal mol}^{-1} \qquad \text{Ans.}$$

(b) $\Delta G^M = RT(x_{Zn} ln a_{Zn} + x_{Cu} ln a_{Cu})$

Since ΔG^M for brass is needed at 1300 K, hence x_{Zn} *and* x_{Cu} should be first calculated at 1300 K by making use of the properties of regular solution.

$$\alpha_{Zn} = \alpha_{Cu} = -1.929, \text{ at } 1200 \text{ K}$$

$$\alpha_1 T_1 = \alpha_2 T_2 \qquad (T_1 = 1200, T_2 = 1300)$$

$$\therefore \alpha_{1200} \times 1200 = \alpha_{1300} \times 1300$$

$$\alpha_{1300} = \alpha_{1200} \times \frac{12}{13} = -1.929 \times \frac{12}{13} = -1.781$$

$$\ln \gamma_{Zn} = \alpha_{1300} . x_{Cu}^2 = -1.781 (0.4)^2 = -0.2849$$

$$\gamma_{Zn} = 0.7520$$

$$a_{Zn} = 0.7520 \times 0.6 = 0.4513$$

$$\ln \gamma_{Cu} = \alpha_{1300} . x_{Zn}^2 = -1.781 (0.6)^2 = -0.6412$$

$$\therefore \gamma_{Cu} = 0.5267$$

$$a_{Cu} = 0.5267 \times 0.4 = 0.2107$$

$$\therefore \Delta G^M = 1.987 \times 1300(0.6 \, ln \, 0.4513 + 0.4 \ln 0.2107)$$

$$= -2842.3 \text{ cal mol}^{-1}$$

Answer Check

$$G^{xs}(1300 \text{ } K) = RT\alpha x_{Zn} x_{Cu} = G^{xs}(1200 \text{ } K)$$
$$= 1.987 \times 1300 \times (-1.781) \, 0.6 \times 0.4 = -1104.1 \text{ cal mol}^{-1}$$
$$= 1.987 \times 1200 \times (-1.929) \, 0.6 \times 0.4 = -1103.9 \text{ cal mol}$$

$$G^{xs}(1300 \, K) = G^{xs}(1200 \, K) = -1104 \text{ cal mol}^{-1}$$

$$\Delta G_{id}^M = RT(x_{Zn} \, ln x_{Zn} + x_{Cu} \, ln x_{Cu})$$
$$= 1.987 \times 1300(0.6 \ln 0.6 + 0.4 \ln 0.4) = -1738.5 \text{ cal mol}^{-1}$$

$$G^{xs} = G^M - \Delta G_{id}^M = -2842.3 - (-1738.5) = -1103.8 \text{ cal mol}^{-1}$$

(c) Difference in change of free energy on dissolution of 1 g atom of Zn at 1200 and 1300 K

$$= \bar{G}_{Zn}(at \text{ } 1200 \text{ } K) - \bar{G}_{Zn}(at \text{ } 1300 \text{ } K)$$

$$= RT \ \ln a_{Zn}(1200) - RT \ln a_{Zn}(1300)$$
$$= 1.987(1200 \ \ln 0.4406 - 1300 \ \ln 0.4513)$$
$$= 1.987[-983.5 - (-1034.3)]$$
$$= 98.9 \text{ cal mol}^{-1} \qquad \text{Ans.}$$

Problem 67 (a) Derive expression for Darken's quadratic formalism.
(b) Classify the following systems in regular and non-regular solutions category.

1. Fe-Mn system at 1600 °C, $\log\gamma^o_{Mn} = 0.1, \alpha_{Fe\text{-}Mn} = 0.1, \ \log\gamma^o_{Fe} = 0.1, \ \alpha_{Mn\text{-}Fe} = 0.1,$
2. Fe-Si system at 1600 °C, $\log\gamma^o_{Si} = -2.75, \alpha_{Fe\text{-}Si} = -3.10, \ \log\gamma^o_{Fe} = -1.64, \alpha_{Si\text{-}Fe} = -0.78.$
3. Fe-Ni system at 1600 °C, $\log\gamma^o_{Ni} = -0.18, \alpha_{Fe\text{-}Ni} = -0.04, \ \log\gamma^o_{Fe} = -0.40, \alpha_{Ni\text{-}Fe} = -0.54.$
4. Fe-Al system at 1600 °C, $\log\gamma^o_{Al} = -1.21, \alpha_{Fe\text{-}al} = -1.21, \ \log\gamma^o_{Fe} = -1.21, \alpha_{Al\text{-}Fe} = -1.21.$
5. Fe-Cu system at 1600 °C, $\log\gamma^o_{CU} = 0.93, \alpha_{Fe\text{-}Cu} = 1.13, \ \log\gamma^o_{Fe} = 1.00, \alpha_{Cu\text{-}Fe} = 1.20.$

Solution

(a) In metallic solutions Raoult's law is strictly obeyed by the solvent designated by 1 in 1–2 system (i.e., when $x_1 \to 1$). In re-examining the available data at finite concentration, Darken has pointed out that an expression of the type $ln\gamma_i = \alpha \ (1 - x_i)^2$ represents departure from Raoult's law at high solute concentration in many metallic systems, hence it may not be applicable to the entire binary system but may be applicable in certain parts of it. With the aid of the Gibbs–Duhem equation: $d \ln \gamma_2 = -\frac{x_1}{x_2} d \ ln \gamma_1$ condition for a strictly regular solution can be derived as:

$$ln\gamma^o_1 = ln\gamma^o_2$$

If the solvent is designated by 1 and the solute by numeral 2, Darken's quadratic formalism is expressed as:

$$ln(\gamma_2/\gamma^o_2) = \alpha_{12}(-2x_2 + x^2_2)$$

For derivation of the expression refer, Sect. 3.12.

(b) The system 1–2, fulfilling the condition: $ln\gamma^o_1 = ln\gamma^o_2$, where the constituents 1 and 2 are the solvent and solute respectively, is categorized as the strictly regular solution.

Since in the system, Fe-Mn: $\log\gamma^o_{Mn} = 0.1 \ = \log\gamma^o_{Fe}$; and in Fe-Al: $\log\gamma^o_{Al} = -1.21 = log\,\gamma^o_{Fe}$, Fe-Mn and Fe-Al belong to the class of strictly regular solution. On this basis Fe-Si, Fe-Ni, and Fe-Cu systems are non-regular in their thermodynamic behavior.

Problem 68 What do you understand by stability and excess stability? Derive expressions for stability, ideal stability and excess stability. Discuss their significance.

Solution

The second derivative of the molar free energy with mole fraction is called stability.

The stability of an ideal solution = RT/x_1x_2.

Excess stability = Stability – Ideal stability

If the stability is anywhere negative, the solution is metastable and a miscibility gap occurs.

Ideal Stability

$$\begin{aligned} G_{id}^M &= RT(x_1 \ln x_1 + x_2 \ln x_2) \\ \frac{\partial G_{id}^M}{\partial x_2} &= RT\left(\frac{\partial x_1}{\partial x_2}\ln x_1 + x_1\frac{\partial \ln x_1}{\partial x_2} + \ln x_2 + x_2\frac{\partial \ln x_2}{\partial x_2}\right) \\ &= RT\left(-\ln x_1 + \frac{x_1}{x_1}\frac{\partial x_1}{\partial x_2} + \ln x_2 + \frac{x_2}{x_2}\frac{\partial x_2}{\partial x_2}\right) \\ &= RT(-\ln x_1 - 1 + \ln x_2 + 1) \\ &= RT(-\ln x_1 + \ln x_2) \end{aligned}$$

$$\begin{aligned} \frac{\partial^2 G_{id}^M}{\partial x_2^2} &= RT\left(-\frac{\partial \ln x_1}{\partial x_2} + \frac{1}{x_2}\frac{\partial x_2}{\partial x_2}\right) \\ &= RT\left[-\frac{1}{x_1}\left(\frac{\partial x_1}{\partial x_2}\right) + \frac{1}{x_2}\right] \\ &= RT\left[\frac{1}{x_1} + \frac{1}{x_2}\right] = \frac{RT}{x_1x_2} \end{aligned} \tag{1}$$

$$G^M = G - (x_1G_1^0 + x_2G_2^0)$$

$$\frac{\partial G^M}{\partial x_2} = \frac{\partial G}{\partial x_2} - \left(\frac{\partial x_1}{\partial x_2}G_1^0 + x_2\frac{\partial G_1^0}{\partial x_2} + G_2^0 + x_2\frac{\partial G_2^0}{\partial x_2}\right) = \frac{\partial G}{\partial x_2} - (-G_1^0 + G_2^0)$$

$$\therefore \frac{\partial^2 G^M}{\partial x_2^2} = \frac{\partial^2 G}{\partial x_2^2} \tag{2}$$

$$\bar{G}_2 = G + (1 - x_2)\frac{\partial G}{\partial x_2}$$

$$\frac{\partial \bar{G}_2}{\partial x_2} = \frac{\partial G}{\partial x_2} + \frac{\partial(1 - x_2)}{\partial x_2}\frac{\partial G}{\partial x_2} + (1 - x_2)\frac{\partial^2 G}{\partial x_2^2}$$

$$= \frac{\partial G}{\partial x_2} - \frac{\partial G}{\partial x_2} + (1 - x_2)\frac{\partial^2 G}{\partial x_2^2} = (1 - x_2)\frac{\partial^2 G}{\partial x_2^2}$$

$$\therefore \frac{\partial^2 G}{\partial x_2^2} = \frac{1}{(1 - x_2)}\frac{\partial \bar{G}_2}{\partial x_2} \quad (3)$$

$$d(1 - x_2)^2 = -2(1 - x_2)dx_2 \qquad \overline{G}_2 = G_2^0 + RT\ln a_2$$

$$\therefore \frac{1}{(1 - x_2)dx_2} = -\frac{2}{d(1 - x_2)^2} \qquad d\overline{G}_2 = RTd\ln a_2$$

$$\therefore \frac{\partial^2 G}{\partial x_2^2} = -\frac{2d\overline{G}_2}{d(1 - x_2)^2} \quad (3a)$$

from equations 2 and 3 we get

$$\therefore \frac{\partial^2 G^M}{\partial x_2^2} = \frac{\partial^2 G}{\partial x_2^2} = -\frac{2d\overline{G}_2}{d(1 - x_2)^2}$$

$$\therefore \text{stability} = \frac{\partial^2 G^M}{\partial x_2^2} = -2RT\frac{d\ln a_2}{d(1 - x_2^2)} = -4.606RT\frac{d\log a_2}{d(1 - x_2^2)} \quad (4)$$

$$\frac{\partial^2 G^{xs}}{\partial x_2^2} = -\frac{2d\overline{G}^{xs}}{d(1 - x_2^2)} = -\frac{2d(RT\ln\gamma_2)}{d(1 - x_2^2)} = -2RT\left[\frac{d\ln\gamma_2}{d(1 - x_2^2)}\right] \quad (5)$$

$$G^{xs} = x_1\overline{G}_1^{xs} + x_2\overline{G}_2^{xs} = x_1 RT\ln\gamma_1 + x_2 RT\ln\gamma_2$$

$$\frac{\partial G^{xs}}{\partial x_2} = \left[\frac{\partial x_1}{\partial x_2}RT\ln\gamma_1 + x_1 RT\frac{\partial\ln\gamma_1}{\partial x_2} + RT\ln\gamma_2 + x_2 RT\frac{\partial\ln\gamma_2}{\partial x_2}\right] \quad (6)$$

$$= -RT\ln\gamma_1 + RT\ln\gamma_2$$

$$x_1 d\ln\gamma_1 + x_2 d\ln\gamma_2 = 0$$

$$\frac{\partial^2 G^{xs}}{\partial x_2^2} = RT\left(-\frac{\partial \ln\gamma_1}{\partial x_2} + \frac{\partial \ln\gamma_2}{\partial x_2}\right) \qquad d\ln\gamma_1 = -\frac{x_2}{x_1} d\ln\gamma_2$$

$$= RT\left(\frac{x_2}{x_1}\frac{\partial \ln\gamma_2}{\partial x_2} + \frac{\partial \ln\gamma_2}{\partial x_2}\right) \qquad \therefore \frac{\partial \ln\gamma_1}{\partial x_2} = -\frac{x_2}{x_1}\frac{\partial \ln\gamma_2}{\partial x_2}$$

$$= RT\frac{\partial \ln\gamma_2}{\partial x_2}\left(1 + \frac{x_2}{x_1}\right) \qquad (7)$$

$$= RT\frac{\partial \ln\gamma_2}{\partial x_2} \cdot \frac{1}{x_1} \qquad d(1-x_2)^2 = -2(1-x_2)dx_2$$

$$= RT\frac{\partial \ln\gamma_2}{\partial x_2}\frac{1}{(1-x_2)} \qquad (1-x_2)dx_2 = -\frac{1}{2}d(1-x_2)^2$$

$$= -2RT\frac{d\ln\gamma_2}{d(1-x_2)^2} \qquad or\ \frac{1}{(1-x_2)dx_2} = -\frac{2}{d(1-x_2)^2}$$

$$= -4.606RT\frac{d\ln\gamma_2}{d(1-x_2)^2}$$

It is thus seen that the excess stability is readily found by multiplying the slope of a curve $ln\gamma_2$ *vs* $(1 - x_2)^2$ by the factor 4.606 RT. Stability and excess stability are properties of the system rather than of a component; alternatively, they may be regarded as joint properties of both components. In the terminal region if the slope $dlog\gamma_2/d(1 - x_2)^2$ is substantially constant, the excess stability is substantially constant.

Problem 69 The following solubility of oxygen in 100 g of silver at 107 °C have been measured.

$p_{O_2}(mm\ Hg)$	128	488	760	1203
oxygen dissolved cm^3/100 g Ag	81.5	156.9	193.6	247.8

(a) Show whether these observations agree with Sievert's law for the solubility of oxygen in silver.
(b) How much oxygen does 100 g of silver absorb at 1075 °C from air?
(c) What pressure of air corresponds to one atmosphere of O_2 with respect to the solubility of oxygen in silver at 1075 °C?

Solution

(a) Sievert's constant, $k_s = \frac{\text{solubility}}{\sqrt{p_{O_2}}}$

$$k_s = \frac{81.5}{\sqrt{128}} = 7.2037$$

$$k_s = \frac{156.9}{\sqrt{488}} = 7.1025$$

$$k_s = \frac{193.6}{\sqrt{760}} = 7.0226$$

$$k_s = \frac{247.8}{\sqrt{1203}} = 7.1444$$

Almost constant values of k_s demonstrate that solubility of oxygen in silver is proportional to the square root of the partial pressure of oxygen in equilibrium with silver. Hence, the observations agree with Sievert's law. Average value of $k_s = 7.1183$.

(b) In air, $p_{O_2} = 0.21 \text{ atm} = 0.21 \times 760 \text{ mm Hg}$

$\therefore$ solubility = $k_s \times \sqrt{0.21 \times 760} = 7.1183 \times \sqrt{1596} = 89.93$ cc. Ans.

(c) $p_{air} = \frac{1}{0.21}\left(p_{\text{atm}} = 1,\ p_{O_2} = 0.21\right)$
$= 4.7619$ atm Ans.

Problem 70 At 1540 °C liquid iron dissolves 0.04% nitrogen in equilibrium with nitrogen gas at one atmospheric gas pressure and 0.23% oxygen in equilibrium with oxygen gas at one atmosphere gas pressure. At that temperature nitrogen pentoxide gas was passed over liquid iron such that equilibrium was attained with fully dissociated gas at a net pressure of one atmosphere. What is the nitrogen and oxygen content of the melt?

Solution

$$N_2O_5(g) = N_2(g) + \frac{5}{2}O_2(g) = 2[N] + 5[O]$$

$$N_2 \text{ gas} : O_2 \text{ gas}$$

$$1 \text{ mol} : \frac{5}{2} \text{ mol}$$

$$2 \text{ mol} : 5 \text{ mol}$$

$$2 \text{ vol} : 5 \text{ vol}$$

$$\therefore p_{N_2} = \frac{2}{7}\text{atm} \quad \text{and} \quad P_{O_2} = \frac{5}{7}\text{atm}$$

$$\text{for } \frac{1}{2}N_2 = [N], \qquad S_N = K_N\sqrt{p_{N_2}}$$

$$0.04 = K_N\sqrt{1}$$

$$\therefore K_N = 0.04$$

$$S_N'(\text{from dissociated } N_2O_5) = K_N\sqrt{p_{O_2}}$$

$$=0.04\sqrt{\frac{2}{7}}=0.02138\% \qquad \textit{Ans.}$$

$$\frac{1}{2}O_2=[O], \qquad S_O=K_O\sqrt{p}$$

$$0.23=K_O\sqrt{1} \quad \therefore K_O=0.23$$

$$\begin{aligned} S'_O(\text{from dissociated } N_2O_5) &= 0.23\sqrt{5/7} \\ &= 0.19439\% \qquad \text{Ans.} \end{aligned}$$

Problem 71 Flow and Chipman found that 100 g of copper at 1100 °C and one atmosphere pressure dissolves 258 cm^3 (STP) of sulfur dioxide. At the same temperature how much SO_2 will be dissolved in 100 g of copper from a gas mixture at one atmosphere pressure consisting of 64% SO_2 and 36% He (by volume)? Helium is insoluble in copper. SO_2 dissolves atomically in copper in accordance with Sievert's law. Prove that the solubility of SO_2 in copper at the given temperature is proportional to the cube root of the partial pressure of SO_2. Calculate the solubility of SO_2 from the gaseous mixture at 1100 °C.

Solution:

Solubility of SO_2 in molten copper is expressed as per the reaction:

$$SO_2(g)=[S]+2[O] \tag{1}$$

$$K=\frac{a^2_{[O]}a_{[S]}}{p_{SO_2}}=\frac{f^2_{[O]}.x^2_{[O]}.f_{[S]}.x_{[S]}}{p_{SO_2}} \tag{2}$$

The reaction can also be represented as

$$SO_2(g)=\frac{1}{2}S_2(g)+O_2(g) \tag{3}$$

$$\frac{1}{2}S_2(g)=[S]$$

$$K_1=\frac{a_{[S]}}{p^{1/2}_{S_2}}$$

$$\therefore p_{S_2}=\left[\frac{a_{[S]}}{K_1}\right]^2=\frac{f^2_{[S]}.x^2_{[S]}}{K_1^2}=\left[\frac{x_{[S]}}{K'_1}\right]^2 \tag{4}$$

$$O_2=2[O]$$

$$K_2=\frac{a^2_{[O]}}{p_{O_2}}$$

$$\therefore p_{O_2} = \frac{a^2_{[O]}}{K_2} = \frac{f^2_{[O]}.x^2_{[O]}}{K_2} = \frac{x^2_{[O]}}{K'_2} \tag{5}$$

From Eq. 3

$$\frac{p_{O_2}}{p_{S_2}} = 2 \tag{6}$$

On combining Eqs. 4, 5 and 6, we get:

$$\frac{p_{O_2}}{p_{S_2}} = 2 = \frac{\frac{x^2_{[O]}}{K'_2}}{\left[\frac{x_{[S]}}{K'_1}\right]^2}$$

$$\text{or} \quad x_{[O]} = K' x_{[S]} \tag{7}$$

$$\text{i.e.} \quad x_{[O]} \propto x_{[S]}$$

On substituting (7) in (2) we get

$$K = \frac{f_{[S]}.x_{[S]}.f^2_{[O]}.x^2_{[O]}}{p_{SO_2}} = \frac{f_{[S]}.x_{[S]}.f^2_{[O]}.K'^2.x^2_{[S]}}{p_{SO_2}} = \frac{K''.x^3_{[S]}}{p_{SO_2}}$$

$$\text{or} \quad x^3_{[S]} = \frac{K}{K''} p_{SO_2}$$

$$\text{or} \quad x^3_{[S]} \propto p_{SO_2}$$

$$\text{i.e.} \quad x_{[S]} \propto p_{SO_2}{}^{1/3}$$

Thus the solubility of SO_2 in copper at the given temperature is proportional to the cube root of the partial pressure of SO_2 in equilibrium with molten copper.

Solubility of SO_2 in 100 g copper is 258 cm^3 at 1 atm pressure

$$\therefore 258 = \mathrm{K}\left(p_{SO_2}\right)^{1/3} = \mathrm{K}(1)^{1/3}$$
$$\therefore \mathrm{K} = 258$$

In the gaseous mixture, $p_{SO_2} = 0.64$ atm

$$\therefore \text{solubility} = 258(0.64)^{1/3} = 164.78 \ \mathrm{cm^3/100 \ g \ Cu} \qquad \text{Ans.}$$

Problem 72 Show that in the composition range where solute obeys Henry's law the solvent obeys Roult's law:

Solution

For the solute, B in a binary solution A–B according to Henry's law:

$$a_B = f_B . x_B \quad \text{i.e.,} \quad a_B \propto x_B$$

$$lna_B = lnf_B + ln\, x_B$$

$$\text{or} \qquad dlna_B = dlnx_B \tag{1}$$

According to Gibbs–Duhem equation for the system: A–B, we have:

$$x_A RT\; dlna_A + x_B RT\; dlna_B = 0$$

$$\text{or} \qquad dlna_A = -\frac{x_B}{x_A} dlna_B \tag{2}$$

$$= -\frac{x_B}{x_A} dlnx_B \qquad \text{(From Eq.1)}$$

$$= -\frac{x_B}{x_A} . \frac{dx_B}{x_B} = -\frac{dx_B}{x_A} = \frac{dx_A}{x_A} = dlnx_A$$

on integration, we get:

$$lna_A = lnx_A + \ln(constant)$$

by definition $a_i = 1$ when $x_i = 1$ and hence, the integration constant is unity. Thus over the composition range in which the solute B obeys Henry's law, the solvent obeys Rault's law.

Problem 73 What are different standard states? Under what condition alternative standard states are used? How will you relate activities in different standard states?

Calculate the free energy change when the standard state of manganese is transferred from pure liquid state to infinitely dilute wt% solution of manganese in iron at 1627 °C, melting point of Mn = 1245 °C, at wt of Mn = 54.94, Fe = 55.85. Assume ideal behaviour of the solution.

Solution

Different standard states are as follows:

(1) Raoultion standard state: pure substance standard state

$$a_i^R = x_i$$

$$x_i \to 1 \text{ (for an ideal solution) } a_i^R = x_i$$

for nonideal solution: $a_i^R = \gamma_i.\ x_i$, where γ_i is activity coefficient.

(2) Henrian standard state- applicable in infinitely dilute solution i.e. component at infinite dilution as standard state.

(i) atom fraction standard state
$a_i^H = x_i$ where Henry's law is obeyed ($x_i \to 0$)
In case of deviation from Henrian behaviour: $a_i^H = f_i.\ x_i$, (f_i is Henrian activity coefficient)

(ii) wt% standard state
In this case 1 wt% (1%) of the component is taken as the standard state

$$a_i^{wt\%} = wt\%i \text{ (limit wt\%} \to 0 \text{ for Henriyan solution activity coefficient} = 1)$$

Solution

$$\mathrm{T} = 1627 + 273 = 1900\ K, \mathrm{M_{Mn}} = 54.94, \mathrm{M_{Fe}} = 55.85, \gamma_{Mn}^0 = 1$$

Mn (pure substance standard state) → Mn (dilute wt% standard state)

$$\Delta G^0 = RT \ln \frac{a_{Mn}^R}{a_{Mn}^{wt\%}}$$

$$= RT \ln \left[\frac{\gamma^0 .x_{Mn}}{wt\%Mn}\right] = RT \ln \left[\frac{\gamma^o M_{Fe}}{100 M_{Mn}}\right]$$

$$= 8.314 \times 1900 \times \frac{1 \times 55.85}{100 \times 54.94}$$

$$= -72455 \text{ J mol}^{-1} \quad \text{Ans.}$$

$$x_{Mn} = \frac{wt\%Mn / M_{Mn}}{\frac{wt\%Mn}{M_{Mn}} + \frac{100 - wt\%Mn}{M_{Fe}}}$$

$$= \frac{wt\%Mn M_{Fe}}{100 M_{Mn}}$$

wt% Mn is negligible compared to 100

$$\therefore \frac{x_{Mn}}{wt\%Mn} = \frac{M_{Fe}}{100 M_{Mn}}$$

Problem 74 Starting with the relationships: $\mu_i = \mu_i^o + RTlnf_i\xi_i = \mu_i^{o'} + RTlnf_i'\xi_i' = \mu_i^{o''} + RTln\gamma_i x_i$, develop expressions for the Gibbs free energy change associated with the transfer from one standard state to another. Derive equations for the infinitely dilute solutions.

Solutions

$$\mu_i = \mu_i^o + RTlnf_i\xi_i = \mu_i^{o'} + RTlnf_i'\xi_i' = \mu_i^{o''} + RTln\gamma_i x_i$$

(i) standard state I (ξ_i)$\longrightarrow$ standard state II (ξ_i')

$$\Delta G_1^o = \mu_i^{o'} - \mu_i^o = -RTln\left[\frac{f_i'\xi_i'}{f_i\xi_i}\right]$$

For the infinitely dilute solutions assuming the following reference state $f_i \longrightarrow 1.0$ *as* $\xi_i \longrightarrow 0$ and $f_i' \longrightarrow 1.0$ *as* $\xi_i' \longrightarrow 0$ we get:

$$\Delta G_1^o = -RTln\left[\frac{\xi_i'}{\xi_i}\right]$$

(ii) standard state I (ξ_i)$\longrightarrow$ standard state III (x_i)

$$\Delta G_2^o = \mu_i^{o''} - \mu_i^o = -RTln\left[\frac{\gamma_i x_i}{f_i\xi_i}\right]$$

For the infinitely dilute solutions assuming the following reference state $\gamma_i \rightarrow \gamma_i^o$ *as* $x_i \rightarrow 0$ and $f_i \longrightarrow 1.0$ *as* $\xi_i \longrightarrow 0$ we get:

$$\Delta G_2^o = -RTln\left[\frac{\gamma_i^o x_i}{\xi_i}\right]$$

(iii) standard state II (ξ_i')$\longrightarrow$ standard state III (x_i)

$$\Delta G_3^o = \mu_i^{o''} - \mu_i^{o'} = -RTln\left[\frac{\gamma_i x_i}{f_i'\xi_i'}\right]$$

For the infinitely dilute solutions assuming the following reference state $\gamma_i \rightarrow \gamma_i^o$ *as* $x_i \rightarrow 0$ and $f_i' \longrightarrow 1.0$ *as* $\xi_i' \longrightarrow 0$ we get:

$$\Delta G_3^o = -RTln\left[\frac{\gamma_i^o x_i}{\xi_i'}\right]$$

Problem 75 Calculate the Gibbs free energy change for the following changes in standard state at 1600 °C.

(a) Hydrogen in liquid iron from (i) wt% to ppm (ii) mole fraction to ppm. Hydrogen in liquid iron follows Henry's law.
(b) Silicon in liquid iron from (i) mole fraction (pure Si as reference state) to wt% (ii) mole fraction (pure Si as reference state) to mole fraction (iii) mol% to wt% $\left[\gamma_{Si}^o = 0.0011\right]$.

Solution

(a) (i) Hydrogen in liquid iron: $[H_{wt\%}] \rightarrow [H_{ppm}]$

$$\Delta G^o = \mu^o_{H\ ppm} - \mu^o_{H\ wt\%} = -RT\ ln\left[\frac{f_H^{ppm}.H_{ppm}}{f_H^{wt\%}.H_{wt\%}}\right]$$

(ii) $[H_{mole\ fr}] \rightarrow [H_{ppm}]$

$$\Delta G^o = \mu^o_{H\ ppm} - \mu^o_{H\ mole\ fr} = -RT\ ln\left[\frac{f_H^{ppm}.H_{ppm}}{f_H.x_H}\right]$$

(b) (i) Si (x, pure Si) → Si (wt%)

$$\Delta G^o = -RT\ ln\left[\frac{f_{Si}^{wt\%}.wt\%Si}{\gamma_{Si}.x_{Si}}\right]$$

For the infinitely dilute solution of Si in liquid iron, $f_{Si}^{wt\%} \rightarrow 1.0\ as\ wt\%Si \rightarrow 0\ and\ \gamma_{Si} \rightarrow \gamma^o_{Si}\ as\ x_{Si} \rightarrow 0$ and $x_{Si} = \frac{wt\%Si.M_{Fe}}{100.M_{Si}}$

$$\therefore \Delta G^o = -RT\ ln\left[\frac{wt\%Si}{\gamma^o_{Si}.x_{Si}}\right]$$

$$= -RT\ ln\left[\frac{100.M_{Si}}{\gamma^o_{Si}.M_{Fe}}\right]$$

$$= -8.314 \times 1873 \times ln\frac{100 \times 28.09}{0.0011 \times 55.85} = -167.1\ \text{kJ mol}^{-1}$$

(ii) Si (x, pure Si) → Si (x, mole fr. std.)

$$\Delta G^o = -RT\ ln\left[\frac{f_{Si}^{mole\%}.x_{Si}}{\gamma_{Si}.x_{Si}}\right]$$

In case of infinitely dilute solution of Si in liquid iron, $f_{Si}^{mole\ fr.} \rightarrow 1.0\ as\ x_{Si} \rightarrow 0\ and\ \gamma_{Si} \rightarrow \gamma^o_{Si}\ as\ x_{Si} \rightarrow 0$

$$\therefore \Delta G^o = -RT\ ln\left[\frac{1}{\gamma^o_{Si}}\right]$$

$$= -8.314 \times 1873 \times ln\left[\frac{1}{0.0011}\right] = -106.1\ \text{kJ mole}^{-1}$$ Ans.

(iii) Si (mol %) → Si (wt%)

$$\Delta G^o = -RT\ ln\left[\frac{f_{Si}^{wt\%}.wt\%Si}{f_{Si}^{wt\%}.mol\%Si}\right] = -RT\ ln\left[\frac{wt\%Si}{mol\%Si}\right]$$

In case of infinitely dilute solution of Si in liquid iron, $f_{Si}^{wt\%} \rightarrow 1.0\ as\ wt\%Si \rightarrow 0$ and $f_{Si}^{mol\%} \rightarrow 1.0$ as mol%$Si \rightarrow 0$

$$\text{mole\%Si} = \left[\frac{\frac{wt\%Si}{M_{Si}}}{\frac{wt\%Si}{M_{Si}} + \frac{100-wt\%Si}{M_{Fe}}}\right] \times 100 = \frac{wt\%Si.M_{Fe}}{M_{Si}}$$

(since wt % Si is negligible as compared to wt % Fe)

$$\therefore \Delta G^o = -RT\ ln\left[\frac{M_{Si}}{M_{Fe}}\right] = -8.314 \times 1873 \times ln\left[\frac{28.09}{55.85}\right]$$

$$= 10.7\ \text{kJ mol}^{-1} \quad \text{Ans.}$$

Problem 76 Liquid copper containing silicon as an impurity is allowed to come into equilibrium with oxygen at a partial pressure of 10^{-14} atm. Calculate the minimum silicon content of the melt at 1200 °C assuming that copper will not oxidize under these conditions.

Given: melting point, Cu: 1083 °C, Si: 1410 °C

Molecular weight, Cu: 63.54, Si: 28.09
$\gamma_{Si}^o = 0.006$ (standard state: pure liquid Si at 1200 °C)
Si(l) + O_2(g) = SiO_2(s), $\Delta G^o = -897050 + 168.7\ T$ J mol^{-1}(298 – 1410 °C)

Solution

$$\Delta G^o \quad \text{at} \quad 1473\ \text{K} = -897050 + 168.7 \times 1473 = -648555\ \text{J mol}^{-1}$$

$$\Delta G^o = -RT \ln K = -648555$$

At equilibrium,

$$K = 9.988 \times 10^{22}$$

$$\text{and}\ K = \frac{a_{SiO_2}}{a_{Si(l)} \cdot p_{O_2}} \quad \text{since SiO}_2 \text{ is a solid}, a_{SiO_2} = 1$$

$$\therefore a_{Si(l)} = \frac{1}{K \cdot p_{O_2}} = \frac{1}{9.988 \times 10^{22} \times 10^{-14}}$$

$$= 1.001 \times 10^{-9}$$

As x_{Si} is very small Henrian behavior can be assumed, hence we can write: $a_{Si} = \gamma_{Si}^o \cdot x_{Si}$

$$\therefore x_{Si} = \frac{a_{Si}}{\gamma^o_{Si}} = \frac{1.001 \times 10^{-9}}{0.006} = 1.669 \times 10^{-7}$$

$$wt\%Si = \frac{x_{Si} \times 100 \times M_{Si}}{M_{Cu}} = \frac{1.669 \times 10^{-7} \times 100 \times \times 28.09}{63.54} = 7.378 \times 10^{-6} \text{Ans.}$$

Problem 77 (M. Nagamori, 1979–1980, Personal communication, Department of Metallurgical Engineering, University of Utah, Salt Lake City)

Crude lead bullion containing Sn, Bi, and Cu is melted at 800 °C and oxidized by adding litharge slag. From the given data show that only tin can be removed successfully by drossing. Assuming that the reaction products are pure solid oxides of Pb, Sn, Bi, and Cu, calculate the minimum Sn content that can be achieved by this softening process. How are Bi and Cu eliminated from the crude lead? Given that:

1. $Pb(l) + ½O_2(g) = PbO(s), \Delta G^o_1 = -53300 + 25.70\ T$ cal
2. $Sn(l) + O_2(g) = SnO_2(s), \Delta G^o_2 = -140180 + 51.52\ T$ cal
3. $2Bi(l) + 3/2O_2(g) = Bi_2O_3(s), \Delta G^o_3 = -141050 + 69.94T$ cal
4. $2Cu(s) + ½\ O_2(g) = Cu_2O(s), \Delta G^o_4 = -40500 - 3.92\ T\ \log T + 29.5\ T$ cal

atomic weights: Pb- 207.2, Sn- 118.7,

and at 800 K : $\gamma^0_{Sn}(in\,Pb) = 6.8, \gamma^0_{Cu}(in\,Pb) = 10\ and\ \gamma^0_{Bi}(in\,Pb) = 0.39$

Solution

Assuming that the crude lead is covered with both PbO and MO (M = Sn, Bi, Cu). Interactions between M in the crude lead and the litharge (PbO) slag results in the following reactions:

$$PbO(s) + M(inPb) = MO(s) + Pb(l)$$

$$\text{At 1073 K,} \quad \Delta G^0_1(PbO) = -25724 \text{ cal}$$

$$\Delta G^0_2(SnO_2) = -84899 \text{ cal}$$
$$\Delta G^0_3(Bi_2O_3) = -66004 \text{ cal}$$
$$\Delta G^0_4(Cu_2O) = -21594 \text{ cal}$$

For the softening reactions:

$$2PbO + Sn = SnO_2 + 2Pb, \Delta G^0_5 = -84899 - 2(-25724) = -33451,$$
$$K_5 = 6.52 \times 10^6$$
$$3PbO + 2Bi = Bi_2O_3 + 3Pb, \Delta G^0_6 = -66004 - 3(-25724) = +11168,$$
$$K_6 = 5.31 \times 10^{-3}$$
$$PbO + 2Cu - Cu_2O + Pb, \Delta G^0_7 = -21594 - (-25724) = +4130,$$
$$K_7 = 1.44 \times 10^{-1}$$

Tin

($a_{SnO_2} = 1 = a_{Pbo}$; both SnO_2 and PbO are solids,
$a_{Pb} \approx 1$, (crude lead contains more than 99% Pb)

$$K_5 = \frac{a_{SnO_2} \cdot a_{Pb}^2}{a_{Pbo}^2 \cdot a_{Sn}} = \frac{1}{a_{Sn}} = 6.52 \times 10^6$$

$$a_{Sn} = \frac{1}{6.52 \times 10^6} = 1.5 \times 10^{-7} = \gamma_{Sn}^0 \cdot x_{Sn} = 6.8.x_{Sn}$$

$$x_{Sn} = 2.3 \times 10^{-8} (\text{i.e., concentration of Sn in Pb is extremely low})$$

At infinitely dilute concentration of the solute Sn, wt % Sn is negligible as compared to the wt % of Pb ~ 100% and according to Equation 3.85,

$$\text{wt\%Sn} = \frac{x_{Sn} \cdot 100.M_{Sn}}{M_{Pb}}$$

$$\therefore \text{wt\%Sn} = \frac{x_{Sn} \times 100 \times 118.7}{207.2}$$

$$= \frac{2.3 \times 10^{-8} \times 100 \times 118.7}{207.2} = 1.3 \times 10^{-6}$$

Thus, the attainable lowest limit of tin in lead is 1×10^{-6} % . Ans.

Bismuth

$$K_6 = \frac{a_{Bi_2O_3} \cdot a_{Pb}^3}{a_{Pbo}^3 \cdot a_{Bi}^2} = \frac{1}{a_{Bi}^2} = 5.31 \times 10^{-3}$$

$$a_{Bi} = 13.7 \gg 1$$

This means that Bi content of the crude lead tries to become as high as 13.7 (a_{Bi}). But the crude lead cannot take up this much of bismuth. In other words, the elimination of Bi from the crude lead by oxidation is not possible. Conversely, Pb will get oxidized preferentially prior to the oxidation of bismuth.

Copper

$$K_7 = \frac{a_{Cu_2O} \cdot a_{Pb}}{a_{Pbo} \cdot a_{Cu}^2} = \frac{1}{a_{Cu}^2} = 0.144$$

$$a_{Cu} = 2.63 = \gamma_{Cu}^0 \cdot x_{Cu} = 10 \cdot x_{Cu}$$

$$x_{Cu} = 0.263$$

This means Cu content in Pb tries to become as high as 0.263(=x_{Cu}). Since the Cu content in the initial crude lead is less than this, Pb will get oxidized preferentially prior to the oxidation of Cu from the Pb-Cu liquid.

In view of the above problems calcium is added to eliminate Bi from Pb by forming Bi_2Ca_3. Sulfur is added to eliminate Cu by forming Cu_2S which is skimmed off from the surface of sulfur-treated Pb alloys.

Problem 78 For removal of zinc, a liquid Cd-Zn alloy is treated with excess solid CdO at 427 °C according to the reaction: Zn (l) + CdO (s) = Cd (l) + ZnO (s). If solid ZnO and solid CdO exist as separate phases without any appreciable mutual solubility, calculate the equilibrium composition of the alloy coexisting with these oxides. The free energy of formation of solid ZnO and solid CdO at 427 °C are -278236 J mol^{-1} and -184514 J mol^{-1}, respectively. The activity coefficients of zinc and cadmium for the pure metals as reference states are given as:

$$log\,\gamma_{Zn} = 0.87(1 - x_{Zn})^2 - 0.30(1 - x_{Zn})^3 \quad \text{(Atomic weight of Zn} = 65.39)$$
$$log\,\gamma_{Cd} = 0.42(1 - x_{Cd})^2 + 0.30(1 - x_{Cd})^3 \quad \text{(Atomic weight of Cd} = 112.41)$$

Solution:
The free energy change for the reaction: Zn (l) + CdO (s) = Cd (l) + ZnO (s) is obtained as:

$$\Delta G_r^o = \Delta G^o(ZnO) - \Delta G^o(CdO)$$
$$= -278236 - (-184514) = -93722\ \text{J} = -\text{RT} \ln \text{K}$$

$$K = \frac{a_{Zno} \cdot a_{Cd}}{a_{Cdo} \cdot a_{Zn}} = \frac{a_{Cd}}{a_{Zn}} \quad (\text{as ZnO and CdO are solids, } a_{Zno} = 1 = a_{Cdo})$$

$$\ln K = \frac{\Delta G_r^o}{-RT} = \frac{-93722}{-8.314 \times 700} = 16.104$$
$$K = 9.86 \times 10^6$$

$$\therefore \frac{a_{Cd}}{a_{Zn}} = 9.86 \times 10^6 = \frac{\gamma_{Cd} \cdot x_{Cd}}{\gamma_{Zn} \cdot x_{Zn}}$$

$$\log \gamma_{Cd} - \log \gamma_{Zn} = 6.994 - \log\left(\frac{x_{Cd}}{x_{Zn}}\right) \tag{1}$$

Substituting the given values of activity coefficients of Cd and Zn in equation (1) we get:

$$0.42x_{Zn}^2 + 0.30x_{Zn}^3 - 0.87(1 - x_{Zn})^2 + 0.30(1 - x_{Zn})^3 = 6.994 - \log\left(\frac{1 - x_{Zn}}{x_{Zn}}\right)$$

On simplification we get:

$$0.45x_{Zn}^2 + 0.84x_{Zn} - 7.564 + \log\left(\frac{1 - x_{Zn}}{x_{Zn}}\right) = 0$$

Since $x_{Zn}$is very small x_{Zn}^2 *and* $0.84x_{Zn}$ can be neglected

$$\therefore \log\left(\frac{1 - x_{Zn}}{x_{Zn}}\right) = 7.564$$

$$\therefore \left(\frac{1 - x_{Zn}}{x_{Zn}}\right) = 3.66 \times 10^7$$

$$\text{or } 1 - x_{Zn} = 3.66 \times 10^7 \times x_{Zn}$$

$$\text{or } x_{Zn} = 2.37 \times 10^{-8}$$

$$wt\%Zn = \frac{x_{Zn} \times 100 \times M_{Zn}}{M_{Cd}} = \frac{2.37 \times 10^{-8} \times 100 \times 65.39}{112.41} = 1.38 \times 10^{-6} \text{Ans.}$$

Problem 79 From experimental measurements of the equilibria between H_2-H_2O gas mixtures pure silica and silicon dissolved in liquid iron, the free energy accompanying the transfer of standard state from pure silicon to the infinitely dilute, wt.% solution of silicon in iron that is

Si (pure, 1) = Si (wt % dil. in Fe) has been expressed as $\Delta G^0 = -28500 - 5.8\,T$ cal mol^{-1}. At 1600 °C, the activity coefficient of silicon in iron, relative to pure silicon as the standard state is 0.0014 at 1 atomic% Si. Calculate the activity coefficient of silicon, relative to the wt% standard state at 0.5 wt % Si concentration.

Solution

$$\Delta G^o = -28500 - 5.8\,T = -28500 - 5.8 \times 1873 = -39360\,\text{cal}$$

$$\text{at\%Si} = 0.01 = \frac{wt\%\text{Si}/28.09}{wt\%\text{Si}/28.09 + 100 - wt\%\text{Si}/55.85}$$

$$\text{wt\%Si} = 0.50$$

$$\Delta G^0 = RT \ln\left[\frac{\gamma_B^0 M_A}{100 M_B}\right] = \text{RT } \ln \frac{\gamma_{Si}^0 M_{Fe}}{100 M_{Si}} = RT \ln \frac{\gamma_{Si}^0 .55.85}{100 \times 28.07} = -39360$$

$$\therefore \gamma_{Si}^0 = 0.00128$$

This is the activity coefficient of Si at infinite dilution, relative to pure Si as the standard state.

Hence the activity coefficient of Si at 1 at% Si, relative to pure Si as the standard state is calculated as follows:

$$a_{Si}(pure) = \gamma_{Si}.x_{Si} = 0.0014 \times 0.01 = 0.000014$$

The activity of Si at this concentration, relative to infinitely dilute wt% solution of Si in Fe as the standard state is related as:

$$\frac{a_{Si}^R}{\gamma_{Si}^0.x_{Si}} = \frac{a^H}{x_{Si}} = \frac{a_{Si}^{wt\%}}{wt\%Si}$$

$$\therefore \frac{a_{Si}(pure)}{a_{Si}(dilwt\%)} = \frac{\gamma_{Si}^0.x_{Si}}{wt\%Si} = \frac{\gamma_{Si}^0 M_{Fe}}{100 M_{Si}}$$

$$\therefore a_{Si}(dil\,wt\%) = \frac{a_{Si}(pure).100M_{Si}}{\gamma_{Si}^0\ M_{Fe}}$$

$$= \frac{0.000014 \times 100 \times 28.09}{0.00128 \times 55.85} = 0.55$$

$$a_{si}^{wt\%} = f_{Si} \cdot wt\%Si$$

$$\therefore f_{Si} = \frac{a_{Si}^{(wt\%)}}{wt\%Si} = \frac{0.55}{0.50} = 1.1 \qquad Ans.$$

Problem 80 The Raoultian activity coefficient of Al at infinite dilution, γ_{Al}^0 in liquid Fe-Al alloys is reported to be 0.063 at 1600 °C. Calculate the standard free energy of formation of $Al_2O_3(s)$ at 1600 °C for each of the three standard states for solution of Al in Fe. Given that

$$M_{Al} = 26.98 \quad \text{and} \quad M_{Fe} = 55.85$$

(1) $2Al(l) + 3/2O_2(g) = Al_2O_3(s)\Delta G_f^o = -1682927 - 323.24\ T\ J$
(2) $2Al(1, H) + 3/2O_2(g) = Al_2O_3(s)$
(3) $2Al(1, wt\%) + 3/2O_2(g) = Al_2O_3(s)$

Solution

$$\frac{a_B^R}{x_B\gamma_B^0} = \frac{a_B^H}{x_B} = \frac{a_B^{wt\%}}{wt\%B}$$

$$\frac{a_i^R}{a_i^H} = \gamma_i^0; \qquad \frac{a_B^{wt\%}}{a_B^H} = \frac{wt\%B}{x_B} = \frac{100M_B}{M_A}$$

$$\text{And} \quad \frac{a_B^{wt\%}}{a_B^R} = \frac{wt\%B}{x_B\gamma_B^0} = \frac{100M_B}{\gamma^0 M_A}$$

$$\text{Raoultion} \rightarrow \text{Henrian}: \quad \Delta G = RT \ln \frac{a_B^R}{a_B^H} = RT \ln \gamma_B^0$$

$$\text{Raoultion} \quad \rightarrow wt\%: \quad \Delta G = RT \ln \frac{a_B^R}{a_B^H} = RT \frac{\gamma^0 x_B}{wt\%B} = RT \ln\left[\frac{\gamma_B^0 M_A}{100 M_B}\right]$$

$$\text{Henrian} \quad \rightarrow wt\%: \ \Delta G = RT \ln \frac{a_B^H}{a_B^{wt\%}} = RT \ln \frac{x_B}{wt\%B} = RT \ln\left[\frac{M_A}{100 M_B}\right]$$

(1) $2Al(1) + \frac{3}{2}O_2(g) = Al_2O_3 - 1682927 - 323.24\ T\ J$

$\Delta G_f^o = -1682927 + 323.24 \times 1873 = -1077498.5\ J$ Ans.

Now (2) free energy change required for 2Al (l, H) $+ \frac{3}{2}O_2(g) = Al_2O_3(s)$

(2a) $2Al(1) = 2Al(1, H)$

reaction 2 is equivalent to (1) – 2(a)

$\therefore$ $\Delta G_{R \rightarrow H}$ for reaction (2a)

$$\Delta G_{R \rightarrow H} = 2RT \frac{a_{Al}^R}{a_{Al}^H} = 2RT \ln \gamma_{Al}^0$$

$\therefore$ free energy change for reaction (2) $= \Delta G_f^o - \Delta G_{R \rightarrow H} (= \Delta G_{Al}^H)$

i.e.

$$\begin{aligned} \Delta G_{Al_2O_3}^H &= \Delta G_f^o - 2\Delta G_{Al}^H \\ &= \Delta G_f^o - 2RT \ln \gamma_{Al}^0 \\ &= -1077498 - 2 \times 8.314 \times 1873 \ln(0.063) \\ &= -1077498 - (-86102) = -991396\ J \quad \text{Ans.} \end{aligned}$$

(3) 2Al (1 wt%) $+ \frac{3}{2}O_2(g) = Al_2O_3(s)$

(3a) $2Al(1) = 2Al(1wt\%)$

For $Al(l) = Al(1wt\%)$:

$$\Delta G_{Al}^{wt\%} = RT \ln \frac{a^R}{a^{wt\%}} = RT \ln \frac{\gamma_{Al}^0 \cdot x_{Al}}{wt\%Al} = RT \ln\left[\frac{\gamma M_{Fe}}{100 M_{Al}}\right]$$

Equation 3 is equivalent to 1 – 3a

$$\begin{aligned} \therefore \Delta G^{wt\%}_{Al_2O_3} &= \Delta G^0_{Al_2O_3} - 2G^{wt\%}_{Al} \\ &= \Delta G^0_{Al_2O_3} - 2RT \ln\left[\frac{\gamma^0_{Al}\cdot M_{Fe}}{100M_{Al}}\right] \\ &= -1077498 - 2\times 8.314\times 1873 \ln\left[\frac{0.063\times 55.85}{100\times 26.98}\right] \\ &= -870631 \text{ J} \quad \text{Ans.} \end{aligned}$$

Problem 81 Vanadiam melts at 1720 °C. The Raoultian activity coefficient of vanadium at infinite dilution in liquid iron at 1720 °C is 0.069. Calculate the free energy change accompanying the transfer of standard state from pure solid vanadium to the infinitely dilute wt% solution of V in pure iron at 1620 °C.

Given that: heat of fusion of V 4800 cal g atom^{-1}

$$M_{Fe}\ 55.85 \quad \text{and} \quad M_V\ 50.95.$$

Solution

$$T^f_V = 1993K \quad \gamma^0 = 0.069 \text{ at } 1893 \text{ K}.$$

$$\Delta S^f = \frac{\Delta H^f}{T^f} = \frac{4800}{1993} = 2.4084 \text{ cal deg}^{-1} \text{ mol}^{-1}$$

(1) Free energy of fusion of vanadium (s → 1) at the operating temperature 1893K.

$$\Delta G' = \Delta H - T\Delta S = 4800 - 1893\times 2.4084 = 240.9 \text{ cal mol}^{-1}$$

We calculate ΔG′ for fusion because the melting point of V is more than the operating temperature (at which V is in solid state).

(2) V(pure 1) → V(wt%)

$$\begin{aligned} \Delta G'' &= RT \ln\left(\frac{\gamma^0_V M_{Fe}}{100 M_V}\right) = 1.987\times 1983 \ \ln\left(\frac{0.069\times 55.85}{100\times 50.95}\right) \\ &= -27033 \text{ cal mol}^{-1} \end{aligned}$$

∴ For energy change V (pure solid → V infinitely dilute wt% solution)

$$\begin{aligned} \Delta G &= \Delta G' + \Delta G'' \\ &= 240.9 - 27033 \\ &= -26792.1 \text{ cal mol}^{-1} \quad \text{Ans.} \end{aligned}$$

Problem 82 J. F. Elliott, 1980 – 1981, Personal communication, Department of Materials Science and Engineering, Massachusetts Institute of technology, Cambridge, USA.

Establish Wagner's reciprocality relationship.

Solution
Wagner's reciprocality relationship:

$$\varepsilon_2^3 = \varepsilon_2^3$$

Consider the partial molar free energy

$$\overline{G_i} = \left(\frac{\partial G}{\partial n_i}\right) \quad \text{and} \quad \overline{G_j} = \left(\frac{\partial G}{\partial n_j}\right)$$

$$\therefore \frac{\partial^2 G}{\partial_{n_i} \partial_{n_j}} = \frac{\partial \overline{G_i}}{\partial n_j} = \frac{\partial \overline{G_j}}{\partial n_i}$$

$$\overline{G_i^M} = RT \ln a_i \quad \text{and} \quad \left.\frac{\partial G}{\partial n_i}\right|_{n_j \neq n_i, T, P} = \mu_i$$

$$\mu_2 = \left(\frac{\partial G}{\partial n_2}\right)_{n_1, n_3, T, P} \quad \text{and} \quad \mu_3 = \left(\frac{\partial G}{\partial n_3}\right)_{T, P, n_1, n_2}$$

$$\therefore \left.\frac{\partial \mu_2}{\partial x_3}\right|_{x_2} = \left.\frac{\partial \mu_3}{\partial x_2}\right|_{x_3} \qquad (a_2 = x_2 \gamma_2)$$

$$\therefore \mu_2 = \mu_2^0 + RT \ln a_2 = \mu_2^0 + RT \ln x_2 + RT \ln \gamma_2$$
$$= \mu_2^0 + RT \ln x_2 + RT\left(\ln \gamma_2^0 + \varepsilon_2^2 x_2 + \varepsilon_2^3 x_3\right)$$

$$\therefore \left.\frac{\partial \mu_2}{\partial x_3}\right|_{x_2} = RT\,\varepsilon_2^3 \quad \text{similarly} \left.\frac{\partial \mu_3}{\partial x_2}\right|_{x3} = RT\,\varepsilon_3^2$$

$$\text{since} \quad \left.\frac{\partial \mu_2}{\partial x_3}\right|_{x_2} = \left.\frac{\partial \mu_3}{\partial x_2}\right|_{x3}$$

$\therefore \varepsilon_2^3 = \varepsilon_3^2$ The required Reciprocality Relationship

Problem 83 Calculate the activity of sulfur in a pig iron containing 0.05%S, 1.0% Si, 5.0%C and 2.0% Mn at 1600 °C

Given $e_S^S = -0.028$, $e_S^{Si} = 0.066$, $e_S^C = 0.114$ and $e_S^{Mn} = -0.025$

Solution:
In Fe-S-Si-C-Mn system:

$$\begin{aligned}\log f_S &= e_S^S\%S + e_S^{Si}\%Si + e_S^C\%C + e_S^{Mn}\%Mn\\ &= -0.028\times 0.05 + 0.066\times 1.0 + 0.114\times 5 + (-0.025)x2\\ &= -0.0014 + 0.066 + 0.570 - 0.05 = 0.5846\\ f_S &= 3.84\\ a_S &= f_S\times wt\% = 3.84\times 0.05 = 0.192 \quad \text{Ans.}\end{aligned}$$

Problem 84 Liquid iron can dissolve 0.045% N under one atmosphere nitrogen gas pressure at 1600 °C. Calculate the partial pressure of nitrogen over molten steel containing 0.005% N and 5.0% Cr at 1600 °C. Given that: $e_N^N = 0, e_N^{Cr} = -0.045$.

Solution
The equation for the dissolution of nitrogen gas in steel can be written as

$$\frac{1}{2}N_2(g) = [N]_{\text{insteel}}$$

$$K = \frac{[a_N]}{\sqrt{p_{N_2}}} = \frac{f_N[\%N]}{\sqrt{p_{N_2}}}$$

$$e_N^N = 0\,\text{i.e.}\,e_N^N = \frac{\log f_N}{\%N} = 0$$

$$\therefore\ \log f_N = 0, \text{or} f_N = 1$$

$$K = \frac{f_N[\%N]}{\sqrt{p_{N_2}}} = \frac{1\times 0.045}{1} = 0.045$$

Steel containing Cr steel: Fe-N-Cr system

$$\begin{aligned}\log f_N &= e_N^N\cdot\%N + e_N^{Cr}\cdot\%Cr\\ &= 0\times 0.005 + (-0.045\times 5) = -0.225\end{aligned}$$

$$f_N = 0.596$$

$$K = \frac{f_N[\%N]}{\sqrt{p_{N_2}}}$$

$$\sqrt{p_{N_2}} = \frac{f_N[\%N]}{K} = \frac{0.596\times 0.005}{0.045} = 0.066$$

$$p_{N_2} = 0.066^2 = 0.0044\ \text{atm} \quad \text{Ans.}$$

Problem 85 A pure manganese alloy containing 0.70 wt% Mn is brought to equilibrium at 1335 °C with a gas mixture of H_2S and H_2 in the ratio: $\frac{pH_2S}{p_{H_2}} = 5.7\times 10^{-4}$. On analysis the equilibrated alloy specimen is found to contain 0.0032 wt% S. If the activity coefficient of sulfur (relative to the infinitely dilute wt% standard state) is unity in dilute binary Fe-S alloys, calculate the interaction coefficient for the effect of manganese on the activity coefficient of sulfur at 1335 °C. Given that:

$$H_2(g) + [S]\left(wt^{\%}\%\text{in Fe}\right) = H_2S(g)$$

$$\Delta G^o = -9870 + 939T\,\text{cal}$$

Solution

For the reaction under question the value of K is given by

$$\log K = \frac{9870}{4.575T} - \frac{9.39}{4.575} = -0.718\,\text{at}\,1608\,K$$
$$K_{1608} = 0.191$$

Assuring the gas behaves ideally:

$$K = \frac{p_{H_2S}}{p_{H_2}[as]} = \frac{p_{H_2S}}{p_{H_2}[wt\%S.f_S]}$$
$$= \frac{5.7\times10^{-4}}{0.0032\times f_s}$$
$$\therefore f_s = \frac{5.7}{0.0032\times0.191} = 0.933$$

In Fe-Mn-S system: $e_S^S = 0$, (since $f_S = 1$ in Fe-S system)

$$\log f_S = \%Se_S^S + \%Mne_S^{Mn} = \%S\times0 + \%Mne_S^{Mn} = \%Mne_S^{Mn}$$
$$e_S^{Mn} = \frac{\log f_S}{\%S} = \frac{\log(0.933)}{0.7} = -0.043 \qquad \text{Ans.}$$

Problem 86 The ratio p_{H_2S}/p_{H_2} in equilibrium with iron containing 0.04 wt % S and 1.2 wt % C is 1.4×10^{-4} at 1600 °C. In dilute Fe-S alloys the interaction parameter: e_S^S is – 0.028, relative to the infinitely dilute wt % standard state of S in iron. Find the effect of carbon on the activity coefficient of sulfur in the ternary alloy. The free energy change accompanying the transfer of standard state from sulfur gas to the infinitely dilute, wt % solution in iron is expressed as:

$$\frac{1}{2}S_2(g) = [S](\text{wt\%in Fe}), \Delta G_1^o = -31520 + 5.27\,T\,\text{cal}$$

$$\text{Given that:}\quad H_2(g) + \frac{1}{2}S_2 = H_2\,S(g), \Delta G_2^o = -21580 + 11.805\,T\,\text{cal}$$

Solution

The free energy change for the reaction under question: H_2(g) + [S] (wt% in Fe) = H_2S (g) is obtained subtracting ΔG_1^o from ΔG_2^o.

$$\Delta G^o = 9940 + 6.535\,T\,\text{cal}$$

$$K = \frac{p_{H_2S}}{p_{H_2}[a_S]} = \frac{p_{H_2S}}{p_{H_2}[wt\%S.f_S]} = \frac{1.4\times10^{-4}}{0.04\times f_S}$$

$$\Delta G^o = -RT\ln K = 9940 + 6.535\,T$$

$$K = 2.58\times10^{-3}\,\text{at}\,1873\ \text{K}$$

$$\therefore f_S = \frac{1.4\times10^{-4}}{0.04\times2.58\times10^{-3}} = 1.357$$

$$\log f_S = 0.132$$

In Fe-S-C system, we have:

$$\begin{aligned}
&\log f_S = \%S.e_S^S + \%C.e_S^C\\
&0.132 = 0.04\times(-0.028) + 1.2e_S^C\\
&1.2e_S^C = 0.132 + 0.00112 = 0.13312\\
&e_S^C = \frac{0.13312}{1.2} = \frac{1.3312}{12} = 0.11 \qquad \textit{Ans.}
\end{aligned}$$

Problem 87 At 1600 °C pure iron dissolves 0.045 wt % nitrogen in equilibrium with nitrogen gas at one atmosphere nitrogen gas pressure and 0.0025 wt % hydrogen in equilibrium with hydrogen gas at one atmosphere hydrogen gas pressure. At that temperature ammonia was passed over steel melt such that equilibrium was attained with dissociated ammonia at a net pressure of one atmosphere. What is the nitrogen and hydrogen contents of the melt containing 1 wt % C, 3 wt % Ti and 4 wt % V at 1600 °C.

Given that: $e_N^N = 0$, $e_N^C = 0.13$, $e_N^{Ti} = -0.063$, $e_N^V = -0.09$,

$$e_H^H = 0, \quad e_H^C = 0.06, \quad e_H^{Ti} = -0.08, \quad e_H^V = -0.007$$

Solution

Dissociation of ammonia: $NH_3(g) = \frac{1}{2}N_2 + \frac{3}{2}H_2$

$$\therefore \frac{p_{N_2}}{p_{H_2}} = \frac{1/2}{3/2} = \frac{1}{3}$$

$$\begin{aligned}
&p_{H_2} + p_{N_2} = 1\,\text{atm}\\
&\therefore p_{N_2} = \frac{1}{4}\,\text{atm}, \quad p_{H_2} = \frac{3}{4}\,\text{atm}
\end{aligned}$$

Since Henry's law is obeyed

$$f_N = 1 \; as \, e_N^N = 0 \quad e_N^N = \frac{\log f_N}{\%N} = 0 \quad \therefore \; \log f_N = 0 \;\text{ or }\; f_N = 1$$

$$f_H = 1 \; as \, e_H^H = 0 \quad e_H^H = \frac{\log f_H}{\%H} = 0 \quad \therefore \; \log f_H = 0 \;\text{ or }\; f_H = 1$$

$$\frac{1}{2}N_2 = [N]$$

$$K_N = \frac{a_N}{p_{N_2}^{1/2}} = \frac{f_N[wt\%N]}{\sqrt{p_{N_2}}}$$

$$= \frac{1 \times 0.045}{\sqrt{1}} = 0.045$$

Activity coefficient changes in steel, that is in Fe-N-C-Ti-V system

$$\log f_N = \%N\, e_N^N + \%C\, e_N^C + \%Ti\, e_N^{Ti} + \%V\, e_N^V$$

$$= 0 + 1 \times 0.13 + 3(-0.063) + 4(-0.09)$$

$$= 0.13 - 0.189 - 0.36 = 0.419$$

$$f'_N = 0.381$$

$$\%\text{ N (in steel)} = \frac{K_N\sqrt{p_{N_2}}}{f'_{N_2}}$$

$$= \frac{0.045 \times \sqrt{1/4}}{0.381} = \frac{0.0225}{0.381} = 0.059 \text{ Ans}$$

$$\frac{1}{2}H_2 = [H]$$

$$K_H = \frac{a_H}{\sqrt{p_{H_2}}} = \frac{f_H \cdot [wt\%H]}{\sqrt{p_{H_2}}} = \frac{1 \times 0.0025}{1} = 0.0025$$

in steel

$$\log f'_H = \%H\, e_H^H + \%C\, e_H^C + \%Ti\, e_H^{Ti} + \%V\, e_H^V$$

$$= 0 + 1 \times (-0.06) + 3(-0.08) + 4(-0.007)$$

$$= -0.06 - 0.24 - 0.028 = -0.328$$

$$f'_H = 0.4698$$

$$\%H = \frac{K_H\sqrt{p_{H_2}}}{f'_H} = \frac{0.0025\sqrt{3/4}}{0.4698}$$
$$= \frac{0.0025 \times 1.732}{2 \times 0.4698}$$
$$= 0.0046\% \quad \text{Ans.}$$

Problem 88

(a) Nitrogen dissolution in molten iron obeys Sievert's law. How will you obtain activity coefficient of nitrogen in Fe-N-X solution if the solubility of nitrogen in pure iron and in Fe-X alloys are experimentally determined.

(b) Solubility of nitrogen in pure molten iron is given by the expression:

$$\log\,[wt\%N]_{Fe} = -\frac{285}{T} - 1.21$$

In a ternary Fe-N-C melt at 1550 °C the solubility of nitrogen was found to be 0.0368 wt % at carbon concentration of 0.54%. Estimate e_N^C and e_C^N assuming $e_N^N = 0$

Atomic weights: Fe 55.85, N 14, C 12.

Solution

(a) In Fe-N system $\quad N_2\ (g) = 2\ [N]\ (\text{in Fe})$

$$K_{Fe-N} = \frac{a_N^2}{p_{N_2}} = \frac{x_N^2 . f_N^2}{p_{N_2}}$$

In Fe-N-X alloys $K_{Fe\text{-}N\text{-}X} = \dfrac{x^2_{N(Fe-N-X)} f^2_{N(Fe-N-X)}}{p_{N_2}}$

At one atmosphere pressure under equilibrium, we have

$$K_{Fe-N} = K_{Fe-N-X}$$

$$\therefore f^2_{N(Fe-N-X)} = \frac{x^2_{N(Fe-N)}}{x^2_{N(Fe-N-X)}}$$

$$or f_{N(Fe-N-X)} = \frac{x_{N(Fe-N)}}{x_{N(Fe-N-X)}}$$

In wt %, $f_{N(Fe-N-X)} = \dfrac{wt\%N_{(Fe-N)}}{wt\%N_{(Fe-N-X)}}$

(b) Fe-N-C system at T^f = 1823 K

In ternary system wt % N = 0.0368

$$\log\,(wt\%N)_{Fe} = -\frac{285}{1823} - 1.21 = -1.3663$$
$$\therefore\ (wt\%N)_{Fe} = 0.043 = a_N^H$$
$$a_N^H = f_N\,wt\%N$$
$$\therefore f_{N(Fe\text{-}N\text{-}C)} = \frac{a_N^H}{wt\%N} = \frac{0.043}{0.0368} = 1.168$$
$$\log f_{N(Fe\text{-}N\text{-}C)} = wt\%N\,e_N^N + wt\%C\,e_N^C$$
$$\log(1.168) = 0.0368(0) + 0.54.e_N^C$$
$$e_N^C = \frac{0.0678}{0.54} = 0.125$$
$$\therefore\ e_C^N = \frac{M_C}{M_N}e_N^C = \frac{12}{14} \times 0.125 = 0.107 \qquad Ans.$$

Problem 89 A liquid Fe-P alloy containing 0.65 wt % phosphorus is brought to equilibrium at 1600 °C with a gas mixture of H_2O and H_2 in the ratio $H_2O/H_2 = 4.94 \times 10^{-2}$. On analysis the equilibrated alloy was found to contain 0.0116 wt % oxygen. The activity coefficient of oxygen, f_O^0 relative to the infinitely dilute wt % standard state in Fe-O binary alloys at 1600 °C is represented as

$$\log f_O^0 = -0.2\,[wt\%O]$$

Calculate the interaction coefficient for the effect of phosphorus on the activity coefficient of oxygen at 1600 °C if the free energy change for the reaction.

H_2 (g) + [O] (wt % in Fe) = H_2O (g) is –5020 cal mol^{-1}

Solution

$$\Delta G = -RT\ln K$$
$$\ln K = -\frac{-5020}{1.987 \times 1873} = \frac{5020}{1.987 \times 1873} = 1.34886$$
$$K = 3.853$$

$$K = \frac{p_{H_2O}}{p_{H_2}[a_O]} = \frac{p_{H_2O}}{p_{H_2}\left[f_O^P.wt\%O\right]} = \frac{0.0494}{f_O^P.0.0116} = 3.853$$
$$\therefore f_O^P = \frac{0.0494}{3.853 \times 0.0116} = 1.10526$$

Fe-P-O system for the effect of P on O

$$\log f_O^P = \%O\,e_O^O + \%P\,e_O^P$$

$$0.043465 = 0.0116(-0.2) + 0.65\, e_O^P$$
$$0.65\, e_O^P = +0.00232 + 0.043465 = 0.04578$$
$$\therefore e_O^P = \frac{0.0467}{0.65} = 0.0704 \qquad Ans.$$

Problem 90 A liquid iron alloy containing 0.20% Si is in equilibrium with pure silica at 1600 °C.

(a) Calculate the equilibrium oxygen content of this alloy.
(b) What would be the equilibrium oxygen content of the alloy on addition of 5 wt% Cr to the liquid iron? The silica content is maintained the same.

Given: $SiO_2(l) = [Si]\ (wt\%) + 2[O]\ (\text{wt}\%),\ logK = -\frac{30400}{T} + 11.58$

$$e_O^O = -0.20,\ e_O^{Si} = -0.23,\ e_O^{Cr} = -0.04.$$

Solution
(a) $SiO_2(l) = [Si]\ (wt\%) + 2[O]\ (\text{wt }\%)$

$$K = \frac{\left[a_{Si}^{wt\%}\right].\left[a_O^{wt\%}\right]^2}{a_{SiO_2(l)}} = [\%Si].[\%O]^2$$

$$\text{at } 1873\ K, logK = -\frac{30400}{1873} + 11.58 = \log\left\{[\%Si].[\%O]^2\right\}$$
$$\text{or } \log[\%Si] + 2\,log[\%O] = -16.23 + 11.58 = -4.65$$
$$\therefore 2\log[\%O] = -4.65 - \log[\%Si] = -4.65 - \log 0.20 = -3.95$$
$$[\%\text{O}] = 0.011 \quad \text{Ans.}$$

(b) In Fe-Si-O alloy

$$\log f_O = \%O.e_O^O + \%\text{Si}.e_O^{Si}$$
$$= 0.011(-0.20) + 0.20(-0.23) = -0.0482$$
$$\therefore f_O = 0.895$$
$$wt\%O = \frac{[a_O]}{f_O} = \frac{0.011}{0.895} = 0.012 wt\%$$

In Fe-O-Si-Cr melt

$$\log f_O = \%O.e_O^O + \%\text{Si}.e_O^{Si} + \%Cr.e_O^{Cr}$$
$$= 0.012(-0.20) + 0.20(-0.23) + 5(-0.04) = -0.2484$$

$\therefore f_O = 0.5644$

$$wt\%O = \frac{[a_O]}{f_O} = \frac{0.011}{0.5644} = 0.0195 \quad \text{Ans.}$$

Problem 91 Calculate the oxygen content of a Fe-O-C melt containing 0.1% C at 1550 °C which is in equilibrium with an atmosphere having (a) p_{CO} = 0.1 atm and (*b*) 0.001 atm.

Given that: $CO(g) = [\%C] + [\%O]$, $\log K = -\frac{1168}{T} - 2.07$

$$e_C^C = 0.14, e_C^O = -0.34, e_O^O = -0.20, e_O^C = -0.45.$$

Solution

(a) Fe-O-C melt at 1823 K, %C = 0.1 and p_{CO} = 0.1 atm

For the reaction: CO (g) = [%C] + [%O], $K = \frac{[a_C].[a_O]}{p_{CO}} = \frac{[\%C.f_C].[\%O.f_O]}{p_{CO}}$

$$\therefore \log K = \log\left[\frac{[\%C.f_C]\cdot[\%O]\cdot f_O}{p_{CO}}\right] = -\frac{1168}{1823} - 2.07 = -2.7107$$

$$\therefore \log[\%O] = \log p_{CO} - \log[\%C] - \log f_C^{wt\%} - \log f_O^{wt\%} - 2.7107 \quad (1)$$

$$\log f_C = \%C.e_C^C + \%O.e_C^O = 0.1 \times 0.14 + \%O.(-0.34) = 0.014 - 0.34.\%O$$

$$\log f_O = \%O.e_O^O + \%C.e_O^C = \%O.(-0.20) + 0.1.(-0.45) = -0.2.\%O - 0.045$$

Substituting the values of $\log f_C$ *and* $\log f_O$ in Eq. (1) we get:

$$\begin{aligned}\log[\%O] &= \log(0.1) - \log(0.1) - 0.014 + 0.34.\%O + 0.2.\%O + 0.045 - 2.7107 \\ &= 0.54.\%O - 2.6797\end{aligned} \quad (2)$$

If the effect of interaction cefficient is neglected, we get:

$$\log[\%O] = \log(0.1) - \log(0.1) - 2.7107 = -2.7107$$

$[\%O] = 1.947\times^{-3}$ as first approximation,

Substituting this value in Eq. (2) we get,

$$\begin{aligned}\log[\%O] &= 0.54.\%O - 2.6797 = 0.54(1.947\times^{-3}) - 2.6797 = -2.6786 \\ [\%O] &= 0.0021 \quad \text{Ans.}\end{aligned}$$

(b) when p_{CO} = 0.001 atm, Eq. 2 becomes:

$$\log[\%O] = 0.54.\%O - 4.6797 \quad (3)$$

When interaction coefficient is neglected,

$$\begin{aligned}\log[\%O] &= \log p_{CO} - \log[\%C] - 2.7107 = \log(0.001) - \log(0.1) - 2.7107 \\ &= -3-(-1) - 2.7107 = -4.7107 \\ \therefore [\%O] &= 1.95 \times 10^{-5} \text{as first approximation}\end{aligned}$$

Substituting this value in Eq. 3, we get:

$$\log[\%O] = 0.54\left(1.95 \times 10^{-5}\right) - 4.6797 = -4.6797$$

$$\therefore [\%O] = 2.1 \times 10^{-5} \text{Ans.}$$

Problem 92 Calculate the oxygen content of an iron melt containing 0.25% C and 0.5% Si which is in equilibrium with pure CO gas at 1 atm at 1550 °C. Assume that silica does not form.

Given: [%C] + [%O] = CO(g), $\Delta G^o = -22400 - 39.6$ T J

$$[\%\text{Si}] + 2[\%O] = SiO_2(s), \Delta G^o = -594100 + 2300.0T\ J$$

$$e_C^C = 0.14, e_C^O = -0.45, e_C^{Si} = 0.18$$

$$e_O^O = -0.20, e_O^C = -0.34.e_O^{Si} = -0.23$$

Solution

$$[\%\text{C}] = 0.25, [\%\text{Si}] = 0.5, \text{and T} = 1823\ \text{K}$$

For the reaction: [%C] + [%O] = CO(g),

$$\Delta G^O(at\,1823K) = -22400 - 39.6 \times 1823 = -94590.8\ \text{J} = -\text{RT}\ln\text{K}$$

and $K = \frac{p_{CO}}{[a_C].[a_O]} = \frac{p_{CO}}{[f_C.\%C].[f_O.\%O]}$

$$\therefore \ln K = -\frac{-94590.8}{8.314 \times 1823} = ln\left[\frac{p_{CO}}{[f_C.\%C].[f_O.\%O]}\right]$$

As $p_{CO} = 1$ atm, $\ln\left[\frac{1}{[f_C\%C]\cdot[f_O\%O]}\right] = 6.241$

$$\text{or} \quad -\log[\%C] - logf_C - \log[\%O] - logf_O = 2.71 \quad (1)$$

$$\begin{aligned}\log f_C &= \%c \cdot e_C^C + \%O \cdot e_C^O + \%Si.e_C^{Si} \\ &= 0.25 \times 0.14 + \%O \cdot (-0.45) + 0.5 \times 0.18 \\ &= 0.125 - 0.45 \cdot [\%O)\end{aligned} \quad (2)$$

$$\begin{aligned}\log f_O &= \%O \cdot e_O^O + \%C.e_O^C + \%Si.e_O^{Si} \\ &= [\%O] \cdot (-0.20) + 0.25.(-0.34) + 0.5.(-0.23) \\ &= -0.2 - 0.2 \cdot [\%O]\end{aligned} \quad (3)$$

Since [%O] is very small, it can be neglected in Eqs. 2 and 3:

$\therefore$ *log* $f_C = 0.125$ and *log* $f_O = -0.2$

From Eq.1, we get:

$$\begin{aligned}\log[\%O] &= -\log[\%C] - \log f_C - \log f_O - 2.71 \\ &= -\log[0.25] - 0.125 - (-0.2) - 2.71 = -2.033 \\ \therefore [\%O] &= 0.0093 \qquad \text{Ans.}\end{aligned}$$

Problem 93 Evaluate the equilibrium constant K for the reaction: CO (g) + H_2O (g) = CO_2 (g) + H_2 (g) at each temperature from the following data:

1. FeO (s) + CO (g) = Fe (s) + CO_2 (g)
2. FeO (s) + H_2 (g) = Fe (s) + H_2O (g)

T, 0 °C	log K_1	log K_2
600	−0.046	−0.479
700	−0.172	−0.375
800	−0.272	−0.302
900	−0.344	−0.226
1000	−0.402	−0.175

Solution:

The reaction under question: CO (g) + H_2O (g) = CO_2 (g) + H_2 (g) is equivalent to reaction 1–2, hence, $\Delta G_r^o = \Delta G_1^o - \Delta G_2^o$

$$\Delta G_r^o = -RT \ln K$$

$$\Delta G_1^o = -RT \ln K_1$$

$$\Delta G_2^o = -RT \ln K_2$$

$$\therefore \Delta G_r^o = -RT \ln K = -RTlnK_1 - (-RTlnK_2)$$

$$RT \ln K = RTlnK_1 - RTlnK_2$$

$\therefore \ln K = \ln K_1 - \ln K_2$

or $\quad \log K = \log K_1 - \log K_2$

$$K = \frac{p_{CO_2} p_{H_2}}{p_{CO} p_{H_2O}}, \quad K_1 = \frac{a_{Fe} p_{CO_2}}{a_{FeO} p_{CO}} = \frac{p_{CO_2}}{p_{CO}}, \quad K_2 = \frac{p_{H_2O}}{p_{H_2}} \quad (a_{Fe} = 1 = a_{FeO})$$

$$\therefore K = \frac{K_1}{K_2}$$

T, 0 °C	log K_1	log K_2	log K = log K_1 – log K_2	K	ΔG° cal mol^{-1}
600	−0.046	−0.479	0.433	2.710	−1729.4
700	−0.172	−0.375	0.203	1.5959	−903.6
800	−0.272	−0.302	0.030	1.0715	−147.3
900	−0.344	−0.226	−0.118	0.7621	+ 633.2
1000	−0.402	−0.175	−0.227	0.5929	+ 1322.0

Problem 94 A CO-CO_2 gas mixture is brought into equilibrium at one atmosphere pressure with a steel containing 0.65 wt% carbon at 1200 K. The resulting gas after equilibration analyzed 95% CO and 5% CO_2 by volume. Calculate the activity of carbon in the steel. What would be the expected analysis of CH_4-H_2 gas mixture if equilibrated with the steel at the same temperature and pressure?

Given that: (1) $2CO(g) = CO_2(g) + C(s), \Delta G_1^o = -40800 + 41.7T$ cal

(2) $CH_4(g) = C(s) + 2H_2(g), \Delta G_2^o = 21600 - 26.2T$ cal

Solution

For the given Eq. (1), T = 1200 K, $p_{CO_2} = 0.05$ atm, $p_{CO} = 0.95$ atm,

$$2CO(g) = CO_2(g) + C(s), \Delta G_1^o = -40800 + 41.7T \text{ cal}$$

$$K = \frac{[a_C].p_{CO_2}}{p_{CO}^2}$$

$$\therefore [a_C] = \frac{K.p_{CO}^2}{p_{CO_2}}$$

$$\begin{aligned} \Delta G_1^o &= -40800 + 41.7 \times 1200 = 9240 \text{ cal} \\ &= -\text{RT} \ln \text{K} \end{aligned}$$

$$\ln \text{K} = -\frac{9240}{1.987 \times 1200} = -3.6752$$

$$\text{K} = 0.02075$$

$$\therefore\ [a_C] = \frac{0.02075 \times (0.95)^2}{0.05} = 0.375 \quad \text{Ans.}$$

(3) $CH_4(g) = C(s) + 2H_2(g), \Delta G_2^o = 21600 - 26.2\, T$ cal

$$\Delta G_2^o = 21600 - 26.2 \times 1200 = -9840 = -RT\, lnK$$

$$K = 61.98$$

$$K = \frac{p_{H_2}^2 \cdot [a_C]}{p_{CH_4}} = \frac{p_{H_2}^2 \times 0.375}{p_{CH_4}}$$

$$\frac{p_{H_2}^2}{p_{CH_4}} = \frac{K}{0.0375} = \frac{61.98}{0.0375} = 165.28$$

$$\text{or } p_{H_2}^2 = 165.28.p_{CH_4} \qquad (3)$$

$$\text{and } p_{CH_4} + p_{H_2} = 1 \qquad (4)$$

From Eqs, 3 and 4, we get:

$$p_{H_2}^2 = 165.28(1 - p_{H_2})$$

$$\text{or } p_{H_2}^2 + 165.28 p_{H_2} - 165.28 = 0$$

$$p_{H_2} = 0.9948 \text{ atm}$$

$$\therefore\ p_{CH_4} = 0.0052 \text{ atm}$$

Hence, the expected analysis of CH_4-H_2 gas mixture would be 99.48% hydrogen and 0.52% methane by volume. Ans.

Problem 95 Determine the maximum temperature to which molybdenum can be heated without formation of MoO_2 in an atmosphere of a gas mixture containing 69% hydrogen and 31% steam by volume.

Given that:

(1) $Mo(s) + O_2(g) = MoO_2(s), \Delta G_1^o = -140100 - 4.6T \log T + 55.85\, T$ cal

(2) $H_2(g) + \frac{1}{2}O_2(g) = H_2O(g), \Delta G_2^o = -58900 + 13.1\, T$ cal

Solution

The reaction under question is: $Mo(s) + 2H_2O(g) = MoO_2(s) + 2H_2(g)$ (3)

$$K_3 = \frac{a_{MoO_2}.p_{H_2}^2}{a_{Mo}.\, p_{H_2O}^2} = \frac{p_{H_2}^2}{p_{H_2O}^2} = \frac{0.69^2}{0.31^2} = 4.9542$$

$$ln\ K_3 = 1.6002$$

$$\Delta G_3^o = \Delta G_1^o - 2\Delta G_2^O = -22300 - 4.6T \log T + 29.65\,T \text{ cal}$$
$$= -RT \; ln \; K_3 = -1.987 \times T \times 1.6002 = -3.18\,T$$

Or $\quad 32.83\,T - 4.6\,T\,logT - 22300 = 0$

$$\text{T} = 1193 \text{ K} = 920°\text{C}$$

Hence, molybdenum can be heated up to 920 °C without formation of MoO_2 Ans.

Problem 96 The following pressures of zinc have been determined for Cu-Zn alloys at 1060 °C.

(a) What is the free energy change when 1 g atom of zinc dissolves in a very large amount of Cu-Zn alloy, in which atom fraction of zinc is 0.3?
(b) Find the difference in the free energy of 1 g atom of zinc in liquid Cu-Zn alloys at atom fractions of 0.45 and 0.15?

x_{Zn}	1.0	0.45	0.30	0.20	0.15	0.10	0.05
p_{Zn} (mm Hg)	3040	970	456	180	90	45	22

Solution

$x_{zn} = 1 = a_{zn}$ (standard state)

Hence, $\quad p^o_{Zn} = 3040 \; mm \; Hg \qquad a_{zn} = \frac{p_{Zn}}{p^o_{Zn}}$

$$\therefore a_{Zn} = \frac{3040}{3040} \quad \frac{970}{3040} \quad \frac{456}{3040} \quad \frac{180}{3040} \quad \frac{90}{3040} \quad \frac{45}{3040} \quad \frac{22}{3040}$$
$$= 1 \quad = 0.319 \quad = 0.15 \quad = 0.0592 \quad = 0.0296 \quad = 0.0148 \quad = 0.007236$$

(a) The change in free energy on addition of 1 g atom of liquid zinc in a very large amount of Cu-Zn alloy is simply the partial molar free energy:

$$\Delta\overline{G}_{Zn} = RT \ln a_{Zn} \qquad (x_{Zn} = 0.3, a_{Zn} = 0.15)$$
$$= 1.987 \times 1333 \times \ln 0.15$$
$$= -5024.8 \text{ cal mol}^{-1} \quad \text{Ans.}$$

(b) The difference in the free energy 1 g atom of zinc in liquid Cu-Zn alloys at atomic fraction of 0.45 and 0.15 $= \Delta\overline{G}_{Zn}(0.45) - \Delta\overline{G}_{Zn}(0.15)$

$$= \text{RT} \ln(0.319) - \text{RT} \ln(0.0296)$$
$$= 1.987 \times 1333[-1.1426 - (-3.5199)] = 6297 \text{ cal mol}^{-1} \qquad \text{Ans.}$$

Problem 97 When 1 mol of argon gas is bubbled through a large volume of an Fe-Mn melt of $x_{Mn} = 0.5$ at 1863 K, evaporation of Mn into the argon causes the mass of the melt to decrease by 1.50 g. The gas leaves the melt at a pressure of 1 atm. Calculate the activity coefficient of Mn in the liquid alloy. Atomic weight of manganese = 54.93.

Given that: $\ln p\,(\text{atm}) = -\frac{33440}{T} - 3.02 \ln T + 37.68$

Solution

First calculate $p^0_{Mn}\ at\ 1863\,K$

$$\begin{aligned}\ln p^0_{Mn}\,(\text{atm}) &= -\frac{33440}{1863} - 3.02 \ln 1863 + 37.68\\ &= -17.9495 - 3.02 \times 7.5299 + 37.68\\ &= -17.95 - 22.74 + 37.68 = -3.01042\end{aligned}$$

$\therefore\ p^0_{Mn} = 0.04927$ atm

Volume of 1 mol of argon at 1863 K $= 22400 \times \frac{1863}{273} = V_s$

$$1.50\text{ g of Mn} = \frac{1.5}{54.93}\text{ mole}$$

$$\therefore\ \text{volume of} 1.5\ g\ of\ Mn\ at\ 1863\,K = \frac{1.5}{54.93} \times 22400 \times \frac{1863}{273} = V_g$$

$$\begin{aligned}p_{Mn} &= \frac{V_g}{V_s}\\ &= \frac{1.5 \times 22400 \times 1863 \times 273}{54.93 \times 273 \times 22400 \times 1863}\\ &= \frac{1.5}{54.93}\end{aligned}$$

$$\begin{aligned}\therefore\ a_{Mn} &= \frac{p_{Mn}}{p^o_{Mn}} = \frac{1.5}{54.93 \times 0.04927}\\ &= 0.5542\end{aligned}$$

$$\therefore\ \gamma_{Mn} = \frac{a_{Mn}}{x_{Mn}} = \frac{0.5542}{0.5} = 1.108 \qquad \text{Ans.}$$

Problem 98 On passing 100 litres of pure nitrogen gas at one atmospheric pressure and 27 °C over the M-Hg alloy containing 70 at% M, the weight of the mercury trapped was found to be 0.75 mg. If the vapour pressure of M is negligible as compared to mercury, calculate the activity coefficient of mercury in the alloy at 27 °C.

Given: $\log p\,(mm\,Hg) = -\frac{3305}{T} - 0.795 \log T + 10.36$ Atomic weight of mercury = 200.6

Solution

Vapor pressure of pure mercury (p^o) at 27 °C (300 K) is calculated as:

$$logp\ (mm\ Hg) = -\frac{3305}{300} - 0.795\log(300) + 10.36 = -2.6259$$

$$p^o_{Hg} = 0.002366 = 23.66 \times 10^{-4} mm\ Hg$$

Volume of 0.75 mg of mercury $= \frac{0.75}{200.6} \times 22.4 = 0.084\ cc\ at\ 300K$
Volume of nitrogen at 300 K = 100 litres

$$\therefore p_{Hg}\ \text{over the alloy} = \frac{0.084}{100 \times 1000} \times 760 = 6.365 \times 10^{-4}\text{mmHg}$$

Hence $a_{Hg} = \frac{p_{Hg}}{p^o_{Hg}} = \frac{6.365 \times 10^{-4}}{23.66 \times 10^{-4}} = 0.269$

$$\therefore \gamma_{Hg} = \frac{a_{Hg}}{x_{Hg}} = \frac{0.269}{0.30} = 0.893\ \ Ans.$$

Problem 99 0.004 g of gold is lost from the Knudsen cell orifice of radius 0.015 cm in 20 h at 1100 °C. Calculate the vapor pressure of gold. Atomic mass of gold is 197.

Solution:

$$p = \frac{m}{tA}\left[\frac{2\pi RT}{M}\right]^{\frac{1}{2}} \qquad \text{t} = 20 \times 3600\ \text{sec}, \quad \text{A} = \pi r^2$$

$$= 0.02256\frac{m}{tA}\left[\frac{T}{M}\right]^{1/2}$$

$$= \frac{0.02256 \times 0.004 \times 7}{20 \times 3600 \times 22 \times (0.015)^2}\left[\frac{1373}{197}\right]^{1/2}$$

$$= 4.7 \times 10^{-6}\ \text{atm} \qquad \text{Ans.}$$

Problem 100 At 500 °C the emf of the galvanic cell: Cd (l)/LiCl-KCl + 5 wt % $CdCl_2$/CdTe (s) + Te (l) was found to be 485 mV. If the temperature coefficient of the emf at this temperature was −0.2417 mV K^{-1}, calculate the free energy and heat of formation of CdTe at 500 °C. Mark the positive and negative electrodes and write the cell reaction.

Solution

Cell: −Cd(1)/LiCl-KCl + 5wt % $CdCl_2$/CdTe(s) + Te(l)+

cell reactions

LHS: Cd → Cd^{++} + 2e, generation of electrons, hence higher activity of liquid Cd serves as negative electrode

RHS: Te + 2e → Te^{--}, consumption of electrons, hence positive electrode

The overall cell reaction: Cd (l) + Te (l) = CdTe (s)

Free energy of formation, $\Delta G^o = -\ nFE$

n = 2, F = 23066 calV^{-1} g equiv^{-1}, E = 0.485 V

$$\therefore \Delta G^0 = -2 \times 23066 \times 0.485 = -22374 \text{ cal mol}^{-1}$$

Heat of formation, $\Delta H^0 = \Delta G^0 + T\Delta S^0$

$$\begin{aligned}
&= -\mathrm{nFE} + \mathrm{T(nFdE/dT)} \\
&= -\mathrm{nF(E - TdE/dT)} \\
&= -2 \times 23066\left[0.485 - 773 \times \left(-0.2417 \times 10^{-3}\right)\right] \\
&= -46132[485 + 773 \times 0.2417] \times 10^{-3} \\
&= -46132(485 + 186.8) \times 10^{-3} \\
&= -46.132 \times 671.8 \\
&= -30990 \text{ cal mol}^{-1} \qquad \text{Ans.}
\end{aligned}$$

Problem 101 What is the mode of conduction in calcia stabilized zirconia (CSZ) electrolyte? The variation of free energy of formation of FeO and Cu_2O with temperature is expressed as-

$$\begin{aligned}
\Delta G^o(\mathrm{FeO}) &= -62050 + 14.95\,\mathrm{T\ cal} \\
\Delta G^o(\mathrm{Cu_2O}) &= -46700 - 3.92\,\mathrm{T}\log\mathrm{T} + 34.1\,\mathrm{T\ cal}
\end{aligned}$$

What would be the open circuit emf of the cell at 1000 °C based on CSZ elecltrolyte using appropriate electrodes of metal-metal oxide mixtures of FeO and Cu_2O? Write the cell reaction.

Solution

When divalent CaO is dissolved in tetravalent ZrO_2, oxygen ions O^{2-} vacancies are generated. These vacancies start moving at high temperature under the influence of potential gradient. This results in a conduction which is predominantly ionic.

The electronic hole conduction is a function of oxygen pressure. At relatively low oxygen pressure, n-type or conduction due to electron results while at high pressure p-type or electron hole conduction occurs. Only within certain critical oxygen pressure range, which varies slightly with temperature, ionic conductivity is obtained with oxides like ZrO_2 and ThO_2.

$$\begin{aligned}
\Delta G^o_{1273}(\mathrm{FeO}) &= -62050 + 14.95 \times 1273 = -43018 \text{ cal} \\
\Delta G^o_{1273}(\mathrm{Cu_2O}) &= -46700 - 3.92 \times 1273 \log 1273 + 34.1 \times 1273 = -18784 \text{ cal}
\end{aligned}$$

Thus FeO is more stable i.e. Fe-FeO electrodes give low p_{O_2} hence should be a negative electrode (by convention located on LHS of the cell).

Therefore, the cell arrangement would be –Fe-FeO/CSZ/Cu, Cu_2O+

The cell reaction: Fe + Cu_2O = 2Cu + FeO

$$\Delta G^o{}_{reaction} = \Delta G^0_{FeO} - \Delta G^0_{Cu_2O} = = -43018 - (-18784) = -2424\,\text{cal}$$

$$\Delta G^0 = -\text{n F E}$$

$$\therefore \text{E} = -\frac{\Delta G^0}{nF} = -\frac{-2424}{2 \times 23066} = 0.5253\ V$$

$$= 525.3\ mV$$

The cell reaction LHS: $O^{2-} = \frac{1}{2}O_2(g) + 2e \quad -ve$

RHS $\frac{1}{2}O_2 + 2e = O^{2-} \quad +ve$

Problem 102 In the following galvanic cell: Ni-NiO/CSZ/Cu-Cu_2O, at 800 °C the emf was measured to be 271 mV. If the standard free energy of formation of Cu_2O at 800 °C is -21.57 kcal mol^{-1}, find out the standard free energy of formation of NiO at this temperature.

Solution
The cell reaction: Cu_2O + Ni = NiO + 2Cu

$$\Delta G^0_r = \Delta G^0_{NiO} - \Delta G^0_{Cu_2O} = -nFE$$

$$\begin{aligned}\therefore \Delta G^0_{NiO} &= -nFE + \Delta G^0_{\text{Cu}_2\text{O}}\\ &= -2 \times 23066 \times 0.271 - 21570\\ &= -12490 - 21570 = -34060\ \text{cal}\\ &= -34.06\,\text{k cal mol}^{-1} \qquad \text{Ans.}\end{aligned}$$

Problem 103 The following data were found for the cell: Pb(1)/LiCl-$PbCl_2$/Pb-Bi (1) at 700 K.

(a) Write the cell reaction and calculate $\Delta\overline{G}_{Pb}$, $\Delta\overline{H}_{Pb}$, *and* $\Delta\overline{S}_{Pb}$.
(b) Calculate the activities of lead for each composition.
(c) Comment on the behavior of the Pb-Bi system.

x_{Pb}	0.848	0.720	0.600	0.496	0.414	0.328	0.230	0.111
E(mV)	5.32	11.48	19.29	27.82	35.94	45.40	59.76	86.15
$10^6 \times dE/dT(VK^{-1})$	7.4	14.4	20.8	27.8	37.6	46.4	64.4	102.0

Solution
(a) Cell arrangement: Pb(1)/LiCl-$PbCl_2$/Pb-Bi (l)

Half cell reactions:
LHS: Pb $\rightarrow$ $Pb^{++} = 2\,e$
RHS: $(Pb\text{-}Bi)^{++} + 2\,e \rightarrow Pb\text{-}Bi$
The overall cell reaction: Pb $\rightarrow$ (Pb-Bi),

(b) $$\Delta\overline{G}_{Pb} = RT\ \ln\ a_{Pb} = -nFE = \Delta\overline{H}_{Pb} - T\Delta\overline{S}_{Pb}$$
$$= \overline{G}_{Pb} - G^0_{Pb} = \overline{\mu}_{Pb} - \mu^0_{Pb}$$

$$\ln\ a_{Pb} = -\frac{nF}{RT}\cdot\frac{E(mv)}{1000} \qquad F = 96500 J\,V^{-1}\,g\,eqv^{-1}$$
$$= -\frac{2\times 96500\times E}{1.987x4.18\times 700\times 1000} = -0.033163\ E$$

When $x_{Pb} = 0.848$, $\ln\ a_{Pb} = -0.033163 \times 5.32 = -0.176425$;

$a_{Pb} = 0.838$

In order to obtain $\Delta\overline{H}$ and $\Delta\overline{S}$, $\Delta\overline{G}$ must be calculated.

$$\Delta\overline{G}_{Pb} = -nFE$$

$$\text{at}\quad x_{Pb} = 0.848,\ \Delta\overline{G}_{Pb} = -2\times 96500\times\frac{5.32}{1000} = -1030$$

$$\Delta\overline{S}_{Pb} = -\frac{\partial\Delta\overline{G}_{Pb}}{\delta T} = +nF\frac{dE}{dT} = +2x96500\times 7.4\times 10^{-6} = 1.43\ \text{J K}^{-1}\,\text{g atom}^{-1}$$

$$\therefore\ \Delta\overline{H}_{Pb} = \Delta\overline{G} + T\Delta\overline{S} = -1030 + 700\times 1.43 = -1030 + 1001$$
$$= -29\,\text{J g atom}^{-1}\quad \text{Ans.}$$

Following the above procedure the results for the entire composition is presented in the given below:

x_{Pb}	0.848	0.720	0.600	0.496	0.414	0.328	0.230	0.111
$-\Delta\overline{G}_{Pb}$ J g atom^{-1}	1030	2220	3720	5360	6950	8770	11530	16620
$-\Delta\overline{S}_{Pb}$ JK^{-1}g atom^{-1}	1.43	2.78	4.02	5.36	7.26	8.95	12.43	19.66
$\Delta\overline{H}_{Pb}$ J g atom^{-1}	29	270	910	1610	1870	2500	2830	2880
a_{Pb}	0.838	0.683	0.527	0.398	0.303	0.223	0.138	0.057

(c) The solutions in the system Pb-Bi exhibit negative deviation from ideal Raoult's law because $a_{Pb} < x_{Pb}$ (i.e. $\gamma_{Pb} < 1$) at all the compositions. This means forces of attraction between Pb – Bi is stronger as compared to Pb-Pb and Bi-Bi attractions.

Problem 104 An electrolytic cell is set up in which one electrode is liquid cadmium and the other is liquid Cd-Pb alloy. The electrolyte is fused LiCl-KCl-$CdCl_2$ mixture. When the atom fraction of cadmium in the alloy is 0.45, the emf is 0.0125 mV at 480 °C. Write the cell arrangement with marking the negative and positive electrodes. Calculate the activity coefficient of cadmium in the alloy at this temperature. Comment on the behaviour of the solution.

Solution

(i) The cell arrangement: –Cd(l)/LiCl-KCl-$CdCl_2$/Cd-Pb (l)+

(ii) $\overline{G}_{Cd} = RT \ln a_{Cd} = -nFE$

$$\log a_{Cd} = -\frac{nFE}{4.575T} = -\frac{2 \times 23066 \times 0.0125}{4.575 \times 753} = -0.1674$$

$$a_{Cd} = 0.68$$

$$\gamma_{Cd} = \frac{a_{Cd}}{x_{Cd}} = \frac{0.68}{0.45} = 1.51$$

(iii) Since γ is more than 1, Cd-Pb system shows positive deviation from Raoult's law. This means Cd-Pb interaction is weaker as compared to Cd-Cd and Pb-Pb interactions.

Problem 105 The emf of the galvanic cell Cd (l)/KCl-NaCl-$CdCl_2$/Cd-Sn (l), $x_{Cd} = 0.258$ has been found to be 0.0324 V at 483 °C.

(a) Calculate $\overline{G}_{Cd} - G^0_{Cd}$ (in cal) *at* 483 °C.

(b) What is the activity of cadmium in the alloy, relative to pure cadmium as standard state?

(c) Calculate the vapor pressure of cadmium over the alloy. The vapor pressure of pure cadmium is given below as a function of temperature.

$$\log \mathrm{p(mmHg)} = -5819\,\mathrm{T}^{-1} - 1.257 \log \mathrm{T} + 12.287$$

(d) At this concentration does the system Cd-Sn show positive or negative deviation from Raoult's law?

Solution

(a) $\overline{G}_{Cd} - G^0_{Cd} = RT \ln a_{Cd} = -nFE = -1.987 \times 23066 \times 0.0324 = -1494.6\ \mathrm{cal\ mol^{-1}}$

(b) $\ln a_{Cd} = -\frac{nFE}{RT} = \frac{-2 \times 23066 \times 0.0324}{1.987 \times 756} = -0.9950$

$$\therefore a_{Cd} = 0.3697 \quad \text{Ans.}$$

(c)
$$\begin{aligned} \log \mathrm{p(mmHg)} &= -5819\,\mathrm{T}^{-1} - 1.257 \log \mathrm{T} + 12.287 \\ &= -5819/756 - 1.257 \times \log 756 + 12.287 \\ &= 0.9716 \end{aligned}$$

$$\begin{aligned} \therefore p^o_{Cd} &= 9.37\ \text{mm Hg} \\ p_{Cd} &= p^o_{Cd} \cdot a_{Cd} = 9.37 \times 0.3697 = 3.46\ \text{mm Hg} \quad \text{Ans.} \end{aligned}$$

(d) $\gamma_{Cd} = \frac{a_{Cd}}{x_{Cd}} = \frac{0.3697}{0.258} = 1.433$

Since γ_{Cd} is greater than unity, Cd-Sn system shows positive deviation from Raoult's law.

Problem 106 An electrolytic cell is set up in which one electrode is liquid thallium, the other liquid thallium-lead alloy. The electrolyte is a fused LiCl-KCl-TlCl mixture. When the mole fraction of thallium in the alloy is 0.20, the emf is 115.2 mV at 438 °C.

(a) Calculate the activity and activity coefficient of thallium in the alloy.
(b) Calculate the vapor pressure of thallium in the alloy.

Given that: Tl (l) = Tl (g), $\Delta G^o = 42530 + 4.95\, T \log T - 40.6\, T$

Solution

(a) $\Delta \overline{G}_{Tl} = RT \ln a_{Tl} = -nFE$

$$\ln a_{Tl} = -\frac{nFE}{RT} = -\frac{1 \times 23066 \times 115.2}{1000 \times 1.987 \times 711} = -1.8809$$

$$a_{Tl} = 0.152 \qquad \text{Ans.}$$

$$\gamma_{Tl} = a_{Tl}/x_{Tl} = 0.152/0.20$$

$$= 0.76 \qquad \text{Ans.}$$

(b) Tl (l) = Tl (g)

$$K = \frac{p^o_{Tl}}{a_{Tl}} = \frac{p^o_{Tl}}{1} = p^o_{Tl} \quad (a_{Tl} = 1)$$

$$\Delta G^o = 42530 + 4.95T \log T - 40.6T = -RT \ln K = -RT \ln p^o_{Tl}$$

$$RT \ln p^o_{Tl} = -42530 - 4.95T \log T + 40.6T$$

$$= -42530 - 4.95 \times 711 \times \log(711) + 40.6 \times 711$$

$$= -23700$$

$$\therefore \ln p^o_{Tl} = \frac{-23700}{1.987 \times 711} = -16.771$$

$$\text{or } p^o_{Tl} = 5.2 \times 10^{-8} \text{ atm}$$

$$= 3.95 \times 10^{-5} \text{ mm Hg}$$

$$p_{Tl} = p^o_{Tl} \times a_{Tl} = 3.95 \times 10^{-5} \times 0.152$$

$$= 6 \times 10^{-6} \text{ mm Hg} \qquad \text{Ans.}$$

Problem 107 If the activities of salts AY_2 and BX (where A and B are, respectively, divalent and univalent cations and X and Y are univalent anions) in a salt mixture: AY_2-BX, are given as $a_{AY_2} = x_{A^{2+}} . x^2_{Y^-}$ and $a_{BX} = x_{B^+} .\ x_{X^-}$, where $x_{A^{2+}}, x_{Y^-}, x_{B^+}$ and x_{X^-} are ionic fractions. Construct an activity (a_{AY_2}, a_{BX} and a_{BY}) vs. mole fraction ($x_{AY_2} = 1.0$ to $x_{BX} = 1.0$) diagram for the Temkin ideal salt mixtures.

Solution

The formation of the salt BY in equilibrium with A^{2+}, B^{+}, X^{-} and Y^{-} ions present in the fused salt mixture may be considered because B^{+} *and* Y^{-} are monovalent. Following the procedure for the derivation of activities of AY_2 and BX in Sect. 3. 13.1, the activity of BY can be obtained in terms of their ionic fractions.

$$\overline{G}_{BY}^{M} = \left(\frac{\partial \Delta G^M}{\partial n_{BY}}\right)_{n_A, n_Y} = RT \ln a_{BY} \tag{1}$$

$$\text{and} \left(\frac{\partial \Delta G^M}{\partial n_{BY}}\right)_{n_A, n_Y} = \left(\frac{\partial \Delta G^M}{\partial n_B}\right)_{n_A, n_Y, n_X} + \left(\frac{\partial \Delta G^M}{\partial n_Y}\right)_{n_A, n_Y, n_X}$$

$$= RT \ln x_B + RT \ln x_Y = RT \ln x_B x_Y \tag{2}$$

$$\therefore a_{BY} = x_B \cdot x_Y$$

Since, we want to calculate the activities of AY_2, BX and BY in the salt mixture, according to Temkin rule, we can write:

$$\begin{aligned} a_{BX} &= x_B \cdot x_X \\ &= \left(\frac{n_B}{n_A + n_B}\right)\left(\frac{n_X}{n_X + n_Y}\right) \\ &= \left(\frac{n_{BX}}{n_{AY_2} + n_{BX}}\right)\left(\frac{n_{BX}}{n_{BX} + 2n_{AY_2}}\right) = x_{BX}\left(\frac{n_{BX}}{n_{BX} + 2n_{AY_2}}\right) \end{aligned}$$

Dividing numerator and denominator by $(n_{AY_2} + n_{BX})$ we get:

$$a_{BX} = x_{BX}\left[\frac{\frac{n_{BX}}{n_{AY_2} + n_{BX}}}{\frac{n_{BX}}{n_{AY_2} + n_{BX}} + 2\left(\frac{n_{AY_2}}{n_{AY_2} + n_{BX}}\right)}\right] = x_{BX}\left[\frac{x_{BX}}{x_{BX} + 2x_{AY_2}}\right] = \left(\frac{x_{BX}^2}{1 + x_{AY_2}}\right) \tag{3}$$

Making use of Eq. 3 we calculate and list the value of a_{BX} for different values of x_{AY_2}:

x_{AY_2}	0	0.05	0.1	0.2	0.3	0.4	0.5	0.6	0.7	0.8	0.9	1.0
a_{BX}	1	0.860	0.736	0.533	0.377	0.257	0.167	0.10	0.053	0.022	0.005	0

Similarly, $a_{BY} = x_B \cdot x_Y$

$$\begin{aligned} &= \left(\frac{n_B}{n_A + n_B}\right)\left(\frac{n_Y}{n_X + n_Y}\right) \\ &= \left(\frac{n_{BX}}{n_{AY_2} + n_{BX}}\right)\left(\frac{2n_{AY_2}}{n_{BX} + 2n_{AY_2}}\right) = x_{BX}\left(\frac{2n_{AY_2}}{n_{BX} + 2n_{AY_2}}\right) \end{aligned}$$

Dividing numerator and denominator by $(n_{AY_2} + n_{BX})$ we get:

$$a_{BY} = x_{BX}\left[\frac{\frac{2n_{AY_2}}{n_{AY_2}+n_{BX}}}{\frac{n_{BX}}{n_{AY_2}+n_{BX}} + 2\left(\frac{n_{AY_2}}{n_{AY_2}+n_{BX}}\right)}\right] = x_{BX}\left[\frac{2x_{AY_2}}{x_{BX} + 2x_{AY_2}}\right] = x_{BX}\left(\frac{2x_{AY_2}}{1 + x_{AY_2}}\right) \quad (4)$$

Making use of Eq. 4 we calculate and list the value of a_{BY} for different values of x_{AY_2}:

x_{AY_2}	0	0.1	0.2	0.3	0.4	0.5	0.6	0.7	0.8	0.9	1.0
a_{BY}	0	0.164	0.267	0.323	0.343	0.333	0.300	0.247	0.178	0.095	0

$$\begin{aligned} \text{and } a_{AY_2} &= x_A \cdot x_Y^2 \\ &= \left(\frac{n_A}{n_A + n_B}\right)\left(\frac{n_Y}{n_X + n_Y}\right)^2 \\ &= \left(\frac{n_{AY_2}}{n_{AY_2} + n_{BX}}\right)\left(\frac{2n_{AY_2}}{n_{BX} + 2n_{AY_2}}\right)^2 = x_{AY_2}\left(\frac{2n_{AY_2}}{n_{BX} + 2n_{AY_2}}\right)^2 \end{aligned}$$

Dividing numerator and denominator by $(n_{AY_2} + n_{BX})$ we get:

$$a_{AY_2} = x_{AY_2}\left[\frac{\frac{2n_{AY_2}}{n_{AY_2}+n_{BX}}}{\frac{n_{BX}}{n_{AY_2}+n_{BX}} + 2\left(\frac{n_{AY_2}}{n_{AY_2}+n_{BX}}\right)}\right]^2 = x_{AY_2}\left[\frac{2x_{AY_2}}{x_{BX} + 2x_{AY_2}}\right]^2 = \left[\frac{4x_{AY_2}^3}{(1 + x_{AY_2})^2}\right] \quad (5)$$

Making use of Eq. 5 we calculate and list the value of a_{AY_2} for different values of x_{AY_2}:

x_{AY_2}	0	0.05	0.1	0.2	0.3	0.4	0.5	0.6	0.7	0.8	0.9	1.0
a_{AY_2}	1	0.0005	0.003	0.022	0.064	0.131	0.222	0.338	0.475	0.632	0.808	1.0

Variation of activity of AY_2, BX and BY with composition is shown in the figure below.

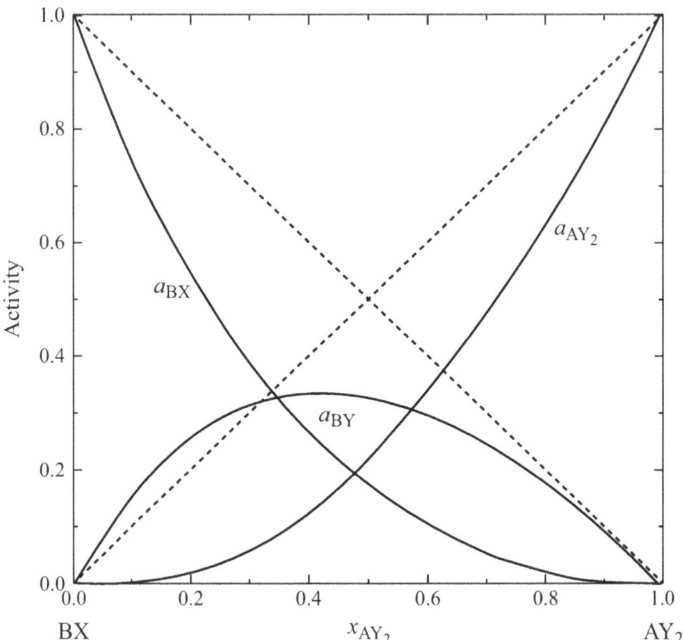

Activity (a_{AY_2}, a_{BX}, a_{BY}) versus mole fraction (x_{AY_2} = 1.0 to x_{BX} = 1.0) diagram

Problem 108 Calculate the activities of $CaCl_2$ and KBr in a salt mixture containing 40 mol % KBr.

Solution

Based on the above derivation we can write expressions for the activity of $CaCl_2$, KBr and KCl in $CaCl_2$-KBr salt mixture as:

$$a_{CaCl_2} = \left[\frac{4x^3_{CaCL_2}}{\left(1 + x_{CaCl_2}\right)^2}\right], a_{KBr} = \left(\frac{x^2_{KBr}}{1 + x_{CaCl_2}}\right) \text{and}\, a_{KCl} = x_{KBr}\left(\frac{2x_{CaCl_2}}{1 + x_{CaCl_2}}\right)$$

In the given problem $x_{KBr} = 0.4$ and $x_{CaCl_2} = 0.6$

$$a_{CaCl_2} = \left[\frac{4 \times (0.6)^3}{(1 + 0.6)^2}\right] = 0.3375$$

$$a_{KBr} = \left(\frac{(0.4)^2}{1 + 0.6}\right) = 0.10$$

$$a_{KCl} = 0.4\left(\frac{2 \times 0.6}{1 + 0.6}\right) = 0.30 \quad \text{Ans.}$$

Problem 109 Calculate the activity of NaCl, NaBr and KCl in a fused salt mixture: NaCl-KBr containing 40 mol % NaCl.

Solution
According to the Temkin rule the fused salt mixture contains Na^+, K^+, Cl^- and Br^- ions:

$$\begin{aligned} a_{NaCl} &= x_{Na^+} \cdot x_{Cl^-} \\ &= \left(\frac{n_{Na^+}}{n_{Na^+} + n_{K^+}}\right)\left(\frac{n_{Cl^-}}{n_{Cl^-} + n_{Br^-}}\right) \\ &= \left(\frac{n_{NaCl}}{n_{NaCl} + n_{KBr}}\right)\left(\frac{n_{NaCl}}{n_{NaCl} + n_{KBr}}\right) = x_{NaCl} \cdot x_{NaCl} = x_{NaCl}^2 \end{aligned}$$

Similarly, $a_{KBr} = x_{KBr}^2$

$$\begin{aligned} a_{NaBr} &= x_{Na^+} \cdot x_{Br^-} \\ &= \left(\frac{n_{Na^+}}{n_{Na^+} + n_{K^+}}\right)\left(\frac{n_{Br^-}}{n_{Br^-} + n_{Cl^-}}\right) \\ &= \left(\frac{n_{NaCl}}{n_{NaCl} + n_{KBr}}\right)\left(\frac{n_{KBr}}{n_{KBr} + n_{NaCl}}\right) = x_{NaCl} \cdot x_{KBr} \end{aligned}$$

Similarly, $a_{KCl} = x_{NaCl} \cdot x_{KBr}$

In the given mixture: $a_{NaCl} = x_{NaCl}^2 = 0.4^2 = 0.16$

$$a_{KBr} = x_{KBr}^2 = 0.6^2 = 0.36$$

and $\quad a_{NaBr} = x_{NaCl} \cdot x_{KBr} = 0.4 \times 0.6 = a_{KCl} = 0.24 \quad$ Ans.

Further Reading and Consultation

1. O. Kubaschewski, C.B. Alcock, *Metallurgical Thermochemistry*, 4th edn. (Pergamon, Oxford, 1967)
2. J. Mackowiak, *Physical Chemistry for Metallurgists* (American Elsevier Publishing Co. Inc, New York, 1966)
3. R.H. Parker, *An Introduction to Chemical Metallurgy*, 2nd edn. (Pergamon, Oxford, 1978)
4. D.R. Gaskell, *Introduction to the Thermodynamics of Materials*, 4th edn. (Taylor & Francis, New York, 2003)
5. L.S. Darken, R.W. Gurry, *Physical Chemistry of Metals* (McGraw-Hill Co. Ltd, London, 1953)
6. C. Bodsworth, A.S. Appleton, *Problems in Applied Thermodynamics* (Longmans, London, 1965)
7. I. Barin, O. Knacke, O. Kubaschewski, *Thermochemical Properties of Inorganic Substances (Supplement)* (Springer, New York, 1977)

Appendixes

Appendix: A – Recommended Values of Physical Constants

Physical constant	Symbol	Value
Acceleration due to gravity	g	$9.81\ ms^{-2}$
Avogadro number	N	6.02252×10^{23} molecules mol^{-1}
Boltzmann constant	k	$1.38054 \times 10^{-23}\ J\ K^{-1}$
Faraday constant	F	96494 Coulombs mol^{-1} $= 23066\ cal\ V^{-1}\ mol^{-1}$
Gas constant	R	$1.987\ cal\ K^{-1}\ mol^{-1}$ $= 8.3143\ J\ K^{-1}\ mol^{-1}$ $= 0.08206$ liter atm $K^{-1}\ mol^{-1}$
Planck constant	h	$6.6256 \times 10^{-34}\ Js$

M. Shamsuddin, *Thermodynamic Measurement Techniques*, The Minerals, Metals & Materials Series, https://doi.org/10.1007/978-3-031-47118-6

Appendix: B – SI Units and Conversion Factors

Physical Quantity	SI unit (a)	cgs unit (b)	conversion factor (from b to a)
Length	m	cm	10^{-2}
Mass	kg	g	10^{-3}
Time	s	s	1
Volume	m^3	cm^3	10^{-6}
Density	kg m^{-3}	g cm^{-3}	10^3
Force	N = kg m s^{-1} = J m^{-1}	dyne	10^{-5}
Pressure	Nm^{-2} = kg $m^{-1}s^{-2}$ = J m^{-3}	dyne cm^{-2}	10^{-1} (1 atm = 760 mm Hg = 760 torr = 1.013 bar = 1.033×10^4 Kg m^{-2} = 1.013×10^5 N m^{-1} = 1.013×10^5 Pascal)
Surface tension	Nm^{-1} = Jm^{-2}	dyne cm^{-1}	10^{-3}
Energy, work	J = kg m^2s^{-2}	cal	4.184
		erg	10^{-7} (1 J = 0.102 kg.m)
Molar free energy, enthalpy	J mol^{-1}	cal mol^{-1}	4.184
Molar entropy, heat capacity	J $K^{-1}mol^{-1}$	cal K^{-1} mol^{-1}	4.184
Concentration	mol m^{-3}	mol l^{-1}	10^3
	mol dm^{-3}	mol l^{-1}	1
Molality	mol kg^{-1}	mol kg^{-1}	1
Ionic strength	mol kg^{-1}	mol kg^{-1}	1
	mol m^{-3}	mol l^{-1}	10^3
	mol dm^{-3}	mol l^{-1}	1

Index

M. Shamsuddin, *Thermodynamic Measurement Techniques*, The Minerals, Metals & Materials Series, https://doi.org/10.1007/978-3-031-47118-6

H

I

J

K

L

M

N

O

P

The manufacturer's authorised representative in the EU is Springer Nature Customer Service Centre GmbH, Europaplatz 3, 69115 Heidelberg, Germany. If you have any concerns regarding our products, please contact ProductSafety@springernature.com

Printed and bound by CPI Group (UK) Ltd, Croydon, CR0 4YY

07/07/2026

02160920-0003